Gas
Installation
Technology

GAS INSTALLATION TECHNOLOGY

R. D. Treloar

Colchester Institute

Blackwell
Publishing

Blackwell Publishing Offices:
Blackwell Publishing Ltd, 9600 Garsington Road, Oxford OX4 2DQ, UK
  Tel: +44 (0) 1865 776868
Blackwell Publishing Inc., 350 Main Street, Malden, MA 02148-5020, USA
  Tel: +1 781 388 8250
Blackwell Publishing Asia Pty Ltd, 550 Swanston Street, Carlton, Victoria 3053, Australia
  Tel: +61 (0) 3 8359 1011

First published 2005 by Blackwell Publishing Ltd

Library of Congress Cataloging-in-Publication Data

Treloar, Roy.
  Gas installation technology / R. D. Treloar. – Ist ed.
    p. cm.
  Includes index.
  ISBN 1-4051-1880-6 (pbk. : alk. paper)
1. Gas-fitting.   2. Gas appliances–Installation.   I. Title.

TH6810.T74 2005
696′.2—dc22   2004019624

ISBN 10: 1-4051-1880-6
ISBN 13: 978-14051-1880-4

A catalogue record for this title is available from the British Library

Set in 10/12 pt Sabon
by TechBooks
Printed and bound in Cornwall, UK
by TJ International Ltd

For further information on Blackwell Publishing, visit our website:
www.blackwellpublishing.com

Contents

Contents

Introduction

This book aims to cover all areas of the gas industry that the operative is likely to encounter. It has purposely grouped together all the various aspects of gas work to include natural gas and LPG with domestic and commercial installations. This may seem strange to some, but it overcomes the problem of repeating topics and cross-checking and allows it all to be contained within the one book.

Much of the work undertaken by the gas engineer follows the same generic principles. The gas principles and processes that are used every day and the type of installation have little or no bearing on the way in which the work is completed. There are variances, hence the need sometimes to identify specific installation types; this has been done throughout the book, where necessary.

The book is not designed to be read from cover to cover and the reader will invariably need to dip into it to retrieve information on a specific problem or interest. A particular topic of interest can be found in one of the following ways:

- **First,** by referring to the Contents page, which identifies the subject areas of the book.
- **Second,** by referring to the Index (p. 439) and choosing a term related to the subject in question.

Good luck in your chosen profession.

Roy Treloar

Acknowledgements

I would like to thank the following organisations and companies for permission to reproduce photographs and extracts from tables.

British Standards Institution Sales & Customer Services (for various tables throughout)
389 Chiswick High Road, London W4 4AL
Telephone: 020 8996 9001
email: cservices@bsi-global.com

Institute of Gas Engineers and Managers (for various tables and the Certificates on p. 181)
Charnwood Wing, Holywell Park, Ashby Road, Loughborough, Leicestershire LE11 3GR
Telephone: 01509 282728
email: general@igem.org.uk

CORGI (for sample card on p. 3 and labels on p. 277)
1 Elmwood, Chineham Business Park, Basingstoke RG24 8WG
Telephone: 01256 37220
http://www.corgi-gas.com

LP Gas Association
Pavilion 16, Headlands Business Park, Salisbury Road, Ringwood, Hampshire H24 3PB
Telephone: 01425 461612
http://www.lpga.co.uk

Society of Laundry Engineers & Allied Trades
Suite 7, Southernhay, 207 Hook Road, Chessington, Surrey KT9 1HJ
Telephone: 020 8391 2266
http://www.sleat.co.uk

Alpha Cookers UK Ltd
Unit 5, Station Road Industrial Estate, Station Road, Thatcham, Berkshire RG19 4QY
Telephone: 01635 876266
http://www.alpha-cookers.co.uk

Duomo UK Ltd
Units 5–6, Judge Court, North Bank, Berryhill Industrial Estate, Droitwich
  Worcestershire WR9 9AU
Telephone: 01905 797989
http://www.duomo.co.uk

Hobart UK
Southgate Way, Orton, Southgate, Peterborough PE2 6GN
Telephone: 07002 101101
http://www.hobartuk.com

Marco Beverages Systems
Telephone: 0207 2744577
http://www.marco-tse.co.uk

Miele Company Ltd
Fairacres, Marcham Road, Abingdon, Oxon OX14 1TW
Telephone: 01235 554455
http://www.miele.co.uk

Mobile Gas Direct Ltd
Unit 1, 459 Uttoxeter Rd, Blythebride, Stoke on Trent ST11 9NT
Telephone: 01782 394444
http://www.mobilegas.co.uk

Powrmatic Ltd
Winterhay Lane, Ilminster, Somerset TA19 9PQ
Telephone: 01460 53535
http://www.powrmatic.co.uk

Reznor UK Ltd
Park Farm Road, Park Farm Ind. Estate, Folkestone, Kent CT19 5DR
Telephone: 01303 259141
http://www.reznor.co.uk

Taylor Portway Ltd
52 Broton Drive Trading Estate, Halstead, Essex CO9 1HB
Telephone: 01787 472551
http://www.portwayfires.com

Trackpipe
OmegaFlex Ltd, Apollo House, Desborough Road, High Wycombe, Bucks
  HP11 2QW
Telephone: 0870 286 8585
http://www.omegaflex.com

Truma Ltd
Truma House , Eastern Avenue, Burton on Trent, Staffordshire DE13 OBB
Telephone: 01283 511092
http://www.truma.com

Sugg Lighting Ltd
Sussex Manor Business Park, Gatwick Road, Crawley, West Sussex RH10 9GD
Telephone: 01293 540114
http://www.sugglighting.co.uk

Yorkshire Fittings
PO Box 166, Leeds LS10 1NA
Telephone: 0113 270 1104
http://www.yorkshirefittings.co.uk

Photographic Workshop Colchester
143 Hythe Hill Colchester Essex CO1 1NF
Telephone: 01206 790221

Abbreviations and Acronyms

$\pm$	plus or minus
$<$	less than
$\leq$	less than or equal to
$>$	greater than
$\geq$	greater than or equal to
$\therefore$	therefore
$\sum$	sum total
Ω	ohm
a.c.	alternating current
ACOP	Approved Code of Practice
ACS	Nationally Accredited Certification Scheme for Individual Gas Fitting Operatives
AR	At Risk
ASD	Atmospheric Sensing Device
BS	British Standard
BS EN	British Standard Europaische Norm (European Standard)
CE	Conformity to Europe (see p. 14)
CO	carbon monoxide
CO_2	carbon dioxide
CORGI	Council of Registered Gas Installers
COSHH	Control of Substances Hazardous to Health
CV	Calorific Value
d.c.	direct current
DFE	Decorative Fuel Effect fire
dm	decimetres
e.m.f.	electromotive force
ECV	Emergency Control Valve
FFD	Flame Failure Device
FSD	Flame Supervision Device
GRP	Glass Reinforced Plastic
H_2O	water
HEL	Higher Explosive Limit
HSE	Health and Safety Executive
Hz	hertz
ID	Immediately Dangerous
IGE/UP	Institute of Gas Engineers/ Utilisation Procedures
IGEM	Institute of Gas Engineers and Managers

ILFE	Inset Live Fuel Effect fire
LCD	Liquid Crystal Display
LEL	Lower Explosive Limit
LFL	Lower Flammable Limit
LPCO	Low Pressure Cut Off
LPG	Liquefied Petroleum Gas
LPGA	LP Gas Association
MIP	Maximum Incidental Pressure
MOP	Maximum Operating Pressure
MPLR	Maximum Permitted Leak Rate
N_2	nitrogen
NCS	Not to Current Standards
NOx	nitrogen dioxide and nitric oxide mix ($NO_2 + NO$)
O_2	oxygen
ODS	Oxygen Depletion System
OPSO	Over Pressure Shut Off
PCB	Printed Circuit Board
PE	polyethylene
PME	Protective Multiple Earthing
p.p.m.	parts per million
PTFE	Polytetrafluoroethylene
PVC	polyvinyl chloride
RIDDOR	Reporting of Injuries Diseases and Dangerous Occurrence Regulations
SAP	Standard Assessment Procedure
SEDBUK	Seasonal Efficiency of Domestic Gas Boiler in the UK
SG	Specific Gravity
TNCS	Terra Neutral Combined Separated (basically Earth and Neutral combined at supply but separated at the building)
UPSO	Under Pressure Shut Off
UV	Ultra Violet

Part 1
The Gas Industry

The Gas Industry

The gas industry has gone through major changes over the last few decades. Prior to and during the early 1960s most gas installation work in the UK was undertaken by British Gas. In 1968 a 22-storey block of flats in Canning Town, East London was devastated by a major gas explosion, which persuaded the industry that a body was needed to oversee this kind of work. As a result, in 1970 a voluntary gas body was formed, called the Confederation of Registered Gas Installers (CORGI).

During the early 1970s, plumbers and heating engineers began to take a greater interest in undertaking gas work, thanks to central heating systems becoming a requirement in the average home. 1972 saw the introduction of the first Gas Safety Regulations, which identified the legal responsibilities to which the installer had to adhere.

With the introduction of the Approved Code of Practice (ACOP) in 1990, gas installers started to take update training and assessment in gas working practices and, by 1991, anyone working in the gas industry for financial gain had to be registered with a Health and Safety Executive (HSE) approved body. The only organisation to date that exists for this purpose emerged from the original voluntary body: CORGI, now called the Council of Registered Gas Installers.

Council of Registered Gas Installers (CORGI)
CORGI was, and still is, tasked with the job of overseeing that the gas industry workforce is adequately trained and competent to undertake gas activities safely. For several years now, all gas engineers have needed to be assessed as competent in the aspect of gas work that they wish to undertake; undertaking any work without this assessment would mean that they are in breach of the law. The assessment that an individual undertakes is called the Nationally Accredited Certification Scheme for Individual Gas Fitting Operatives (ASC) and there are many different assessments. These are listed in the next section. The address is:
CORGI
1 Elmwood
Chineham Business Park
Crockford Lane, Basingstoke RG24 0WG
Telephone: 01256 37220   http://www.corgi-gas.com

Working in the Gas Industry
Today, if you wish to work in the gas industry you need to belong to a company registered with CORGI. Becoming a member is no easy task and the following activities are necessary to gain registration:

1. You will need to gain training through a registered company and be given the opportunity to undertake activities in the type of gas work in which you wish

to be assessed. During this period of employment, you will need to complete a portfolio with authenticated evidence to show that you have undertaken 'on the job' work under the guidance of a CORGI registered installer and completed several installations yourself.

2. You will need to undertake and be able to show evidence that you have completed 'off the job' training at an established training centre.
3. You will need to demonstrate the skills obtained from (1) and (2) above, and pass the written and practical ACS assessment.
4. Only after completing (1)–(3) above can you obtain work with a registered gas business. Alternatively, you may apply to become CORGI registered in your own right.

CORGI Registration

Not every individual gas operative is registered with CORGI in their own right. It is the businesses that are registered. A business, however, must list all the gas engineers that they employ on their list of named gas fitting operatives. As proof that the operative is maintained on the register, CORGI issues a card-type certificate annually. The operative should carry this as proof of competence. On the back of the card is a list of the work categories that the operative is allowed to perform.

Working without CORGI Registration

By careful study of the Gas Safety (Installation and Use) Regulations it will be found that it is possible to undertake gas work, but within a certain degree of limitation. For example, DIY work is an area that may be practised without the need for registration. However, a word of caution, this does mean 'for yourself', or possibly for very close family, and without financial gain. The work must still be completed competently and in compliance with the Gas Regulations and associated industry documents, such as manufacturers' instructions and British Standards. At the time of writing, certain work within factories and mines, etc. were outside the scope of CORGI but, as before and under the Health and Safety at Work Regulations, this must still be completed competently and safely.

Nationally Accredited Certification Scheme for Individual Gas Fitting Operatives (ACS)

In order for gas operatives to undertake work in a particular aspect of the profession they need to have undertaken the appropriate ACS gas assessment in the specific area of work. There are many different assessments that are applicable to domestic, commercial, natural gas and LPG installations; there are also some specialist and service provider assessments. In addition to the list of assessments identified below, there is a range of changeover assessments, providing conversions between the various core assessments e.g. domestic to commercial (CCN1 to COCN1). These are given in the next section.

Specific CORGI Assessment Categories (in alphabetical order)
CABLP1 – Domestic LPG Gas fired mobile cabinet heaters
CBHP1 – Boosters/compressors and high pressure pipes
CCCN1 – Core commercial catering safety
CCLNG1 – Core commercial laundry gas safety
CCLP1B – Core domestic LPG gas safety: Boats, yachts and other vessels
CCLP1EP – Core domestic LPG gas safety: External pipework and gas vessel connections
CCLP1LAV – Core domestic LPG gas safety: Leisure accommodation vehicles (caravans and motor homes)
CCLP1MC – Core domestic LPG gas safety: Mobile cabinet heaters (limited scope)
CCLP1PD – Core domestic LPG gas safety permanent dwellings
CCLP1RPH – Core domestic LPG gas safety: Residential park homes
CCN1 – Core domestic natural gas safety
CCP1 – Commissioning indirect fired commercial plant and equipment
CDGA1 – Commercial direct fired heating appliances
CEN1 – Domestic central heating/hot water boilers (< 70 kW net input)
CESP1 – Natural gas core emergency service provider
CGFE1 – Gas fuelled engines
CGLP1 – Commercial catering vehicle gas generators
CIGA1 – Commercial indirect fired heating appliances
CKHB1 – Domestic range cookers/boilers
CKR1 – Domestic natural gas cookers
CLE1 – Commercial laundry equipment
CMA1 – Specific core meter installations
CMA2LS – Limited core domestic gas safety
CMCALP1 – Commercial LPG mobile catering appliances (boiler rings, bains marie, hot cupboards, toasters)
CMCALP2 – Commercial LPG mobile catering appliances (fryers, fish and chip ranges, grillers, griddles, doughnut fryers)
CMCALP3 – Commercial LPG mobile catering appliances (pressure/expansion boilers, tea urns, cappuccino machines)
CMCALP4 – Commercial LPG mobile catering appliances

CMCLP1 – LPG commercial core mobile catering gas safety

CMET1 – Diaphragm RPD > 6 m³/h and < 107 m³/h

CMET2 – Diaphragm RPD and turbine meters

CMIT1LS – Gas meter instrumentation operatives

COCLP1 – LPG core commercial gas safety

COCN1 – Natural gas core commercial gas safety

COCNPI1LS – Natural gas core commercial gas safety pipework (installer/commissioner)

COMCAT1 – Commercial catering (freestanding stoves/ovens, boiling tables, hot plates, bains marie, hot cupboards)

COMCAT2 – Commercial catering (pressure/expansion boiler, steaming ovens, boiling pans, dishwashers, urns)

COMCAT3 – Commercial catering (deep fat and pressure fryers, brat pans, griddles, under and over fired grills)

COMCAT4 – Commercial catering (fish and chip ranges)

COMCAT5 – Commercial catering (conveyor type ovens and force draught burner appliances)

CORT1 – Commercial radiant heaters

DAH1 – Natural gas domestic ducted air heaters (< 70 kW net)

DFDA1 – Domestic forced draught burning appliances

EFJLP1 – Poly-electro fusion jointing

HTR1 – Domestic gas fires and wall heaters

HTRLP2 – LPG caravan gas fires

HTRLP3 – LPG caravan heaters

HWB1 – Hot water boilers 15–140 kW (swimming pool boilers)

ICAE1 – Commercial first fix appliances and equipment

ICAE1LS – Limited scope commercial first fix appliances and equipment

ICPN1 – Commercial pipework first fix > 28 mm

ICPN1LS – Limited scope commercial pipework first fix

LAU1 – Natural gas domestic laundry appliances

LEI1 – Natural gas leisure and miscellaneous appliances (barbeques, greenhouse heaters, gas lights)

LEILP1 – LPG Leisure & miscellaneous appliances (barbeques, greenhouse heaters, gas lights)

MET1 – Domestic natural gas meters (install/exchange primary and secondary meters)

MET2 – Domestic natural gas meters for service providers (install/exchange primary and secondary meters)

MET3LS – Meter installations gas safety

MET4 – Diaphragm gas meter installations

REFLP2 – LPG caravan gas refrigerators

TPCP1 – Testing and purging low pressure commercial gas pipework > 1 m³

TPCP1A – Testing and purging low pressure commercial gas pipework < 1 m³

VESLP1 – Single LPG gas pipework above and below ground < 0.1 m³

VESLP2 – Multiple LPG gas pipework above and below ground > 0.1 m³

WAHLP1 – LPG warm air heaters in boats

WAT1 – Domestic natural gas instantaneous water heaters

WATLP2 – LPG caravan gas water heaters

Required ACS Assessments and Flowcharts

In order to undertake an assessment in the particular work category that you require, e.g. CEN1 (domestic boilers), you also need to hold the specific core assessment because the core assessment is a prerequisite to all appliance assessments.

There are several core assessment categories and many have similar assessment criteria, therefore it is not always necessary to undertake the complete core for a specific range of appliances as a changeover assessment can be obtained. These include:

CoCATA1 – Changeover COCN1 to CCCN1
CoDC1 – Changeover CCN1 to CCCN1
CoDNCML1 – Changeover CCN1 to CMCLP1
CoDNCO1 – Changeover CCN1 to COCN1
CoDNESP1 – Changeover CCN1 to CESP1
CoLPNG1 – Changeover CCLP1 to CCN1
CoNGLP1 – Changeover CCN1 or COCN1 to CCLP1
CoNGLP1B – Changeover CCN1 to CCLP1B
CoNGLP1LAV – Changeover CCN1 to CCLP1LAV
CoNGLP1PD – Changeover CCN1or COCN1 to CCLP1PD
CoNGLP1RPH – Changeover CCN1 to CCLP1RPH

Importance of Maintaining a Current Core Assessment

Because the core is a mandatory requirement for any further appliance assessment it must always be maintained as a valid and current assessment certificate. Assessments are valid for five years from the date of issue. Should the core certification run out, then the validation of all other certificates received after the date of the core, including changeover cores, cease, even though they may have time to run. As soon as the core category has been re-assessed and the assessment passed the other certificates become valid again.

Individual ACS Requirements

At first sight the vast list of ACS assessment criteria can look quite daunting and it is difficult to choose which assessment to undertake. However, this is can be approached by using the following flowcharts.

Points to consider

- You must hold the specific core for the area of gas work in which you wish to work, e.g. commercial, domestic or LPG and have any prerequisite assessments. *On the flowcharts these are the assessments through which the line flows.*
- If you have a domestic natural gas core and, say, CKR1 you can work on natural gas cookers only. However, if you also have the LPG core, you can also work on LPG cookers. It is the core that denotes the type of installation into which the cooker is installed.

Domestic Appliances in Commercial Premises

Where operatives hold the domestic ACS qualifications for a specific appliance (e.g. CEN1 boilers < 70 kW net) they can work on that appliance in commercial premises, providing the pipework, including a local isolation valve, does not exceed 28 mm in diameter. Should an operative hold the commercial core and wish to work on a domestic type appliance, they do not need to obtain the domestic core to acquire a domestic appliance assessment, such as CEN1 above. However, they cannot work in domestic premises unless they also obtain the domestic core.

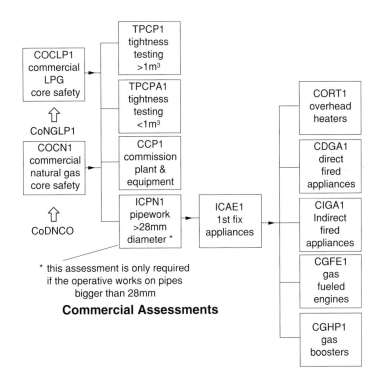

Commercial Assessments

* this assessment is only required if the operative works on pipes bigger than 28mm

Continued over the page:

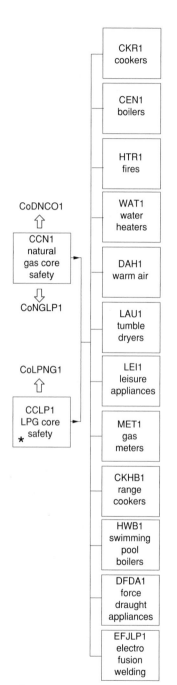

**Domestic Natural Gas
and LPG Assessments**

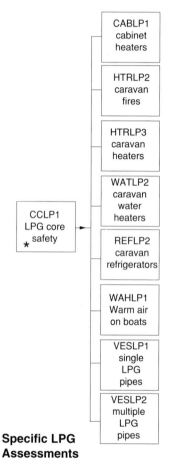

**Specific LPG
Assessments**

* Note CCLP1 is followed by the letters denoting
the core work area applicable e.g.:

PD - permanent dwellings
LAV - leisure accommodation vehicles
RPH - residential part homes
B - boats
EP - external pipework
MC - mobile cabinet heaters (limited scope)

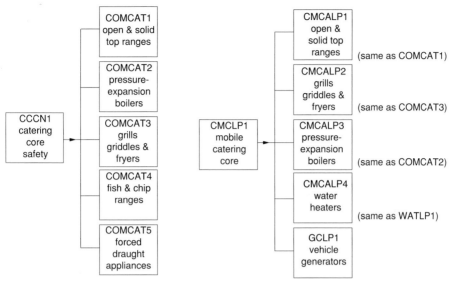

Commercial Catering

Commercial Catering in Mobile Units

CCCN1 catering core safety
- COMCAT1 open & solid top ranges
- COMCAT2 pressure-expansion boilers
- COMCAT3 grills griddles & fryers
- COMCAT4 fish & chip ranges
- COMCAT5 forced draught appliances

CMCLP1 mobile catering core
- CMCALP1 open & solid top ranges (same as COMCAT1)
- CMCALP2 grills griddles & fryers (same as COMCAT3)
- CMCALP3 pressure-expansion boilers (same as COMCAT2)
- CMCALP4 water heaters (same as WATLP1)
- GCLP1 vehicle generators

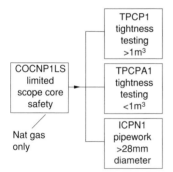

**Limited Scope Operative
(pipe installer/commissioner)**

COCNP1LS limited scope core safety
- TPCP1 tightness testing >1m³
- TPCPA1 tightness testing <1m³
- ICPN1 pipework >28mm diameter

Nat gas only

Commercial Laundry

CCLNG1 core commercial laundry → CLE1 commercial laundry *

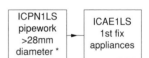

Limited Scope Operative

ICPN1LS pipework >28mm diameter * → ICAE1LS 1st fix appliances

This operative does not make a connection to a live gas supply and is involved in 1st fix pipework and appliances only.

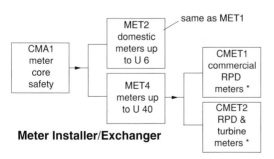

Meter Installer/Exchanger

CMA1 meter core safety
- MET2 domestic meters up to U 6 — same as MET1
- MET4 meters up to U 40
  - CMET1 commercial RPD meters *
  - CMET2 RPD & turbine meters *

CESP1 gas safety core *

Emergency Service Providers

* in addition to the above ICPN1 and TPCP1/1a may be required

Legislation Affecting the Gas Installer 1

Legislation places a mandatory/legal responsibility on the gas operative who must comply with it in order to work within the industry. The piece of legislation that affects the gas engineer most is the Gas Safety (Installation and Use) Regulations. However, this is not the only legislation that needs to be observed. In fact, in this area most actions undertaken as general work activities are affected by some piece of legislation or other.

Referring to the flow diagram opposite, which only represents a very small percentage of the legislation that must be observed, it will be seen that the law is divided into two parts. The first are Acts of Parliament and the second are Regulations, drawn up by government and often following an Act of Parliament. Regulations are usually policed by an authority such as the Health and Safety Executive (HSE), local authorities and their building control officers (BCOs) or water authority inspectors. Acts of Parliament and the Regulations are updated as necessary and care needs to be taken to ensure that the latest version is referred to.

Below the dotted line opposite falls the industry guidance documentation. These documents do not have legal status as such. However compliance with these documents is generally deemed to indicate good working practices and shows that the minimum standards have been maintained. Industry guidance documentation may also form part of a specification for a contract and, as such, non-compliance may lead to a civil action being taken against the gas fitting company.

Criminal and Civil Law

English law is divided into criminal law and civil law and different courts and procedures are followed for each.

Criminal law is penal law involving a crime against the State and is punishable by imprisonment and/or a fine. Action is taken by the police or by such bodies as the HSE or local authority.

Civil law pertains to the rights of private individuals and to the legal proceedings involving those rights. In a civil case the aim of a trial is to establish facts, based on evidence, and to determine liability. Civil law provides for compensation for the injured party, usually in the form of damages or an injunction demanding certain action to be taken.

Examples of the legislation given opposite are described in more detail over the page.

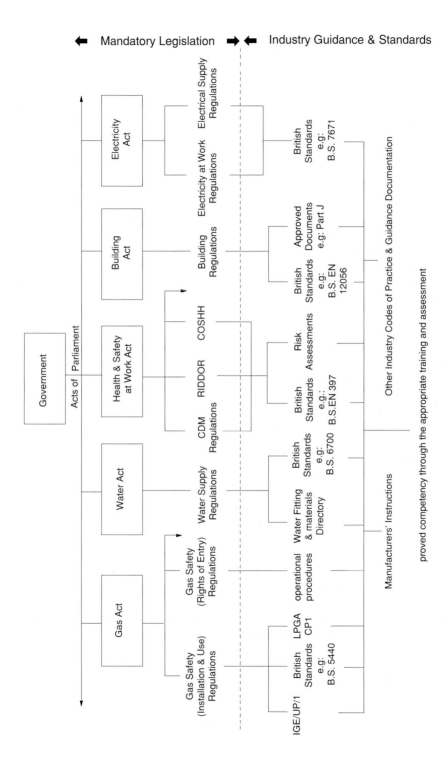

Compliance With the Law

Legislation Affecting the Gas Installer 2

Gas Act

The Gas Act has seen many changes over the years and is responsible for putting in place the following Regulations, to name a few:

- *Gas Safety (Installation and Use) Regulations*
  These are the main regulations that are applicable to the gas fitting operative, but it must be emphasised that they amount to only a few of many regulations that are applicable. The Gas Regulations are described in more detail on page 16.
- *Gas Safety (Rights of Entry) Regulations*
  Under these regulations the gas supplier may, with police assistance if necessary, enter a property to make an installation safe.
- *Gas Safety (Management) Regulations*
  Under these regulations the gas supplier must submit a 'safety case' identifying their procedures. They must operate a full emergency gas service, to include a central telephone emergency number and must investigate any major gas incident, such as a poisoning or gas explosion.

In 1995 this Act was updated to include new licensing provisions enabling competition in the domestic gas market for the sale of gas.

Health and Safety at Work Act

This Act, first issued in 1974, includes a large number of regulations for working operatives. There are possibly hundreds of regulations that fall under this Act. Some examples are cited below.

- *Control of Substances Hazardous to Health (COSHH) Regulations*
  These contain the statutory duties designed to protect operatives and all others from the effects of working with substances that may cause harm to their health. The COSHH Regulations make it a requirement that you maintain a list of all hazardous materials and substances used, and have to hand the necessary protection advice and first aid information.
- *Reporting of Injuries, Diseases and Dangerous Occurrences Regulations (RIDDOR)*
  This regulation applies to the reporting of dangerous and unsafe situations and is covered in depth on page 280.
- *Management of Health and Safety at Work (MHSW) Regulations and Construction (Design and Management) (CDM) Regulations*
  These Regulations place a wide range of duties on employers, contractors, designers and clients, etc. to ensure that health and safety is maintained throughout the construction process. This includes ensuring that adequate risk assessments are carried out.

There are many, many more regulations falling within the Act and the Regulations outlined above. It must be understood that they are only part of the whole picture.

Building Act

In the early days, local Building Acts were applied to specific districts. However, these days they are in the form of Building Regulations. There are differences in England/Wales, Northern Ireland and Scotland. The Isle of Man also has some variations. However, in essence, the Regulations all identify the minimum requirements to be applied to building works as well as aspects affecting safety, energy conservation, etc. The Regulations itself is quite a small document. However, alongside it sits a series of Approved Documents (or Technical Standards, in Scotland), which set out detailed design requirements. Examples of those used in England and Wales include:

- *Approved Document F*: covering ventilation;
- *Approved Document J*: covering heat producing appliances;
- *Approved Document L*: covering conservation of fuel.

Water Act

The earliest Water Acts made provision for local water byelaws. These were made to suit the various local water authority conditions. However, since 1999, all water installation work in England and Wales has had to comply with a set of mandatory *Water Supply Regulations*. As with building controls, Northern Ireland and Scotland have their own variations.

Electrical Safety Quality and Continuity Regulations (Formerly Electricity Act)

The earliest Electrical Acts date back to the late 19th century when electricity was first produced. This Act is responsible for putting in place the following Regulations:

- *The Electrical Supply Regulations*
  This identifies the types of supply that may be used to serve a particular property, etc.
- *The Electricity at Work Regulations*
  These regulations, among other things, identify specific tasks that must be undertaken when working on electrical supply systems. For example, the Regulations states the minimum tests to be carried out when checking an installation and also identifies how the test equipment should be checked for correct operation prior to and after testing.

The effect of legislation does not stop here and many other laws are involved. For example, issued under The Road Traffic Act, the **Carriage of Goods Regulations** identifies the carriage of LPG cylinders greater than 2.5 kg in closed vans. Another piece of legislation to affect the operative is the **Environmental Protection Act**, which aims to protect the environment and requires an operative to have a licence to carry rubbish in the back of their vehicle. It is difficult to keep abreast of new laws; however, ignorance is no defence in the eyes of the law.

Industry Documents and British Standards

In addition to the vast amount of legislation already mentioned, the operative needs to be aware of an array of industry documents that must be followed in order to ensure that an installation conforms to the required standard.

Industry documents include manufacturers' instructions; British Standards (BS); Institution of Gas Engineers and Managers (IGEM) procedures; LP Gas Association (LPGA); and codes of practice, to name a few. As with legislation, these documents are continually being updated and reviewed so it is essential that the latest version of the document be used. Another problem is that one document often overlaps another and invariably they have different or conflicting views. CORGI has identified a hierarchy to documentation used. The first requirement is to follow mandatory legislation; you then need to observe the specific manufacturer's instructions, followed by the British Standards, IGE and LPGA codes of practice. Finally, where assistance is still required, you should follow other industry guidelines such as those of HVCA and CORGI.

Food for Thought
If we looked at the Gas Safety (Installation and Use) Regulations we would find that the law requires an appliance to be installed in accordance with the manufacturer's instructions. The instructions themselves may well indicate that the appliance must be installed in accordance with a relevant BS or industry document. Therefore, in essence, if you fail to install or test an appliance by the methods suggested in the BS, etc., this may well suggest that you are not in compliance with the manufacturer's instructions and therefore the law.

Gas Appliance (Safety) Regulations
This is the standard that all new gas appliances must meet. A supplier must not offer for sale any appliance that does not meet this standard. The provision of this regulation ensures appliances are tested and quality guaranteed and, above all, are safe to use.

To identify that an appliance meets these high standards it is marked with a special logo, no less than 5 mm high, bearing the letters 'CE'. The letters CE stand for the French phrase 'Conformité Europeené', implying that an appliance conforms to European safety legislation, and it is recognised throughout Europe.

CE Mark

Contact details for the following Standards and Guidance Documents are given at the front of this book.

Useful British Standards Applicable to Gas Installation Work
BS EN 498: Dedicated LPG barbecues for outdoor use
BS EN 699: Corrugated metallic flexible hoses for catering appliances
BS EN 1443: Chimneys – general requirements
BS EN 1949: Installation of LPG systems for habitation in leisure accommodation vehicles
BS 1179: Glossary of terms used in the gas industry
BS 1710: Identification of pipelines and services

BS 3016: Pressure regulators and automatic changeover valves
BS 3212: Flexible rubber tubing and hose assemblies for LPG installations
BS 5440 pt. 1 and 2: Installation/maintenance of flues and ventilation $\leq$ 70 kW net
BS 5482 pt. 1, 2 and 3: Installation of LPG in dwellings, caravans and boats
BS 5546: Installation of hot water supplies for domestic purposes using gas $\leq$ 70 kW net
BS 5601: Ventilation and heating of caravans
BS 5864: Installation of gas fired ducted-air heaters $\leq$ 70 kW net
BS 5871 pt. 1, 2 and 3: Installation of gas fires, convector heaters, back boilers and DFE appliances
BS 5854: Flues and flue structures in buildings
BS 5925: Ventilation principles and designing natural ventilation
BS 6172: Installation and maintenance of domestic cooking appliances
BS 6173: Installation of gas catering appliances used in catering establishments
BS 6230: Installation of gas fired forced convection air heaters for commercial heating
BS 6400: Installation of domestic-sized meters $\leq$ 6 m^3/h
BS 6644: Installation of gas-fired hot water boilers between 70 and 2 MW gross
BS 6764: Habitation and stability requirements for leisure accommodation vehicles/homes
BS 6798: Installation of gas-fired boilers $\leq$ 70 kW net
BS 6891: Installation of low pressure gas pipework up to 28 mm in domestic premises
BS 6896: Installation of gas-fired overhead radiant heaters for commercial heating
BS 7624: Installation and maintenance of domestic direct gas-fired tumble dryers $\leq$ 3 kW

Institution of Gas Engineers and Managers (Utilisation Procedure Publications)

IGE/UP/1: Strength testing, tightness testing and purging of commercial installations > 1 m^3
IGE/UP/1A: Strength testing, tightness testing and purging of commercial installations $\leq$ 1 m^3
IGE/UP/1B: Tightness testing and purging of domestic natural gas installations
IGE/UP/2: Gas installation pipework, boosters and compressors in commercial premises
IGE/UP/3: Gas fuelled spark ignition and dual fuel engines
IGE/UP/4: Commissioning of gas fired plant in commercial premises
IGE/UP/6: Application of positive displacement compressors to natural gas fuel systems
IGE/UP/7: Gas installations in timber framed dwellings
IGE/UP/8: Gas installations for caravan holiday/residential homes and permanently moored boats
IGE/UP/9: Natural gas fuel systems to gas turbines and auxiliary fired burners
IGE/UP/10: Installation of gas appliances in commercial premises

Useful LP Gas Association (Codes of Practice publications)

CP1: Installation and maintenance of bulk LPG storage
CP7: Storage of full and empty LPG cylinders/cartridges
CP17: Purging LPG vessels and systems
CP21: Safety checks on LPG appliances in caravans
CP22: LPG piping system design and installation
CP24: The use of LPG cylinders
CP25: LPG central storage and distribution
CP27: The carriage of LPG cylinders in closed vans
CP29: The labelling requirements for commercial LPG cylinders
GN2: A guide to servicing cabinet heaters

Gas Safety (Installation and Use) Regulations

The Gas Safety Regulations, which have been enacted under the Gas Acts, are divided into a number of topics, with each set of regulations having a specific focus. Their subject matter is indicated in the brackets, as follows:

- Gas Safety (Installation and Use) Regulations;
- Gas Safety (Rights of Entry) Regulations;
- Gas Safety (Management) Regulations;
- Gas Safety (Meters) Regulations.

Where this book makes reference simply to the Gas Regulations, this refers to the specific set of regulations dealing with the Installation and Use.

The first set of the Gas Regulations was published in 1972 and it has since undergone many changes. The latest revision is due out in 2005. The Regulations are available either as a Statutory Instrument as laid before Parliament or as a series of 'Approved Documents' as supplied by the HSE.

The Gas Regulations are divided into seven parts as follows:

Part A: General
This part begins by citing the dates on which the new Regulations come into effect. The bulk of this section deals with the general interpretation and application of the regulations and defines specific terms, for example, in relation to any premises. The term 'responsible person' is defined as the occupier or owner of the property and not the gas operative, as is often mistakenly thought. This part also defines those buildings to which the regulations do not apply.

Part B: Gas Fittings – General Provisions
This part deals with the general competency of operatives and/or the working practices that have to be complied with. This section makes reference to employers and self-employed persons being classes of persons approved by the Health and Safety Executive (HSE). The regulations do not state that every operative needs to be approved. However, this comes within the rules of CORGI and all businesses must provide the names of every competent and ACS assessed operative working for them. This section also puts an onus on responsible persons to ensure that those working for them are adequately qualified.

This part does more than define competency, it also goes on to emphasise the need to ensure that gas is not freely discharged from a pipe, and that a naked flame is not used to assist in finding a gas leak.

Part C: Meters and Regulators

This section deals with the positioning and labelling of meters, both primary and secondary. It also deals with the supply regulator and, where applicable, any meter by-pass.

Part D: Installation Pipework

This part of the regulations considers the safe use and location of pipework. It considers bonding and ventilation of voids where necessary. Where pipework is placed within commercial premises it also identifies appropriate marking.

Part E: Gas Appliances

This part of the regulations covers the necessary testing and checks that must be undertaken when working on an appliance and it also includes testing and checking the flue and ventilation. Under Part E, operatives are required to leave the manufacturers' instructions with the consumer, and it is an offence for an operative to take them away on completion of a contract. This ensures that the appliance can be maintained to the specified standards as prescribed by the manufacturer.

Part F: Maintenance

This part of the regulations is quite new and was first introduced in 1998. It places specific duties on the owners of properties within which gas appliances are situated. For example, it requires the landlord of a property to ensure that any appliances installed within their property is inspected for safe operation at no less that 12 monthly intervals. All appropriate records of these checks need to be maintained.

Part F deals mainly with the duties of landlords but it also includes employers and their workplace.

Part G: Miscellaneous

This final part deals with miscellaneous issues, such as how to deal with gas escapes. It also cites regulations that have been revoked or amended.

What is not in the Gas Regulations

It should be remembered that the Regulations do not specify how work should be carried out. They simply identify what should and should not be done. How this is achieved is up to the installer. To assist in this process you need to refer to the various industry documents and manufacturers' instructions, as already mentioned. Nor do the Gas Regulations identify what materials should be used. These may, however, be specified in other legislation, such as the Building Regulations, which also need to be complied with. Particular attention should be made to Approved Document J, which deals with the requirements for flues.

Part 2
Gas Utilisation

Gas: Its Origin

The word 'gas' is derived from the Greek word meaning chaos, possibly because its molecules are continually moving in all directions.

Natural gas is not a new discovery. The use of natural gas is mentioned in China around 900 BC and in about AD 1000 wells were drilled and the gas was piped through bamboo tubes. The discovery of natural gas in Europe was in England in 1659. Unfortunately because of the difficulty of transporting the gas over long distances it remained unused in an economy which was based up coal, oil and electricity. Later a gas was manufactured, known as town gas. It was produced by a process of mixing coal gas and water gas. Coal gas was produced by heating coal to approximately to 1000°C, whereas water gas was produced by passing steam over red-hot coke. This early form of gas had a high proportion of hydrogen (around 45%) and included about 15% carbon monoxide. This resulted in a gas that was both highly explosive and very toxic. However, this gas is now no longer in use, and the gas supplied to a large proportion of the United Kingdom is natural gas.

Natural gas is a gas that has, over millions of years, accumulated beneath the Earth's surface. It is the result of the break down of organic matter in the natural process of degradation. As the gas rises, it becomes trapped by the impervious layers of the Earth as shown in the illustration opposite. There are many types of natural gas, including methane, ethane, propane and butane. However, in general, the term 'natural gas' usually refers to gas that has a high proportion of methane.

Liquefied petroleum gas (LPG) is also found at the oil and gas fields and is obtained from crude oil or as a condensate product of natural gas. A variety of LPGs are produced by the many refineries and the composition varies with the process used. The term LPG applies to the group of hydrocarbon gases that can be liquefied by applying moderate pressures or temperature (see 'Boiling point' on page 22).

Substitute Natural Gas (SNG)
It is possible to manufacturer natural gas, either for adding to an existing supply for peak loads, or as a direct substitute. It is made from other petroleum products, such as LPG or light distillate grade liquid fuels.

Toxicity and Odour
Some gases are toxic or poisonous and inhaling them may result in death. Carbon monoxide (CO) is one such gas and was a constituent of town gas as supplied in the days before conversion to natural gas in the gas distribution main. Fuel gases such as natural gas and propane, etc. are non-toxic as they do not contain any CO, but they can produce it if they are not fully burnt. Natural gases such as methane, propane, etc. are odourless and an odorant, such as diethyle sulphide and ethyl butyl mercaptan, are added at the point of distribution to give the gas a recognisable smell.

Note: CO, when it is produced from natural gas, is an odourless gas.

Constituents of Gases
The natural gases are not commercially used as pure methane or propane, but are mixed with other gases in order to improve their burning quality. The table opposite shows the constituents of the gases.

Typical percentages by volume of the constituents of gases

Constituent	Symbol	Natural gas	LPG:Propane
Methane	CH_4	90.0	–
Ethane	C_2H_6	5.3	1.5
Propylene	C_3H_6	–	12.0
Propane	C_3H_8	1.0	85.9
Butane	C_4H_{10}	0.4	0.6
Nitrogen	N_2	2.7	–
Carbon dioxide	CO_2	0.6	–
		100.0%	100.0%

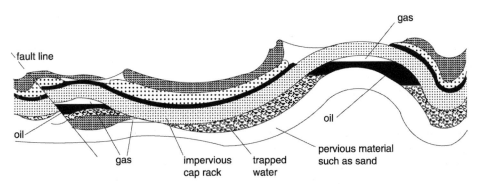

gas

fault line

oil

oil

gas   impervious   trapped   pervious material
      cap rack   water   such as sand

oil

Section through Earth's strata showing Natural Gas traps

Characteristics and Properties of Gases

What is Gas?

All substances are made up of tiny particles, called molecules. In a solid, each molecule is strongly attracted to its neighbouring molecule by what is known as cohesion and there is very little space between each molecule to allow for any movement. In liquids cohesion still exists, but as a result of a change in pressure or temperature, the molecules have more kinetic energy and move about more vigorously, giving the material its fluid properties. The molecule of a gas, on the other hand, have no cohesion and are free from adjoining molecules and can move in any direction.

The three prime gases used in the gas supply industry, namely natural gas and the two types of LPG, have different qualities. The table opposite gives their various characteristics and properties. These values are only a guide as they may vary under different atmospheric conditions. The Quality/Unit column is explained below.

Chemical Formulae

This is the shorthand abbreviation used to identify a material. As mentioned above, substances are made up of molecules, and each molecule in turn is made up of atoms. From the table opposite it can be seen that there is 1 carbon and 4 hydrogen atoms for every 1 methane gas molecule, hence its formula: CH_4. (See previous page for a more accurate list of the constituent parts of natural gas and LPG.)

Boiling point

Gases such as methane (natural gas) are found only in liquid form at extreme temperatures or pressures, far above those in our atmosphere. You can see from the table opposite that, at atmospheric pressure, the temperature would need to be above $-162°C$ before the liquid would change to a gas. This means that $-162°C$ is the boiling point of liquid methane. (Gas vaporisation is explained further on page 26 in the context of LPG.)

Specific Gravity (SG) (also called Relative Density)

This term is used to compare one substance with another. Liquids and solids are compared with water, which has a specific gravity of 1. Those with a specific gravity number less than 1 will float on water, and those with a specific gravity greater than 1 will sink. Gases however, are compared with dry air under the same atmospheric conditions; air has a specific gravity of 1. Therefore, any gas with a specific gravity greater than 1, such as propane, will sink in air; those with a specific gravity less than 1, such as methane, will rise.

Physical characteristics and properties of gases (typical values)

Quality/unit	Natural methane	LPG propane	LPG butane
Chemical formulae	CH_4	C_3H_8	C_4H_{10}
Boiling point	−162°C	−42°C	−2°C
Specific gravity of liquid fuel	−	0.5	0.57
Specific gravity of gas vapour	0.58	1.78	2.0
Gross calorific value	38.5 MJ/m³	95 MJ/m³	121.5 MJ/m³
Gas family	2nd	3rd	3rd
Flammability limits	5–15%	2.3–9.5%	1.9–8.5%
Air/gas ratio	9.81:1	23.8:1	30.9:1
Oxygen/gas ratio	2:1	5:1	6.5:1
Flame speed	0.36 m/s	0.46 m/s	0.38 m/s
Ignition temperature	704°C	530°C	500°C
Max flame temperature	1000°C	1980°C	1996°C
System operating pressure	21 ± 2 mbar	37 ± 5 mbar	28 ± 5 mbar

2 Gas Utilisation

Graham's Law of Diffusion (see page 60) proved that a light gas will mix twice as fast as a gas that is four times its weight. You can experience this when purging a natural gas system, compared with butane or propane installations; you can see from the table above that they are heavier and therefore more difficult to disperse.

Calorific Value (CV) Gross and Net

Gas is sold on the basis of its *gross calorific value*. This figure varies at different times and in different locations due to the gas field supplying the fuel. The gross CV is the amount of heat energy produced by the complete consumption of a known quantity of fuel. The CV for various fuels will be found to differ and one can see, for example, that more heat can be obtained from burning 1 m³ of propane than from burning 1 m³ methane. When the fuel gas is consumed, water is formed as a by-product. When this water condenses from vapour to liquid state, latent heat is given up. This latent heat is not usually available for use as it passes out through the flue system, so it cannot be regarded as part of total heat energy available. Therefore it is not counted when quoting energy efficiencies. Appliance are therefore invariably specified as a net CV.

Gas Family

The amount of heat given off from a burner will be dependant on a number of factors, which can be divided into two groups:

- Group 1 relates to the SG;
- Group 2 depends on the size of the injector and pressure of the gas within the system.

The factors in Group 2 are dependant on the design or adjustment of the appliance, whereas those in Group 1 are dependant on the properties of the gas being supplied. These two characteristics can be linked to give what is known as a 'Wobbe number' that, in turn, is used to define the international family of gases into which the fuel can be categorised. The Wobbe number is calculated as $(CV \div \sqrt{SG})$.

Example Natural gas with a CV of 38.6 and SG of 0.58 would have a Wobbe number of: $38.6 \div \sqrt{0.58} = 50.68$.

Knowing the Wobbe number, we can see from the table below that this is natural gas.

Families of gases

Family	Gas type	Approx. Wobbe no.
1st	Manufactured (town gas)	24–29
2nd	Natural	48–53
3rd	LPG	72–87

Flammability Limits (explosive limits)

If one could open a matchbox full of gas in a room, wait a second, then try to ignite the gas, nothing would happen: there is too much air. Conversely, if you strike a match in a room that is completely full of gas, again no combustion will occur: there will not be enough air. The gas will only ignite if there is a certain percentage range of air/gas mixture; this is referred to as its 'flammability limits'. The table on the previous page gives the lower explosive limit (LEL) for natural gas as 5% and the higher explosive limit (HEL) as 15%, therefore it will only burn within this range (5–15%).

Air/Gas Ratio

As we have seen, air needs to be mixed with the gas in order for it to burn. The flammability limits defined for natural gas suggest a mixture of 5–15% gas. However, how much air is required for all the fuel to be consumed? Some air may be present for combustion to occur, but is it sufficient? If there is insufficient air combustion will be incomplete and CO will be produced (see page 29). This is what is meant by the term 'air/gas ratio'. From the table on the previous page we see that natural gas needs 9.81 volumes of air for complete combustion of 1 volume of gas. (See also page 31, for an explanation of why more air is required for LPG.)

Oxygen/Gas Ratio

This is basically very similar to the air/gas ratio above. However, air is a mixture of gases: approximately 80% nitrogen and 20% oxygen (i.e. one-fifth oxygen). The nitrogen plays no part in the combustion process and so can be ignored here.

Rounding up the numbers, natural gas has an air/gas ratio of 10:1. Only 1/5 of the gas in air is oxygen and we know that 1/5 of 10 is 2. Therefore we can estimate that the required oxygen/gas ratio is 2:1. To put it simply, twice as much oxygen is required than natural gas to ensure complete combustion.

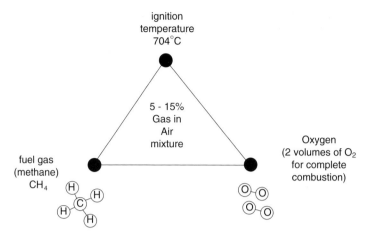

Essentials for Combustion

Flame speed

As fuel burns, the flame travels through the gas/air mixture at a set speed. This can be illustrated by watching a piece of paper burn and observing the speed at which the flame travels. In gases this speed is not so easily observed, but nevertheless the same phenomena occurs. With gases, such as hydrogen, the flame speed is so great it breaks the sound barrier with a loud bang. Burner pressures should be carefully adjusted to ensure a flame speed that keeps the flame at the head of the burner. If the pressure was too great, it would eject the gas faster than it could be consumed, resulting in a flame lift off. On the other hand, insufficient pressure would result in the flame burning back inside the mixing tube.

Ignition Temperature

Fuel will not burn if the ignition temperature is not reached. For example, paper needs to be heated to around 80–100°C before it can be consumed by fire. To quench a fire, the fire brigade pour on water. For this, water at around 6°C is used to cool the solid material. Gases also need to reach a sufficient temperature for combustion and for natural gas we can see from the table on page 23 that this temperature is around 704°C. When the gas is ignited by a match, ignition spark, etc., rapid heat transfer through the fuel soon enables it to reach the required ignition temperature.

System Operating Pressure

The system operating pressure is the pressure that the gas is regulated down to for all domestic dwellings and most commercial installations. The actual gas pressure of the regional grid network supplying the homes would be considerably higher, see page 42.

Liquefied Petroleum Gas (LPG)

There are two main types of liquefied petroleum gas (LPG): butane and propane. They are only used in their pure form for test purposes, and are supplied as commercial butane or commercial propane, which are a mixture of the gas with other gases added to improve the working characteristics.

The key difference between LPG and natural gas is that when a modest pressure is applied to the gas and it becomes a liquid. This makes it possible to store large quantities of the fuel in specially constructed containers.

Gas Vapourisation

In order to convert a liquid to a gas, heat is required. You can see this when you boil a kettle of water to make a cup of tea. The liquid heats up and bubbles and steam rises. When the LPG cylinder on the pressurised LPG vessel valve is opened, the liquid released is subject to atmospheric pressure. It can be seen from the table on page 23 that butane, for example, boils at –2°C. If the valve were opened in a very cold climate, nothing would happen. At a temperature of, say, 16–20°C, on a typically UK day, the heat from the atmosphere would rapidly boil the liquid and the vapour would be drawn off. The greatest transference of heat is from the inside wetted metal surface that is in contact with the liquid gas, so air is free to circulate around the vessel or cylinder. The photographs and illustrations opposite show liquid gas boiling.

When calculating the gas requirements for a particular supply, it is important to remember that it is the size of the vessel in which the LPG is contained and not the volume of liquid that determines the amount of gas that can be generated per hour. This is because larger vessels have a larger wetted surface area.

Container Pressure

The pressure within the vessel containing LPG varies, depending on the type of gas. Typical pressures for propane are 6–7 bar and for butane they are 1–2 bar. The containers are protected from over pressure that could occur due to excessive expansion caused by, for example, exposure to heat by a safety pressure relief valve. This forms part of the outlet valve on a cylinder where a cylinder is used.

The vessels in which the LPG is supplied are never filled to more than 80–87% and a void, called an 'ullage space', is left to allow for any expansion of the liquid caused by changes in atmospheric pressure or temperature. This space also allows for compressed vapour to form above the liquid level in readiness for use.

Vessel used to demonstrate the liquid butane boiling

When the valve is opened the vessel is subjected to atmospheric pressure. As the liquid undergoes the physical change from liquid to gas it boils, taking its heat from the area surrounding the vessel.

Close up photo of the container showing the LPG bubbling as it boils

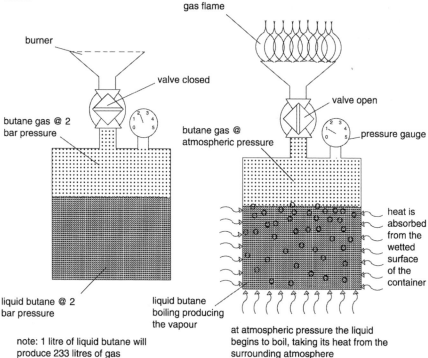

gas flame

burner

valve closed

butane gas @ 2 bar pressure

valve open

butane gas @ atmospheric pressure

pressure gauge

liquid butane @ 2 bar pressure

liquid butane boiling producing the vapour

heat is absorbed from the wetted surface of the container

at atmospheric pressure the liquid begins to boil, taking its heat from the surrounding atmosphere

note: 1 litre of liquid butane will produce 233 litres of gas & 1 litre of liquid propane will produce 274 litres of gas

Inside view of vessel containing LPG

The Combustion Process

Combustion of a fuel involves a chemical reaction; it produces heat as the fuel changes into a new compound, just as heat is produced within a compost heap. A compound is a substance with more than one kind of atom. For example, a hydrocarbon is a compound of carbon (C) and hydrogen (H). The proportion of carbon to hydrogen varies depending on the hydrocarbon: methane has 1 carbon atom to 4 hydrogen atoms (CH_4), whereas propane has 3 carbon atoms and 8 hydrogen atoms (C_3H_8).

Hydrocarbons react with oxygen in the presence of heat to undergo a chemical change, so producing a new by-product. When a piece of wood is burnt, wood being a hydrocarbon, we see the wood slowly disappear leaving nothing but a pile of ash. The process of combustion is misunderstood as being to destroy and consume. In reality, during the combustion process, the wood or gas, etc. is not consumed; it is simply converted to another form.

The Complete Combustion of Methane

Methane (CH_4) will burn when in the presence of oxygen (O_2). Where the supply of oxygen is unlimited it uses 2 volumes of oxygen to every volume of fuel gas supplied. From the following illustration you can see that during combustion the molecular structure of the gases changes into a new form. They are still gases and they still contain the same number and types of atoms. The difference is that these newly formed gases, in this case carbon dioxide (CO_2) and water vapour (H_2O), have new and different qualities. CO_2 and H_2O are both non-toxic gases. This is often expressed as:

$$CH_4 + 2\,O_2 = CO_2 + 2\,H_2O.$$

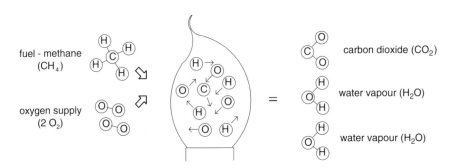

The Combustion Process

Incomplete Combustion

The gas flame is the visual chemical reaction that occurs when the fuel gas ignites with oxygen to form new compounds. As the oxygen and gas diffuse, the transition from the hydrocarbon to the resultant CO_2 and H_2O is not instantaneous. The heat generated by the flame causes the original constituents to break down, slowly forming different compounds, such as alcohols (CH_3OH), aldehydes (HCHO), free carbon (C) and carbon monoxide (CO). In the presence of sufficient oxygen, these hydrocarbons will continue to burn until complete combustion has taken place. However, if the process is interrupted through lack of oxygen, or the flame temperature falls below its ignition temperature, the un-burnt gases will be discharged into the atmosphere. For example, in the combustion of methane, if the available oxygen is restricted by just 5% (i.e. 1.9 volumes O_2 instead of 2), we would begin to see levels of carbon monoxide (CO) being produced and if the O_2 were further restricted, the level of CO would rise:

$$CH_4 + 1.9\,O_2 = 0.05\,CO + 0.9\,CO_2 + 1.9\,H_2O + 0.05\,CH_3OH$$

Dangers of Carbon Monoxide (CO) and the Effects of CO in the Air

CO in the environment can have a devastating effect on human health. For most of us it would be the most dangerous gas we might ever experience. It is often referred to as the 'hidden gas' because it is very hard to detect. However, the effects of CO are quite dramatic and in high concentrations it will kill within a few minutes, as can be seen in the table below.

When we breathe our lungs absorb oxygen. However, carbon monoxide is absorbed more readily into the blood stream than oxygen and when our haemoglobin, or red blood cells, become saturated with CO no oxygen can be absorbed. One of the functions of oxygen is to remove waste matter from our tissues, so without oxygen, our blood rapidly become poisoned.

The effects of CO in the air

CO in air	Saturation of CO in blood	Effects of CO in adults
0.01%	0–15%	Slight headache after 2–3 hours
0.02%	1–30%	Mild headache, feeling sick and dizziness after 2–3 hours
0.05%	30–50%	Strong headache palpitations and nausea within 1–2 hours
0.15%	50–55%	Severe headache, nausea and dizziness within 30 minutes
0.3%	55–60%	Severe headache, nausea, and dizziness within 10 minutes; increased breathing and convulsions, leading to collapse, possible death after 15 minutes
0.6%	70–75%	Severe symptoms within 1–2 minutes, death within 15 minutes
1.0%	85–90%	Immediate symptoms; death will occur within 1–3 minutes

Air Requirements for Combustion

The table below gives an indication of the amount of oxygen required to consume completely one volume of a gas.

Oxygen requirement to provide complete combustion of 1 volume of gas

Fuel		Volume of oxygen
Methane	CH_4	2.0
Ethane	C_2H_6	3.5
Propylene	C_3H_6	4.5
Propane	C_3H_8	5.0
Butane	C_4H_{10}	6.5

To calculate the oxygen requirements and, consequently, the air requirement for complete combustion of a specific fuel gas comprising a mixture of gases, such as natural gas, first list the constituents in order of percentage volume, taken from the table on page 21. Then multiply each percentage by the volume required for the individual fuel to obtain the oxygen required by each individual gas, from the table above. Adding these gives the total amount of oxygen required.

O_2 Requirement for natural gas

Constituent		%		Volume of O_2 required		
Methane	CH_4	90.0	×	2.0	=	180.0
Ethane	C_2H_6	5.3	×	3.5	=	18.55
Propane	C_3H_8	1.0	×	5.0	=	5.0
Butane	C_4H_{10}	0.4	×	6.5	=	2.6
Carbon dioxide	CO_2	0.6	×	0	=	0.0
Nitrogen	N_2	2.7	×	0	=	0.0
		100.0				206.15

Note: No consumption of the N_2 or CO_2 takes place as these are not fuel gases.

Therefore, as can be seen, 100 m³ of natural gas would require 206.15 m³ of O_2. This equates to 1 m³ of natural gas needing 2.062 m³ of O_2.

In round figures, 2 volumes of O_2 are needed to consume 1 volume of natural gas.

In most cases the oxygen supply is taken from the air surrounding the appliance. There is approximately 21% oxygen in the atmosphere, therefore $2.06 \times (100 \div 21)$ = 9.81 m³ of air would be needed for complete combustion of 1 m³ of natural gas.

In round figures, 10 volumes of air are needed to consume 1 volume of natural gas.

O$_2$ requirement for propane gas

Constituent		%	Volume of O$_2$ required		
Ethane	C$_2$H$_6$	1.5	×	3.5 =	5.25
Propylene	C$_3$H$_6$	12.0	×	4.5 =	54.0
Propane	C$_3$H$_8$	85.9	×	5.0 =	429.5
Butane	C$_4$H$_{10}$	0.6	×	6.5 =	3.9
		100.0			492.65

This equates to 1 m^3 of propane needing 4.926 m^3 of O$_2$ for complete combustion.

In round figures, 5 volumes of O$_2$ are needed to consume 1 volume of propane gas.

This means that 4.93 × (100 ÷ 21) = 23.47 m^3 of air are needed for combustion.

In round figures, 24 volumes of air are needed to consume 1 volume of propane gas.

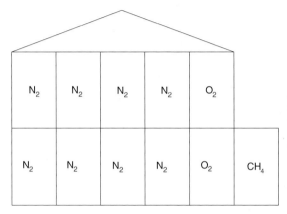

Imagine a house full of air. It would all be required to consume the volume of natural gas in the extension. This same amount of air would consume less than half the gas if it was propane, hence the need for good ventilation.

Lots of Air is Required for Combustion!

Ventilation Requirements for Natural gas and LPG Systems

It will be noted that LPG requires more air for combustion. Yet when calculating the size of ventilation grille needed, for a permanent dwelling, the same size grille is used irrespective of the type of gas used. The size of the ventilation does not need to be larger for the LPG installation because, for the same heat input, less gas is consumed, i.e. the calorific value of natural gas is 38.5 MJ/m^3, whereas that for propane is 95 MJ/m^3.

To obtain 1 kW of heat from natural gas, the following amount of fuel is consumed:

$$1 \text{ kW} \times 3.6 \div \text{CV} = 1 \times 3.6 \div 38.5 = 0.094 \text{ m}^3 \text{ of gas.}$$

To obtain 1 kW of heat from propane gas, the following amount of fuel is consumed:

$$1 \text{ kW} \times 3.6 \div \text{CV} = 1 \times 3.6 \div 95 = 0.038 \text{ m}^3 \text{ of gas.}$$

So although LPG requires more air, less gas is consumed!

Products of Combustion

The products of complete combustion, as previously stated, are carbon dioxide (CO_2) and water vapour (H_2O). Their quantities can be calculated by making a table for each of the fuel gases. *Note:* The percentage by volume for the constituents was given in the previous section.

Products of combustion for natural gas with sufficient O_2 supplied

Constituent percentage by volume %		Ratio of CO_2 and H_2O produced by complete combustion of fuel		Total products of combustion*	
		CO_2/Vol.	H_2O/Vol.	CO_2	H_2O
Methane CH_4	90	1	2	90	180
Ethane C_2H_6	5.3	2	3	10.6	15.9
Propane C_3H_8	1	3	4	3	4
Butane C_4H_{10}	0.4	4	5	1.6	2
CO_2	0.6			0.6	
N_2	2.7				
	100			105.8	201.9

*For example: the total CO_2 for ethane is found thus 5.3 (% vol.) × 2 (CO_2/Vol.) = 10.6; the total H_2O for ethane is found thus 5.3 (% vol.) × 3 (H_2O/Vol.) = 15.9.

Therefore if 1 m³ of natural gas was consumed, the total products of combustion and ultimately the total volume of gases leaving the appliance would be as follows:

Carbon dioxide	1.058
Water vapour	2.019
Nitrogen in gas	0.027
Nitrogen in air (9.81–2.06)*	7.750 *From the previous section.*
	10.854 or approx. 11m³

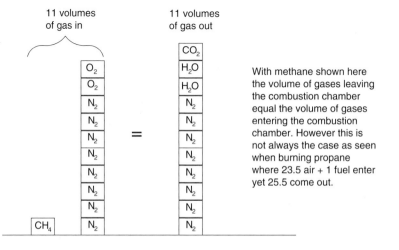

11 volumes of gas in = 11 volumes of gas out

With methane shown here the volume of gases leaving the combustion chamber equal the volume of gases entering the combustion chamber. However this is not always the case as seen when burning propane where 23.5 air + 1 fuel enter yet 25.5 come out.

Products of combustion for natural gas with sufficient O_2 supplied

Constituent percentage by volume %		Ratio of CO_2 and H_2O produced by complete combustion of fuel		Total products of combustion*	
		CO_2/Vol.	H_2O/Vol.	CO_2	H_2O
Propane C_3H_8	85.9	3	4	257.7	343.6
Propylene C_3H_6	12	3	3	36	36
Ethane C_2H_6	1.5	2	3	3	4.5
Butane C_4H_{10}	0.6	4	5	2.4	3
	100			299.1	387.1

For 1 m³ of propane gas consumed the total products of combustion and ultimately the total volume of gases leaving the appliance would be as follows:

Carbon dioxide	2.991	
Water vapour	3.871	
Nitrogen in air (23.46–4.926)*	18.534	*From the previous section.
	25.396	or approximately 26 m³

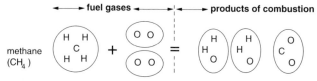

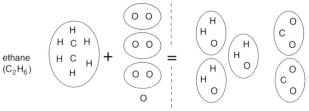

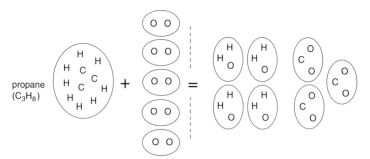

Typical Combustion Equations   note how the number of atoms one side of the equal sign equals those on the other

The Gas Flame

The gas flame can have many different shapes and colours, each designed for a particular purpose; however these flames generally fall into one of two categories: post-aerated or pre-aerated.

The Post-aerated Flame (non-aerated flame, neat flame or luminous flame)
A flame in which no air has been mixed with the gas prior to combustion. This type of flame is rarely used in modern gas supply systems because of its unstable characteristics.

The Pre-aerated Flame (aerated flame)
This is the type of flame and burner design used in most modern gas appliances. Air is drawn in by natural draught or using a fan to mix with the gas prior to combustion. The premixed air and gas at this stage is called primary air and it makes up some 40–50% of the air that is needed The additional air that is needed to support complete combustion is obtained from the air that surrounds the flame and is known as secondary air. Forced draught pre-mix burners draw all the air that they need prior to the combustion process and no secondary air supply is required.

Burners that rely on natural draught draw the air into the mixing chamber. The flow of gas is restricted as it passes through an injector, forcing it to flow with increased velocity, and so creating a negative pressure at the primary air-port, which causes it to draw in the required air.

Both the post-aerated and pre-aerated flames produce the same amount of heat output per volume of gas consumed. However, the temperature of the pre-aerated flame is much higher due to the smaller intensity of the flame. When all the air is forced, as with the pre-mix burner, a smaller intensely hot flame results.

The Flame Front
The gas–air mixture flows through the mixing chamber and is finally discharged at the burner ports where it is ignited. The point where the unburnt gas ends and the actual flame begins is referred to as the 'flame front'. At this point the gas speed equals the flame speed. The shape of the flame front varies slightly depending on the design of burner. It is often cone shaped because the friction of the gas against the burner sides slows the gas, so that it travels faster at the centre, pushing the flame front away from the burner head. The primary air effects, the shape of the flame front, and a small amount of air produces a long flame front because the gas maintains a high speed at its centre. On the other hand, when large volumes of primary air are added, the flame front becomes flatter and noisier.

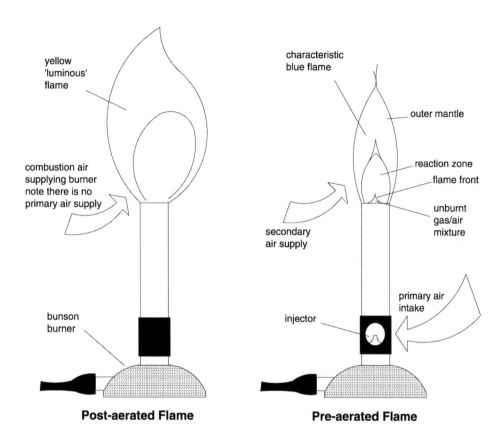

Post-aerated Flame

Pre-aerated Flame

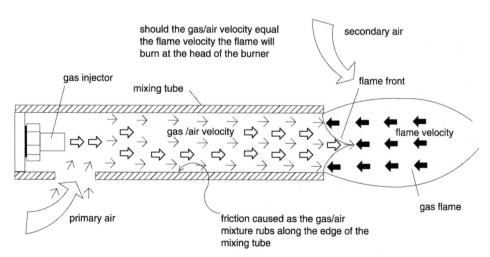

Gas/Air Speed = Flame Speed

Flame Pattern and Characteristics

A well-adjusted flame can be recognised by its colour: a bright blue–green inner cone and an outer flame that is slightly darker with a bluish–purple tinge. Such a flame is stable and there is a constant flow as the gas leaves the burner ports. Problems with a gas flame can often be identified by observing and understanding the flame pattern. The pictures opposite illustrate the effect that a wrongly fitted or blocked injector will have on the flame pattern. Similarly, should the primary air-port become blocked or damaged it will also affect the flame characteristics. In both cases, the amount of primary air intake is affected causing incomplete combustion of the fuel. A greater understanding of these simple observations illustrates the need for servicing and maintenance checks on appliances.

Flame Lift and Lighting Back

The speed at which the gas–air mixture leaves the burner ports should match the flame speed, so producing a flame front at the burner head. However, where there is a difference in gas/air speed and flame speed caused, for example, by incorrect pressure adjustment, the flame will either try to lift away from the head of the burner or lighting back down inside the mixing tube to the injector. The latter is caused by the reduction in gas/air speed to below the flame speed. In both cases incomplete combustion will result. Gases that burn with a slow velocity, such as natural gas and LPG, seldom suffer lighting-back problems. However, the same cannot be said for flame lift, where high gas pressure will soon disrupt the stability of the flame.

Retention Flames

Many burners have very small burner port-holes, referred to as retention ports, adjacent to the main burner ports, through which the gas flow is restricted, resulting in a small stable flame. Should a main burner flame try to lift off from the burner head, these stable retention flames continually re-ignite the gas/air mixture, holding it securely in place. Cold air from the room feeds secondary air to the burner. This air is usually at about 21°C, and it tends to cool the flame below the ignition temperature required (see page 25). The retention flame also helps to increase the temperature of the gas/air mixture at this point, again assisting in keeping the flame stable.

When servicing an appliance, care needs to be taken to maintain the condition of the burner head to ensure that these retention ports remain clear and undamaged.

Flame Chilling

Flame chilling occurs when the gas is cooled below its ignition temperature of between 500 and 700°C (see page 23). This leads to the flame going out before complete combustion has occurred. It can be caused by flame impingement, where the flame touches the heat exchanger, which although hot, it is cooler than the ignition temperature of the gas. Where flame impingement occurs, extensive sooting or carbon build-up will result.

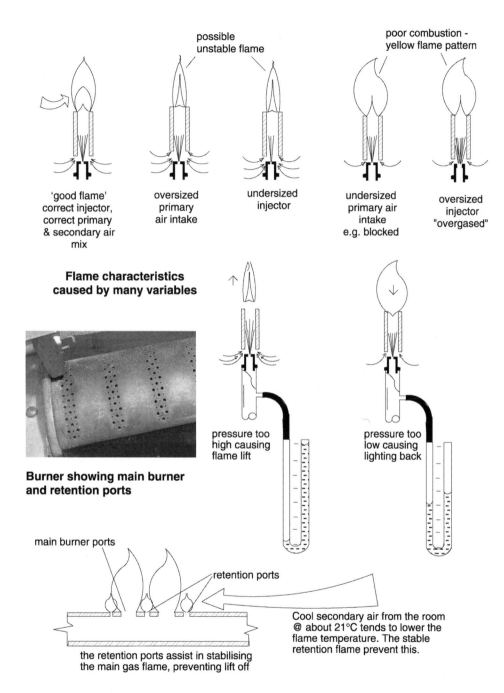

possible
unstable flame

poor combustion -
yellow flame pattern

'good flame'
correct injector,
correct primary
& secondary air
mix

oversized
primary
air intake

undersized
injector

undersized
primary air
intake
e.g. blocked

oversized
injector
"overgased"

**Flame characteristics
caused by many variables**

**Burner showing main burner
and retention ports**

pressure too
high causing
flame lift

pressure too
low causing
lighting back

main burner ports

retention ports

the retention ports assist in stabilising
the main gas flame, preventing lift off

Cool secondary air from the room
@ about 21°C tends to lower the
flame temperature. The stable
retention flame prevent this.

Atmospheric Burners

There are many designs of atmospheric burner, but they all operate with the same basic components: injector, primary air port, mixing tube and burner.

The Injector

This is usually made from brass with a central hole or series of holes through which the gas can pass. The pressure within the gas pipe is a few millibars above atmospheric pressure, so as the gas passes out through the relatively small hole it moves at about 45 m/s, causing air trapped in its wake to be drawn in at the primary air port. The size and design of the injector controls the amount of heat input to an appliance as indicated by the manufacturer, therefore it is essential that the correct injector is selected for a specific burner as an incorrectly sized injector could have fatal consequences. During maintenance or when cleaning an injector, care needs to be taken to ensure that its hole does not become enlarged. For example, should an injector with a 5 mm diameter hole become enlarged to, say, 6 mm it would supply over 40% more gas, which is nearly half as much again as that intended.

Primary Air Port

The primary air port is located at the point where the injector leaves the pipeline, usually to the side or bottom. The size of this hole is generally fixed, however sometimes some form of aeration control is incorporated to control the amount of primary air drawn into the burner by way of an air shutter or throat restrictor.

Mixing Tube

This is the chamber between the primary air port and the burner within which the gas and air can premix prior to combustion. Some mixing tubes, particularly those where there may be a resistance to flow, incorporate a venturi, which provides a more consistent pressure to the burner.

Burner

The burner provides the base on which the flames can burn with the required shape and heat input. The shape of the burner varies depending on the appliance in which it is installed, e.g. round for a cooker and a long box or bar inside a boiler. The burner is designed to produce an even flame distribution and baffles are often put in to serve this purpose. Sometimes, a gauze is incorporated with bar burners to even out internal pressures. This also has the effect of preventing light back, as the flame will not travel back through the gauze. The burner ports are spaced to allow secondary air to reach all the flames, yet must be close enough to allow cross-lighting from an igniter situated at one end. To prevent flame lift, retention ports are sometimes incorporated as was explained on the previous page. Some burners, referred to as a ribbon burners, use alternating strips of corrugated and flat metal.

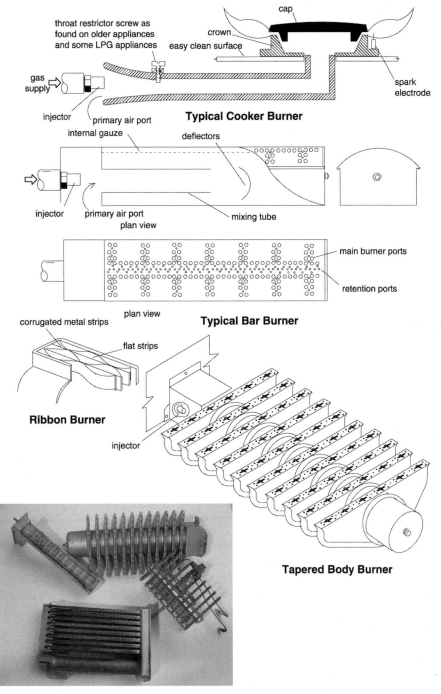

throat restrictor screw as found on older appliances and some LPG appliances

cap

crown

easy clean surface

gas supply

spark electrode

injector

primary air port

Typical Cooker Burner

internal gauze

deflectors

injector

primary air port

plan view

mixing tube

main burner ports

retention ports

plan view

Typical Bar Burner

corrugated metal strips

flat strips

Ribbon Burner

injector

Tapered Body Burner

Selection of Atmospheric Burners

Pre-Mix Burners

These burners, unlike atmospheric burners, do not have a primary and secondary air supply to the flame; all air is premixed prior to combustion. As a result the flame characteristics differ in that no outer mantle can be seen as all the gases are fully consumed by the air supplied through the point of intake. These flames tend to be more compact and, as a result, the temperature of the flame is higher. They can therefore have a smaller combustion chamber without risk of impingement and flame chilling (see page 36). For a commercial application it may be possible that all the air supply could be induced into the burner by raising the pressure of the gas supply. However, a fan that is adjusted to deliver a predetermined flow rate is usually used. The fan may be placed either upstream of the burner, in which case it is called a 'forced draught burner', or downstream, where it would be referred to as an 'induced draught burner'.

Forced Draught Burner

These burners are found typically on industrial appliances and in large domestic boilers. A centrifugal fan is used to blow air past the injectors, which are located centrally upstream of the burner ports. As the gas emerges from the injector it initially premixes with the air before entering the burner ports. Here a flame pattern is created that interacts with the discharge from other burner ports, so preventing flame lift. The position of the burner head in the housing can be adjusted to enable the length of the flame to be adjusted and varies the rate of mixing in order to obtain the best possible combustion efficiency.

Most modern systems package all the components in one unit, called a package burner. However the controls may be fitted on to the pipe that serves the burner, in which case it is called a gas train. This design of burner has an automatic control sequence that prevents unburnt gases from being pre-ignited within the combustion chamber and may follow a typical format as shown on the page opposite.

Induced Draught Burner

With this type of burner the fan serves two purposes: the first is to draw the combustion air into the burner and the second, to remove the products of combustion from the heat exchanger. Gas enters the burner assembly and mixes with the air drawn in by the combustion fan before passing through the burner head.

Tunnel Burners

These are specialist burners, associated with high temperature work such as in a furnace or where rapid heating is required. The gas/air mixture burns within a tunnel of refractory material. The tunnel may form part of the burner design or may be part of the wall furnace itself. The flame is directed towards whatever is to be heated.

Note: At any time, if the correct sequence is not fully adhered to the burner will shut off the appliance. Some burners will conclude with a post purge as the appliance shuts down.

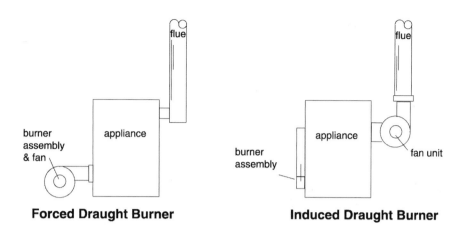

Forced Draught Burner   **Induced Draught Burner**

**Typical Forced Draught/
Pre-Mix Burners**

Stage	Typical ignition sequence for a force draught burner
1	The air-proving device is checked and all contacts confirmed to be in the correct position prior to any air flow.
2	A pre-purge sequence begins and runs for 30 seconds to clear the combustion chamber of any remaining combustion products. Also a check is made for the presence of a flame that, if detected, will cause the burner to go into lockout.
3	A spark is produced at the burner head whereupon the control box confirms its presence for a minimum period of 5 seconds.
4	Pilot gas is introduced and the flame stability is checked for a further 5 seconds, whereupon the spark would go out.
5	Main gas burner valve opens and full ignition takes place, the pilot gas terminates and flame monitoring continues throughout the run period.

*At any time where the correct sequence is not fully adhered to the burner will shut the appliance off. Some burners will conclude with a post purge as the appliance shuts down.

Gas Pressure and Flow

As previously stated, natural gas in the United Kingdom comes from underground gas fields beneath the North Sea. The gas pressure there is in the region of 100 bar. Large volumes of gas are distributed through a national gas transmission system and pass into the regional grid network at pressures ranging from 7 bar down to 30 mbar. It arrives in the customers' property at pressures of 21 ± 2 mbar and is finally used at pressures ranging from 3 to 20 mbar at the domestic appliance. The pressures at the various stages of gas distribution are as shown below.

	High pressure	Intermediate pressure	Medium pressure	Low pressure
Maximum	>7 bar	7 bar	2 bar	75 mbar
Minimum	7 bar	2 bar	75 mbar	30 mbar*

*This pressure may be as low as 19 mbar where the supply cannot be maintained.
All pressure <75 mbar is regarded as low pressure.

The pressure of an LPG supply depends on the type of gas being used, i.e. propane or butane, and because it is supplied and stored on site in liquid form its pressure is subject to the ambient pressure and temperature conditions and is therefore variable. However, typical pressures within a propane vessel would be 6–7 bar and for butane 1.5–2 bar. The gas supply within a building is reduced to 37 mbar $\pm$ 5 mbar for propane and 28 $\pm$ 5 mbar for butane.

Gas Flow in Pipes

When the gas within a pipe is static, the pressure is the same throughout its entire length. However, as soon as the gas begins to flow, the pressure falls progressively from the entry point to the exit. This is called 'pressure absorption' or 'pressure loss' (pressure drop). This is a result of the friction that takes place as the gas rubs along the pipe wall. The rate at which the pressure drops is constant, provided that the pipe diameter and material are the same throughout the length. The illustration opposite shows four pressure gauges fitted equidistantly along a 9 m pipe. When no gas is flowing, the pressure throughout the pipe is the 'standing pressure'. However, when the gas valve is opened, the gas pressure readings are seen to drop as the gas progresses through the pipe; this is the 'working pressure'. (See page 46 for a method of taking standing and working pressure readings.)

Factors Affecting Pressure Absorption

Length: With double the length of pipe, the pressure loss is doubled. Conversely if its length is halved, the pressure loss would be halved.

Diameter: The pressure loss is inversely proportional to the fifth power of the pipe diameter (d^5). In simple terms, if you halve the diameter of a pipe, its pressure would increase 32 times.

Material: A smooth pipe wall will generate less friction than a coarse surface.

Quantity of gas: The amount of pressure loss is proportional to the square of the volume of gas flowing (m^3)^2. In simple terms, double the quantity and the pressure loss increases four-fold.

Specific gravity: Double the weight of the gas and the pressure loss will be doubled. Conversely, halve its weight and the pressure loss will be halved.

Photo showing gas flow diminishing due to undersized pipe

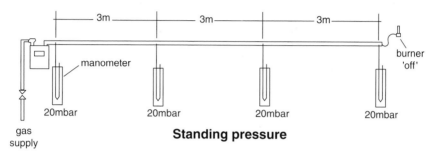

Standing pressure

Working pressure

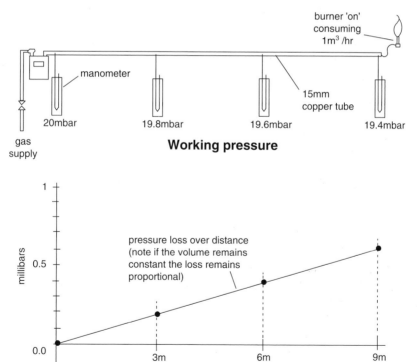

Illustration of 1m^3 of gas flowing through 15mm copper tube showing the pressure absorption (pressure loss)

Gas Pressure Readings

The gas pressure in an installation varies and can be identified in several ways as shown below.

Maximum Incidental Pressure (MIP)
This is the maximum pressure that could be experienced within a gas installation under fault conditions. For example, where a gas booster is installed, this would be the MIP. Alternately it could be where a supply regulator that feeds the installation malfunctions. If the maximum operating pressure of the service enters the installation it becomes the MIP of the supply.

Maximum Operating Pressure (MOP)
This is the maximum pressure experienced within a gas installation under normal operating conditions. This pressure is determined by the regulator that separates the installation from the service. For a typical building using natural gas this pressure would not exceed 30 mbar (See page 42).

Operating Pressure (Working Pressure)
This is the usual pressure experienced within a gas installation under normal operating conditions. For a natural gas installation, at the point of entry to the building or at the meter this would typically be 21 ± 2 mbar. The pressure would progressively drop throughout the length of the system due to pressure absorption or pressure loss, described on the previous page.

Maximum Pressure Absorption Across the Installation (Pressure Loss/Drop)
In order to comply with the appropriate standards, the maximum pressure absorption in a gas installation, after the primary meter, must not exceed 1 mbar for natural gas installations or 2.5 mbar where LPG is to be used. Where these pressures are exceeded the pipe would be considered undersized.

Pressure Absorption Across the Natural Gas Meter
The minimum pressure that the gas supplier must provide at the outlet of the emergency control valve, under peak flow conditions, is 19 mbar. The pressure absorption across the primary meter must not exceed 4 mbar at maximum rate of flow, which would provide a minimum of 15 mbar at the meter outlet. A further maximum of 1 mbar drop could occur across the pipework, leaving only 14 mbar at the appliance – a pressure at which manufacturing standards permit the appliance to operate safely. Ideally, we look for 21 ± 2 mbar at the meter and need to report any shortfall to the supplier, but they may be unable to improve on the pressure of the gas supplied.

Appliance Operating Pressure (Burner Pressure)
This is the pressure at which an appliance is designed to operate. This pressure may be the supply pressure or it may be regulated down to a fixed or variable pressure,

as deemed suitable by the manufacturer. Often appliances are range rated, which requires the regulator to be adjusted up or down to suit a particular gas input.

Standing Pressure

The term 'standing pressure' relates to the pressure in a pipe when no gas is flowing. This pressure would not normally exceed the MOP identified above.

See over for a method of taking pressure readings.

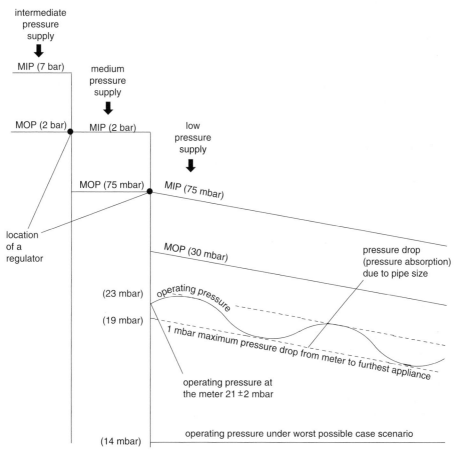

Illustration Showing Different Pressures Within a Pipeline

Quick Reference Guide to Taking Pressures

Standing Pressure Test
- Turn off the supply and connect a manometer to the system at a suitable test point. Ensure the gauge is calibrated to zero.
- 'Slowly' open the supply, allowing gas into the system.
- With no appliances operating the standing pressure can be read. [Maximum pressures: 30 mbar for natural gas, 47 mbar for propane and 38 mbar for butane gas.]
- Close supply, remove the manometer and reseal the test point.
- Turn on the supply and test the joint with leak detection spray.

Meter Regulator Outlet Working Pressure
- Turn off the supply and connect a manometer to the test point at the meter. Ensure the gauge is calibrated to zero.
- 'Slowly' open the supply, allowing gas into the system.
- Operate an appliance to provide a sufficient gas flow of at least 0.5 m^3/h. This would typically be an appliance on 'full' running in excess of 6 kW. [The pressure should be 21 ± 2 mbar for natural gas, 37 ± 5 mbar for propane and 28 ± 5 mbar for butane.]
- Close supply, remove the manometer and reseal the test point.
- Turn on the supply and test the joint with leak detection spray.

Appliance Inlet Working Pressure (Maximum Working Pressure Drop)
- Turn off the supply and connect a manometer to a suitable test point at the appliance, prior to any regulator. Ensure the gauge is calibrated to zero.
- 'Slowly' open the supply, allowing gas to the appliance.
- Turn on this gas appliance and any other gas appliances on the section being tested. [For natural gas the pressure should be no more than 1 mbar below that read at the meter. For LPG the maximum drop should not exceed 2.5 mbar.]
- Close supply, remove the manometer and reseal the test point.
- Turn on the supply and test the joint with leak detection spray.

Appliance Operating Pressure (Burner Pressure)
- Turn off the supply and connect a manometer at the appliance, after any regulating device. Ensure the gauge is calibrated to zero.
- 'Slowly' open the supply and turn on the appliance to allow gas to flow.
- The gas pressure is observed and should be that recommended by the manufacturer.
- Close supply, remove the manometer and reseal the test point.
- Turn on the supply, run appliance, and test the joint with leak detection spray.

Note: When using a 'U' gauge, to ensure accuracy both columns should always be read.

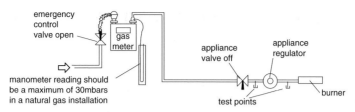

emergency
control
valve open

gas
meter

appliance
valve off

appliance
regulator

manometer reading should
be a maximum of 30mbars
in a natural gas installation

test points

burner

Standing Pressure
(no gas flowing)

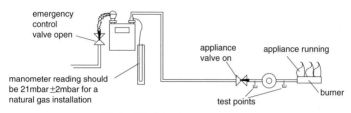

emergency
control
valve open

appliance
valve on

appliance running

manometer reading should
be 21mbar ±2mbar for a
natural gas installation

test points

burner

Meter Working Pressure
(gas flowing)

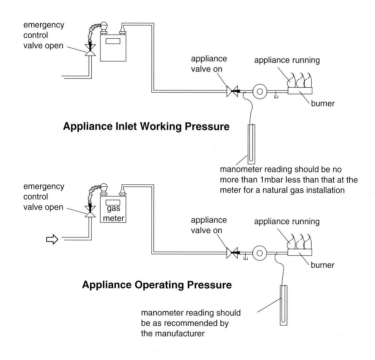

emergency
control
valve open

appliance
valve on

appliance running

burner

Appliance Inlet Working Pressure

manometer reading should be no
more than 1mbar less than that at the
meter for a natural gas installation

emergency
control
valve open

gas
meter

appliance
valve on

appliance running

burner

Appliance Operating Pressure

manometer reading should
be as recommended by
the manufacturer

Estimating a Suitable Pipe Size

Relevant Industry Documents
BS 6891 and IGE/UP/2

Basic pipe sizing is often undertaken using a computer package or a hand-held disc calculator, such as the 'Mears'. However, you can make a simple calculation such as the one illustrated here to give you an indication as to a suitable size. (See also page 54 for an alternative method of pipe sizing.) When pipe sizing. the selected size chosen must take account of the maximum possible flow rate of gas for the appliances on the system, with an allowance for possible future extensions. As explained on the previous page, the gas in the pipe will have a pressure loss. For the system to be considered suitably sized, for a natural gas installation this pressure loss in must not exceed 1 mbar. (For systems using LPG, see page 154.)

The method described here needs some time and consideration, but is quite simple to perform once the concept is understood. It consists of dividing the pipe installation into sections, each of which is to take a proportion of the maximum 1 mbar drop. For example, the system shown below may have several branches but any 'individual run' has no more than four sections. If the maximum drop is not to exceed 1 mbar, then each section must have a pressure loss of less than $(1 \div 4) = 0.25$ mbar.

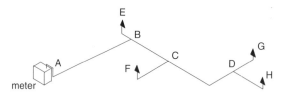

Internal Installation Pipework

Having determined the number of sections, there are four stages to calculating the pipe size for each section. A completed example is shown over the page, each stage being completed as follows:

Stage 1: Determine the total gas flow in m^3 that could pass through the pipe section. This is found by totalling up the gross kilowatt rating of all the appliances and multiplying this total by 0.094 to convert natural gas kW to m^3. For example:

$$17 \text{ kW} \times 0.094 = \underline{1.598 \text{ or } 1.6 \text{ m}^3}$$

Where only the net heat input has been specified, this needs to be first converted to a gross input by simply multiplying the net figure by 1.1.

Note that, at this stage, when the gas flow has been determined, it is possible to establish the size of the meter required.

Stage 2: Find the effective pipe length. This is the total measured length plus an allowance for fittings; this can be found from Table 1 opposite.

Stage 3: Multiply the effective pipe length by the total number of possible sections.

Stage 4: Using tables 2–4 opposite find the appropriate pipe size, ensuring that the maximum length allowed will permit the required gas flow that was calculated in Stage 1.

Table 1 Pressure loss due to fittings

Approx size		Metallic	≤25 mm	40 mm	50 mm	65 mm	80 mm	150 mm
		PE	≤25 mm	55 mm	75 mm	100 mm	140 mm	200 mm
90° elbows and tees			0.5 m	1.0 m	1.5 m	2.0 m	2.5 m	3.0 m
Pulled bends			0.3 m	0.45 m	0.65 m	1.0 m	1.2 m	2.2 m

Table 2 Flow discharge through copper tube to EN1057 and corrugated stainless steel

Pipe size	Max length: allowing 1mbar pressure differential between each end									
	3	6	9	12	15	20	25	30	40	50
10	0.86	0.57	0.5	0.37	0.3	0.22	0.18	0.15		
12	1.5	1.0	0.85	0.82	0.69	0.52	0.41	0.34		
15	2.9	1.9	1.5	1.3	1.1	0.95	0.92	0.88	0.66	0.52
22	8.7	5.8	4.6	3.9	3.4	2.9	2.5	2.3	2.0	1.78
28	18	12	9.4	8.0	7.0	5.9	5.2	4.7	4.0	3.6
35	32	22	17	15	13	11	9.5	8.5	7.2	6.3
42	54	37	29	25	22	19	17	15	13	12
	Discharge in m³/h									

Table 3 Flow discharge through steel tube to BS1387 and 3610

Pipe size	Max length: allowing 1mbar pressure differential between each end									
	3	6	9	12	15	20	25	30	40	50
15	4.3	2.9	2.3	2.0	1.7	1.5	1.4	1.3		
20	9.7	6.6	5.3	4.5	3.9	3.3	2.9	2.6		
25	18	12	10	8.5	7.5	6.3	5.6	5.0	4.1	3.6
32	32	22	17	15	13	11	9.5	8.5	7.2	6.3
40	54	37	29	25	22	19	17	15	13	12
50	110	75	60	50	42	35	32	28	25	22
65	220	150	120	100	92	79	70	63	54	47
80	320	240	180	150	135	115	100	95	80	70
100	650	460	360	300	270	230	200	180	150	140
150	1800	1250	1000	850	750	620	560	500	420	380
	Discharge in m³/h									

Table 4 Flow discharge through polyethylene tube

Pipe size	Max length: allowing 1mbar pressure differential between each end									
	3	6	9	12	15	20	25	30	40	50
20	4.0	2.7	2.1	1.8	1.6	1.3	1.2	1.1	0.9	0.8
25	8.7	5.8	4.6	3.9	3.6	2.8	2.6	2.3	2.0	1.78
32	18	12	9.4	8.0	7.2	6.0	5.4	4.8	4.0	3.6
50	54	37	29	25	22	19	17	15	13	12
63	110	75	60	50	42	35	32	28	25	22
90	280	190	150	125	115	95	85	76	65	60
125	650	460	360	300	270	230	200	180	150	140
	Discharge in m³/h									

Tables such as those above can be found in many industry documents including BS 6891 and IGE/UP/2.

Domestic Natural Gas Pipe Sizing

Example (*using notes from the previous page*)

Looking at the example illustrated opposite you can see that there are a maximum three possible sections: (A to B) + (B to C) + (C to D) *or* (C to F). Therefore the table for pipe sizing could be completed as shown below.

Section	Stage 1	Stage 2	Stage 3	Stage 4
	Find the total gas flow in m³.	**Determine the effective pipe length.**	**Determine the max. pipe length for sizing.**	**Compare length with required flow.***
	Max. kW × 0.094	*Actual length + Fittings*	*Length × No. of sections*	
A–B	All appliances are supplied therefore: $32 + 17.2 + 4.6 = 53.8$ kW × 0.094 = <u>5.06 m³</u>	Measured length = 5 m + 0.5 m is added (from Table 1) for each elbow of which there are four ∴ 4 × 0.5 + 5 m = <u>7 m</u>	∴ 7 × 3 = 21 m Table 2 has no column for 21 m, therefore check the column the next size up, i.e. <u>25 m.</u>	Suggested pipe size: <u>28 mm</u>
B–C	The cooker and fire are supplied therefore: $17.2 + 4.6 = 21.8$ kW × 0.094 = <u>2.05 m³</u>	Measured length = 2 m. There are no fittings. ∴ Effective length = 2 m	∴ 2 × 3 = <u>6 m</u>	Suggested pipe size: <u>22 mm</u>
C–D	Only the cooker, therefore: 17.2 kW × 0.094 = <u>1.62 m³</u>	Measured length = 2 m + 0.5 m is added (from Table 1) for each elbow of which there are two ∴ 2 × 0.5 + 2 m = <u>3 m</u>	∴ 3 × 3 = <u>9 m</u>	Suggested pipe size: <u>22 mm</u>
C–F	Only the fire, therefore: 4.6 kW × 0.094 = <u>0.43 m³</u>	Measured length = 2.5 m + 0.5 m is added (from Table 1) for each elbow or tee of which there is 1 elbow and 1 tee ∴ 2 × 0.5 + 2.5 m = <u>3.5 m</u>	∴ 3.5 × 3 = 10.5 m Table 2 has no column for 10.5 m. You therefore check the column the next size up, i.e. <u>12 m.</u>	Suggested pipe size: <u>12 mm</u>
B–E	Only the boiler, therefore: 32 kW × 0.094 = <u>3.01 m³</u>	Measured length = 1 m + 0.5 m is added (from Table 1) for each elbow or tee of which there is 1 elbow and 1 tee. 2 × 0.5 + 1 m = <u>2 m</u>	∴ 2 × 2 = <u>4 m</u> *Note:* There are only two sections between this pipe and the meter, therefore the max. pressure drop for this section could be increased to 0.5 mbar.	Suggested pipe size: <u>22 mm</u>

*From Table 2 select the maximum length allowed in Stage 3 and run you finger down until you find a volume that will still just allow the volume in Stage 1 to flow.

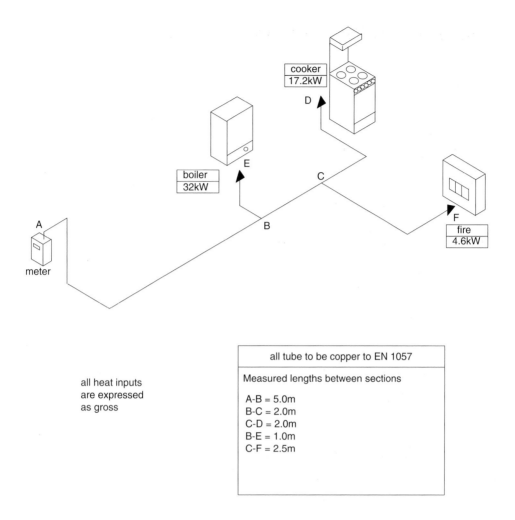

all heat inputs
are expressed
as gross

all tube to be copper to EN 1057
Measured lengths between sections
A-B = 5.0m B-C = 2.0m C-D = 2.0m B-E = 1.0m C-F = 2.5m

A second example for a commercial installation is given over the page.

Note: The same method of calculation can be used for sizing LPG installations. Because of its higher specific gravity and its higher operating gas pressure, LPG requires smaller pipe diameters, therefore an LPG system could easily be converted to natural gas later without fear of any problems of pipe size. Accurate LPG pipe sizing, however, can be obtained by referring to page 154.

Commercial Pipe Sizing

Example (*following the format from previous page*)

Looking at the installation illustrated opposite you can see that there are two sections: (A to B) + (B to C) *or* (A to B) + (C to D). Owing to the size of the system, steel pipe has been selected. Copper pipework may not be used for gas installations exceeding 67 mm in diameter.

The pipe sizing could be completed as shown on the following table.

Section	Stage 1	Stage 2	Stage 3	Stage 4
	Find the total gas flow in m³.	**Determine the effective pipe length.**	**Determine the pipe length for sizing purposes.**	**Compare length with required flow.**[*]
	Max. kW × 0.094	*Actual length + Fittings‡*	*Length × No. of sections*	
A–B	Both appliances are supplied, therefore: 600 + 600 = 1200 kW × 0.094 = <u>112.8 m³</u>	Measured length = 6 m + from Table 1 an amount is added for the tee which the gas will flow round (estimated size 80 mm) ∴ 6 + 2.5 = <u>8.5 m</u>	∴ 8.5 × 2 = 17 m Table 3 shows no column for 17 m. You therefore check column the next size up, i.e. <u>20 m.</u>	Suggested pipe size of <u>80 mm</u> confirmed suitable.
B–C	Only one appliance supplied therefore: 600 kW × 0.094 = <u>56.4 m³.</u>	Measured length = 2 m + from Table 1 an amount is added for the one elbow (estimated size 40 mm) ∴ 2 + 1 = <u>3 m.</u>	∴ 3 × 2 = <u>6 m</u>	Suggested pipe size of 40 mm @ Stage 2 *not suitable.* ∴ **Re-do!**
B–C second estimate	Only one appliance supplied therefore: 600 kW × 0.094 = 56.4 m³	Measured length = 2 m + from Table 1 an amount is added for the one elbow (estimated size 50 mm) ∴ 2 + 1.5 = <u>3.5 m</u>	∴ 3.5 × 2 = 7 m Table 3 shows no column for 7 m. You therefore check column the next size up, i.e. <u>9 m.</u>	Suggested pipe size of <u>50 mm</u> confirmed suitable.
C–D	Only one appliance supplied therefore: 600 kW × 0.094 = 56.4 m³	Measured length = 4 m + from Table 1 an amount is added for the three elbows (estimated size 65 mm) ∴ 4 + (3 × 2) = <u>10 m</u>	∴ 10 × 2 = <u>20 m</u>	Suggested pipe size of <u>65 mm</u> confirmed suitable.

*‡ Note: In Stage 2 an estimate pipe size needs to be made so that an accurate pressure loss due to fittings can be made (see Table 1 on page 49).

*From Table 3, page 49, select the maximum length allowed in Stage 3 and run your finger down to cross-align with the estimated pipe size in Stage 2. The discharge volume must still just allow the volume in Stage 1 to flow.

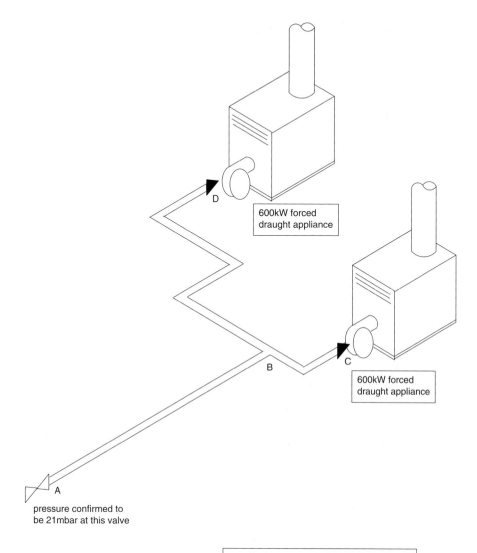

D

600kW forced
draught appliance

C

600kW forced
draught appliance

B

A

pressure confirmed to
be 21mbar at this valve

all heat inputs
are expressed
as gross

all tube to be mild steel to BS 3601
Measured lengths between sections A-B = 6.0m B-C = 2.0m C-D = 4.0m

Determining Existing Losses due to Pipe Size

The previous method of pipe sizing demonstrated a method that could be used in determining the size for a pipe on a new installation. Here we illustrate another method that could be employed. It takes into consideration the total pressure drop, unlike the previous method, which only made an allowance for each section.

This method will prove most suitable for an existing installation where possible pressure losses need to be considered, such as in extending an existing system, or in diagnosing the cause of a problem in obtaining suitable pressures.

Pages 50–51 show a domestic gas installation. In the following example, we assume that this is the existing system with the pipe sizes that we obtained in the worked example.

Method
This is best achieved by the completion of a table, as shown opposite. The stages of each section are as follows.

Stages 1–2: Complete this as described in the previous worked example, dividing the installation into sections, then determining the gas flow and effective pipe length.

Stage 3: Enter the size of pipe installed.

Stage 4: Refer to Table 2 on page 49. Place a straight edge along the size of pipe size installed and run a finger along until the furthest column is reached that still just allows the required gas volume to flow (Stage 1). Enter this maximum length at Stage 4.

For example, section A–B requires 5.04 m³ of gas and has a 28 mm diameter pipe. Therefore, with the straight edge on 28 mm it can be seen that the 25 m column can be selected. The 30 m column would only allow 4.7 m³.

Stage 5: Divide the length just found in Stage 4 into the effective length identified in Step 2 to give the actual pressure loss for the section in question (in mbars).

Finally, the actual loss is found by simply adding together the various losses from the appropriate sections to give a total pressure drop at the various appliances.

Thus:

$$\text{Section A–E (meter to boiler)} = (\text{A to B}) + (\text{B to E})$$
$$= 0.28 + 0.13 = \underline{0.41 \text{ mbar drop}}$$

$$\text{Section A–D (meter to cooker)} = (\text{A to B}) + (\text{B to C}) + (\text{C to D})$$
$$= 0.28 + 0.067 + 0.06 = \underline{0.4 \text{ mbar drop}}$$

$$\text{Section A–F (meter to fire)} = (\text{A to B}) + (\text{B to C}) + (\text{C to F})$$
$$= 0.28 + 0.067 + 0.175 = \underline{0.52 \text{ mbar drop}}$$

Section	Stage 1	Stage 2	Stage 3	Stage 4	Stage 5
	Gas flow (m³) *Max. kW ×* 0.094	**Effective length (m)** *Actual length +* *Fittings*	**Pipe size (mm)**	**Max. length allowing 1 mbar drop (m)**	**Pressure loss for section (mbar)**
A–B	All appliances are supplied therefore: $32 + 17.2 + 4.6 = 53.8$ kW $\times 0.094 = \underline{5.06 \text{ m}^3}$	Measured length 5 m + 0.5 m is added (from Table 1) for each elbow of which there are four $\therefore 4 \times 0.5 + 5 \text{ m} = \underline{7 \text{ m}}$	28 mm	25 m	$7 \div 25 =$ 0.28 mbar
B–C	The cooker and fire supplied therefore: $17.2 + 4.6 = 21.8$ kW $\times 0.094 = \underline{2.05 \text{ m}^3}$	Measured length = 2 m. There are no fittings $\therefore$ effective length = $\underline{2 \text{ m.}}$	22 mm	30 m	$2 \div 30 =$ 0.067 mbar
C–D	Only the cooker therefore: 17.2 kW $\times$ $0.094 = \underline{1.62 \text{ m}^3}$	Measured length = 2 m + 0.5 m is added (from Table 1) for each elbow of which there are two $\therefore 2 \times 0.5 + 2 \text{ m} = \underline{3 \text{ m}}$	22 mm	50 m	$3 \div 50 =$ 0.06 mbar
C–F	Only the fire therefore: 4.6 kW $\times$ 0.094 $= \underline{0.43 \text{ m}^3}$	Measured length = 2.5 m + 0.5 m is added (from Table 1) for each elbow or tee of which there is 1 elbow and 1 tee $\therefore 2\times$ $0.5 + 2.5 \text{ m}$ $= 3.5 \text{ m}$	12 mm	20 m	$3.5 \div 20 =$ 0.175 mbar
B–E	Only the boiler therefore: 32 kW $\times$ 0.094 $= \underline{3.01 \text{ m}^3}$	Measured length = 1 m + 0.5 m is added (from Table 1) for each elbow or tee of which there is 1 elbow and 1 tee $\therefore 2 \times 0.5 + 1 \text{ m}$ $= \underline{2 \text{ m}}$	22 mm	15 m	$2 \div 15 =$ 0.133 mbar

Gas Rates and Heat Input

The *gas rate* refers to the amount of gas used in m³, usually over one hour, whereas the *heat input* refers to the amount of heat generated in kW from burning a certain quantity of fuel.

In order to determine the gas consumption of an appliance, a gas meter is used to measure the amount of gas consumed in one hour in m³. There are several types of gas meter, providing readings in different units. Occasionally one is found reading in cubic metres per hour (m³/h). However, for payment purposes it only registers in either ft³, or m³ and not per hour.

It would be uneconomical to run an appliance for an hour just to determine its gas flow over this period, therefore some kind of estimate is made, depending on the meter type. However it is necessary to let the appliance run for about 10 minutes to allow it to warm up. This allows the injectors to heat up and expand and takes account of the effects of the physical laws relating to gas (see page 60).

Taking the Gas Rate with an Older Meter (reading in ft³)
Run the appliance and observe the test dial, timing how long it takes, in seconds, to consume 1 ft³ of gas. For the smaller U6 meter this would be once round the test dial. For larger meters, 1 ft³ may only be 1/10 of the test dial. For larger commercial appliances, which use large volumes of gas, a reading of, say, 10 ft³ may be taken and the time taken for the gas to be consumed divided by 10, to give the time for 1ft³.

Knowing the time in seconds (t) to burn 1 ft³, and with 3600 s in an hour, it is a simple process to calculate the gas rate per hour by understanding the following sum: $3600 \div t$. Finally to convert ft³/h into m³/h divide by a conversion factor of 35.3.

Example If we find that it takes 39 s to consume 1ft³, then consumption per hour is:

$$3600 \div 39 = 92.3 \text{ ft}^3/\text{h} \div 35.3 = \underline{2.6 \text{ m}^3/\text{h}}$$

Finding the Gas Rate from a Metric Meter (in m³)
Take the meter reading, then take it again after two minutes. The amount of gas used in two minutes is the difference between the two readings (i.e. second reading minus the first). To find the gas rate over an hour in this example of a two-minute test you simply multiply by 30 (as $2 \times 30 = 60$ minutes).

Example Assume the index at the meter reads 07341.926 at the beginning of the two-minute test period and 07341.954 two minutes later. The consumption per hour is:

$$07341.954 - 07341.926 = 0.028 \text{ m}^3 \text{ in } 2 \text{ minutes} \times 30 = \underline{0.84 \text{ m}^3/\text{h}}$$

Finding the Gross Heat Input
The gross heat input is expressed in kilowatt hours and is found by multiplying the gas rate in m³ by the calorific value (CV) then divide by 3.6.

Example To find the heat input where 0.84 m³/h of gas is consumed. (Take the CV to be 38.5.)

$$\therefore 0.84 \times 38.5 \div 3.6 = 8.98\,\text{kWh}$$

(*Note*: The 'h' is often omitted, giving 8.98 kW.)

If we reconsidered the previous example with a CV of 39.1, the heat input would be:

$$0.84 \times 39.1 \div 3.6 = 9.12\,\text{kW}$$

This is only a slight difference, yet it exists, and awareness of it is useful when comparing the heat input of an appliance with that given by the manufacturer. Your result may be based on a different CV from that used by the manufacturer. Typical calorific values can be found on page 23.

Finding the Net Heat Input
The net heat input (see page 23) is found by dividing the gross input by 1.1.

Example Taking the gross input from the previous example as 9.12 kW. The net heat input will be $9.12 \div 1.1 = 8.29\,\text{kW}$.

Having found the heat input, one can check it against the manufacturer's data badge to confirm that the appliance is being supplied with the correct quantity of gas. *Note*: For older appliances this may be indicated as a gross input and for newer appliances it should be recorded as a net input.

Sometimes quick-reference tables are used as illustrated overleaf to determine the gas rate and heat input, however they are limited to smaller sized appliances and take account of only one particular CV.

2 Gas Utilisation

Test Dial of Imperial Meter

Typical Gas Meters

Test Dial of Metric Meter

Quick Reference Tables: Gas Rate/Heat Input

The tables produced here are only suitable for domestic natural gas installations with a calorific value (CV) of 38.5. If the CV increases in value, the figures indicated would reduce. For a more accurate reading see the previous page.

To determine the gas rate/heat input of an appliance, all other appliances in the system need to be turned off and the appliance in question run for about 10 minutes to enable it to warm up. The following procedure then needs to be adopted.

Meters reading in m³, e.g. E6 and G4 Record the gas volume used over a 2-minute period by recording a reading then take a second reading after 2 minutes, deducting the first figure from the second.

Meters reading in ft³, e.g. U6 Observe the test dial on front of the meter and record the time in seconds that it takes to complete a full circle, i.e. 1 ft³.

Gas rate and heat input for E6 type gas meters (average $CV = 38.5$)

2 min gas flow m³	Gas rate m³/h	Gross input (kW)	Net input (kW)	2 min gas flow m³	Gas rate m³/h	Gross input (kW)	Net input (kW)	2 min gas flow m³	Gas rate m³/h	Gross input (kW)	Net input (kW)
0.010	0.30	3.21	2.92	0.074	2.22	23.73	21.58	0.138	4.14	44.26	40.24
0.012	0.36	3.85	3.50	0.076	2.28	24.37	22.16	0.140	4.20	44.90	40.82
0.014	0.42	4.49	4.08	0.078	2.34	25.01	22.74	0.142	4.26	45.54	41.41
0.016	0.48	5.13	4.67	0.080	2.40	25.66	23.33	0.144	4.32	46.18	41.99
0.018	0.54	5.77	5.25	0.082	2.46	26.30	23.91	0.146	4.38	46.82	42.57
0.020	0.60	6.41	5.83	0.084	2.52	26.94	24.49	0.148	4.44	47.46	43.16
0.022	0.66	7.06	6.42	0.086	2.58	27.58	25.08	0.150	4.50	48.11	43.74
0.024	0.72	7.70	7.00	0.088	2.64	28.22	25.66	0.152	4.56	48.75	44.32
0.026	0.78	8.34	7.58	0.090	2.70	28.86	26.24	0.154	4.62	49.39	44.91
0.028	0.84	8.98	8.16	0.092	2.76	29.50	26.83	0.156	4.68	50.03	45.49
0.030	0.90	9.62	8.75	0.094	2.82	30.15	27.41	0.158	4.74	50.67	46.07
0.032	0.96	10.26	9.33	0.096	2.88	30.79	27.99	0.160	4.80	51.31	46.66
0.034	1.02	10.90	9.91	0.098	2.94	31.43	28.58	0.162	4.86	51.95	47.24
0.036	1.08	11.55	10.50	0.100	3.00	32.07	29.16	0.164	4.92	52.59	47.82
0.038	1.14	12.19	11.08	0.102	3.06	32.71	29.74	0.166	4.98	53.24	48.41
0.040	1.20	12.83	11.66	0.104	3.12	33.35	30.33	0.168	5.04	53.88	48.99
0.042	1.26	13.47	12.25	0.106	3.18	33.99	30.91	0.170	5.10	54.52	49.57
0.044	1.32	14.11	12.83	0.108	3.24	34.64	31.49	0.172	5.16	55.16	50.16
0.046	1.38	14.75	13.41	0.110	3.30	35.28	32.08	0.174	5.22	55.80	50.74
0.048	1.44	15.39	14.00	0.112	3.36	35.92	32.66	0.176	5.28	56.44	51.32
0.050	1.50	16.04	14.58	0.114	3.42	36.56	33.24	0.178	5.34	57.08	51.90
0.052	1.56	16.68	15.16	0.116	3.48	37.20	33.83	0.180	5.40	57.73	52.49
0.054	1.62	17.32	15.75	0.118	3.54	37.84	34.41	0.182	5.46	58.37	53.07
0.056	1.68	17.96	16.33	0.120	3.60	38.48	34.99	0.184	5.52	59,01	53.65
0.058	1.74	18.60	16.91	0.122	3.66	39.13	35.58	0.186	5.58	59.65	54.24
0.060	1.80	19.24	17.50	0.124	3.72	39.77	36.16	0.188	5.64	60.29	54.82
0.062	1.86	19.88	18.08	0.126	3.78	40.41	36.74	0.190	5.70	60.93	55.40
0.064	1.92	20.52	18.66	0.128	3.84	41.05	37.32	1.192	5.76	61.57	55.99
0.066	1.98	21.17	19.25	0.130	3.90	41.69	37.91	0.194	5.82	62.22	56.57
0.068	2.04	21.81	19.83	0.132	3.96	42.33	38.49	0.196	5.88	62.86	57.15
0.070	2.10	22.45	20.41	0.134	4.02	42.97	39.07	0.198	5.94	63.50	57.74
0.072	2.16	23.09	21.00	0.136	4.08	43.62	39.66	0.200	6.00	64.14	58.32

Gas rate and heat input for U6 type gas meters (average $CV = 38.5$)

Seconds to burn 1 ft^3	Gas rate m^3/h	Gross Input (kW)	Net Input (kW)	Secs to burn 1 ft^3	Gas rate m^3/h	Gross Input (kW)	Net Input (kW)	Secs to burn 1 ft^3	Gas rate m^3/h	Gross Input (kW)	Net Input (kW)
400	0.255	2.73	2.48	207	0.493	5.27	4.79	77	1.324	14.16	12.87
396	0.258	2.75	2.50	204	0.500	5.34	4.86	76	1.342	14.35	13.04
392	0.260	2.78	2.53	201	0.507	5.42	4.93	75	1.360	14.54	13.22
388	0.263	2.81	2.55	198	0.515	5.51	5.01	74	1.378	14.73	13.40
384	0.266	2.84	2.58	195	0.523	5.59	5.08	73	1.397	14.93	13.58
380	0.268	2.87	2.61	192	0.531	5.68	5.16	72	1.416	15.14	13.77
376	0.271	2.90	2.64	189	0.540	5.77	5.24	71	1.436	15.36	13.96
372	0.274	2.93	2.66	186	0.548	5.86	5.33	70	1.457	15.57	14.16
368	0.277	2.96	2.69	183	0.557	5.96	5.42	69	1.478	15.80	14.37
364	0.280	3.00	2.72	180	0.567	6.06	5.51	68	1.500	16.03	14.58
360	0.283	3.03	2.75	177	0.576	6.16	5.60	67	1.522	16.27	14.80
356	0.286	3.06	2.78	174	0.586	6.27	5.70	66	1.545	16.52	15.02
352	0.290	3.10	2.82	171	0.596	6.38	5.80	65	1.569	16.77	15.25
348	0.293	3.13	2.85	168	0.607	6.49	5.90	64	1.593	17.03	15.49
344	0.296	3.17	2.88	165	0.618	6.61	6.01	63	1.619	17.31	15.74
340	0.300	3.21	2.92	162	0.630	6.73	6.12	62	1.645	17.58	15.99
336	0.304	3.24	2.95	159	0.641	6.86	6.23	61	1.672	17.87	16.25
332	0.307	3.28	2.99	156	0.654	6.99	6.35	60	1.700	18.17	16.52
328	0.311	3.32	3.02	153	0.667	7.13	6.48	59	1.729	18.48	16.80
324	0.315	3.36	3.06	150	0.680	7.27	6.61	58	1.758	18.80	17.09
320	0.319	3.41	3.10	147	0.694	7.42	6.74	57	1.789	19.13	17.39
316	0.323	3.45	3.14	144	0.708	7.57	6.88	56	1.821	19.47	17.70
312	0.327	3.49	3.18	141	0.723	7.73	7.03	55	1.854	19.82	18.02
308	0.331	3.54	3.22	138	0.739	7.90	7.18	54	1.889	20.19	18.36
304	0.335	3.59	3.26	135	0.755	8.08	7.34	53	1.924	20.57	18.70
300	0.340	3.63	3.30	132	0.773	8.26	7.51	52	1.961	20.97	19.06
296	0.345	3.68	3.35	129	0.791	8.45	7.68	51	2.000	21.38	19.44
292	0.349	3.73	3.39	126	0.809	8.65	7.87	50	2.040	21.80	19.83
288	0.354	3.79	3.44	123	0.829	8.86	8.06	49	2.081	22.25	20.23
284	0.359	3.84	3.49	120	0.850	9.08	8.26	48	2.125	22.71	20.65
280	0.364	3.89	3.54	117	0.872	9.32	8.47	47	2.170	23.20	21.09
276	0.370	3.95	3.59	114	0.895	9.56	8.70	46	2.217	23.70	21.55
272	0.375	4.01	3.64	111	0.919	9.82	8.93	45	2.266	24.23	22.03
268	0.381	4.07	3.70	108	0.944	10.09	9.18	44	2.318	24.78	22.53
264	0.386	4.13	3.75	105	0.971	10.38	9.44	43	2.372	25.35	23.05
261	0.391	4.18	3.80	102	1.000	10.69	9.72	42	2.428	25.96	23.60
258	0.395	4.23	3.84	99	1.030	11.01	10.01	41	2.487	26.59	24.18
255	0.400	4.28	3.89	96	1.062	11.36	10.33	40	2.550	27.26	24.78
252	0.405	4.33	3.93	93	1.097	11.72	10.66	39	2.615	27.95	25.42
249	0.410	4.38	3.98	91	1.121	11.98	10.89	38	2.684	28.69	26.09
246	0.415	4.43	4.03	90	1.133	12.11	11.01	37	2.756	29.47	26.80
243	0.420	4.49	4.08	89	1.146	12.25	11.14	36	2.833	30.28	27.54
240	0.425	4.54	4.13	88	1.159	12.39	11.26	35	2.914	31.15	28.32
237	0.430	4.60	4.18	87	1.172	12.53	11.39	34	3.000	32.07	29.16
234	0.436	4.66	4.24	86	1.186	12.68	11.53	33	3.090	33.04	30.04
231	0.441	4.72	4.29	85	1.200	12.83	11.66	32	3.187	34.07	30.98
228	0.447	4.78	4.35	84	1.214	12.98	11.80	31	3.290	35.17	31.98
225	0.453	4.85	4.41	83	1.229	13.14	11.94	30	3.399	36.34	33.04
222	0.459	4.91	4.47	82	1.244	13.30	12.09	28	3.642	38.94	35.40
219	0.466	4.98	4.53	81	1.259	13.46	12.24	26	3.922	41.93	38.13
216	0.472	5.05	4.59	80	1.275	13.63	12.40	24	4.249	45.43	41.30
213	0.479	5.12	4.65	79	1.291	13.80	12.55	22	4.636	49.55	45.06
210	0.486	5.19	4.72	78	1.307	13.98	12.71	20	5.099	54.51	49.56

2 Gas Utilisation

The Physical Laws Relating to Gas

Graham's Law of Diffusion

This law is named after Thomas Graham (1805–1869), who discovered that gases will mix with one another quite readily due to the continuous movement of the molecules. However, the rate at which they mix depends on the specific gravity or density of the gases. Graham made a container, consisting of two separate compartments with a small hole in their dividing wall. He placed a different gas in each of the compartments, one having a higher specific gravity than the other. He found that more faster, lighter molecules passed through the hole than the slower, heavier molecules of the heavier gas. After many experiments he discovered that the rates of diffusion, or mixing, varied inversely as the square root of the density of the gas. Thus:

$$\text{Diffusion rate} \propto 1 \div \sqrt{\text{Density}}$$

Which basically means that a light gas will diffuse twice as fast as a gas four times its density. (See also page 22 relating to specific gravity.)

Boyle's Law

This is named after Robert Boyle (1627–1691) who discovered the relationship between volume and pressure of a gas. He found that the absolute pressure of a given mass of gas is inversely proportional to its volume providing its temperature remains constant (Absolute pressure = Atmospheric pressure (1013 mbar) + gauge pressure).

So:

$$\text{Pressure} \times \text{Volume} = \text{Constant}$$

To put it simply, if the absolute pressure (gauge pressure + atmospheric pressure) on a given quantity of gas decreases, then its volume will increase. So if the absolute pressure is increased four-fold, the volume would be reduced to one quarter. The formula can be redefined as:

$$P_1 V_1 = P_2 V_2$$

where P_1 = original pressure, V_1 = original volume, P_2 = final pressure and V_2 = final volume.

Practical example Suppose that the supply pressure to a building is 80 mbar and the total volume is 1 m³, if the pressure was reduced to 20 mbar the new volume of the gas would be calculated as follows.

$$P_1 V_1 = P_2 V_2, \text{ so } P_1 V_1 \div P_2 = V_2$$
$$\therefore (1013 + 80) \times 1 \div (1013 + 20) = \underline{1.06\,\text{m}^3}\text{an increase of } \underline{6\%}$$

Conversely, if the supply pressure is increased to 800 mbar when a new medium pressure regulator is fitted, supplying the same 1 m³ volume of gas, and also reduced to 20 mbar, there is an interesting result:

$$P_1 V_1 \div P_2 = V_2$$
$$\therefore (1013 + 800) \times 1 \div (1013 + 20) = \underline{1.76\,\text{m}^3}, \text{ an increase of } \underline{76\%}$$

This increase is probably the reason that the gas supplier installed the regulator prior to the meter.

Charles's Law

This is named after Jacques Charles (1746–1823), a French physicist who discovered that all gases increase in volume by the same proportion if heated through the same temperature range, provided the pressure remained constant. This proportion is 1/273 of their volume at freezing point (0°C or 273K) for each 1K rise above 273K. Therefore, the temperature of a volume of gas would need to be increased from 0°C to 273°C in order to double its volume. (*Note:* 1K rise = 1°C rise.)

Where the pressure remains constant Charles's law is expressed as:

$$\text{Volume} \div \text{Temperature} = \text{Constant}$$

In simple terms, if the temperature increases, so does the volume. As with Boyle's law, the formula can be redefined as:

$$V_1 \div T_1 = V_2 \div T_2$$

where V_1 = original volume, T_1 = original temperature, V_2 = final volume and T_2 = final temperature.

Practical example If 1 m³ of gas enters a building from outside where the temperature is 2°C and passes into a building where the temperature is 21°C, the gas would increase in volume by:

$$V_1 \div T_1 = V_2 \div T_2, \text{ so } (V_1 \times T_2) \div T_1 = V_2$$
$$\therefore 1 \times (21 + 273) \div (2 + 273) = \underline{1.07 \text{ m}^3}, \text{ an increase of } \underline{7\%}$$

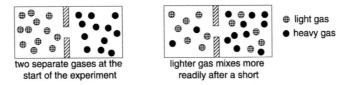

⊕ light gas
● heavy gas

two separate gases at the start of the experiment

lighter gas mixes more readily after a short

Graham's Law of Diffusion

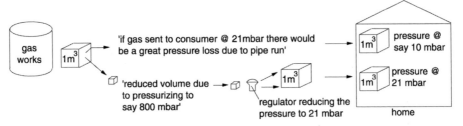

gas works

1m³

'if gas sent to consumer @ 21mbar there would be a great pressure loss due to pipe run'

1m³ pressure @ say 10 mbar

'reduced volume due to pressurizing to say 800 mbar'

1m³ regulator reducing the pressure to 21 mbar

1m³ pressure @ 21 mbar

home

Practical example of Boyle's Law

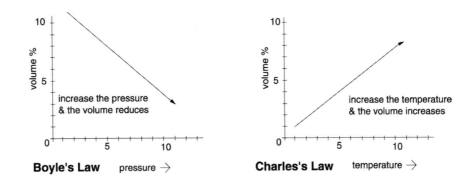

volume %

increase the pressure & the volume reduces

Boyle's Law pressure →

volume %

increase the temperature & the volume increases

Charles's Law temperature →

Measurement of Gas 1

There are many different designs of gas metering devices, each using a different method to record the volume of gas flow. Some record the amount of gas used for billing purposes, whereas others give an indication as to the rate at which the gas is flowing, possibly for test purposes. Gas meters fall into two distinct classes: Displacement and Inferential.

Displacement Meters

These meters record a definite volume by the displacement of gas through one of the following methods:

- a bellows chamber or compartment; or
- spaces between impellers or vanes.

Positive Displacement Meters

This type of meter is a 'unit construction' (U-meter). It is a design that has been around since the 1800s. Since then it has undergone many improvements, but it still operates along the same principles of the early design. The U-type meter uses diaphragms that are alternately inflated and deflated by the gas pressure flowing through. The movements of the diaphragms, via pivoting flag rods, opens and closes two valves located at the top of the meter. These valves control the passage of gas through the four measuring compartments. The diagram opposite illustrates the cycle of opening and closing each compartment.

1. The gas flows into the inner chamber at the front of the meter, and in so doing allows gas to pass out into the system from the adjoining front chamber. As the diaphragm moves, it causes the flag rods to turn, which cause the valves at the top to move.
2. Owing to the movement of the flag rod, the inlet port to the front inner chamber eventually closes, as does the outlet from the outer front chamber and gas can now only pass into the rear outer chamber, causing the gas to flow out from the inner rear chamber.
3. and 4. As the rear inner chamber empties, so the movement of the flag rods continues and eventually the gas is allowed to flow into the front outer chamber and out from the front inner chamber filled during (1) above, and so the process continues.

As the flag rods pivot back and forth, opening and closing the valves, they cause a worm wheel to turn. This, via a gear box, slowly turns the dial to register the quantity of gas used. The diaphragm used in the movement is made from a synthetic fibre called reinforced nitro-rubber. The typical U6 and U40 meters shown in the photograph opposite are two of the many meters that use this design. The number following the U indicates the volume that the meter is capable of passing in m^3/h. This range of meters is from U4 (4 m^3/h) 25 mm (1 inch) connections, through to U160 (160 m^3/h) 100 mm (4 inch) connections.

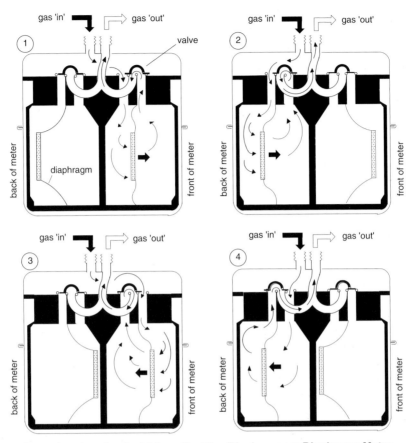

Operational cycle of a 'U' type Positive Displacement Diaphragm Meter

U Type Meters

Type	Badged rating m³/h	ft³/h	Connection type
U4	z	141	25 mm (1") threaded
U6/G4	6	212	25 mm (1") threaded
U10	10	353	32 mm (1¼") threaded
U16	16	565	32 mm (1¼") threaded
U25	25	883	50 mm (2") threaded
U40	40	1412	50 mm (2") threaded
U65	65	2295	65 mm (2½") flanged
U100	100	3530	80 mm (3") flanged
U160	160	5650	100 mm (4") flanged

**Typical 'U' type
Diaphragm Meters**

Measurement of Gas 2

Rotary Displacement Meters

An example of a rotary displacement meter is shown opposite. It measures the gas flow by trapping it between two impellers that rotate in opposite directions. From the diagram of the meter, one can see that the bottom impeller turns anticlockwise, allowing the gas to enter the space between the impeller and the casing. As the impeller reaches the horizontal position the measured quantity of gas is contained between the impeller and case wall. As the impeller continues to turn, the gas discharges through to the outlet. This design of meter ranges in sizes from 43 m³/h to 9063 m³/h, and therefore can measure much greater volumes than the previous type of meter. All rotary displacement meters operate at pressures up to 2 bar and some can operate up to 100 bar.

With both the positive displacement and rotary displacement meter, the moving parts are activated by the gas flowing through the meter. Should the meter seize up, the gas supply will shut off. So if the supply needs to be continuous, a by-pass would need to be fitted or two meters would need to be fitted in parallel, to allow continuity of supply during shut down, or for maintenance purposes.

Inferential Meters

Inferential meters record the gas flow by deducting one known quantity from another. These include:

- the rotation speed of a turbine;
- the measured differential across an orifice;
- the measured difference between static and kinetic pressures;
- the measured temperature changes across a wire.

There are many designs of inferential gas meter, including the turbine meter, orifice meter and Ultrasonic meter.

Turbine Meter

With this design the gas impinges on to specially shaped air-foil blades that rotate at high speeds as the gas flows through. The speed of rotation of the blades is proportional to the velocity and volume of gas passing over them. The meter records both the velocity and volume per hour. The drive from the turbine spindle is transferred by a magnetic coupling to record the flow on to an index mechanism counter. This type of meter could not record the amount of gas consumed because when the gas stops flowing the turbine blades do not stop immediately. They are available in a full range of sizes from 249 m³/h to 6514 m³/h with operating pressure ranging from 7 to 100 bar. With these meters, the gas can still flow freely through, should the vanes stop for any reason. The rotary anemometer is similar to the turbine meter, but has flat aluminium blades. It is usually mounted vertically into a central housing, so that

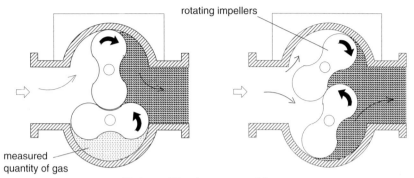

rotating impellers

measured quantity of gas

Rotary Displacement Meter

Rotary Displacement Meter

Turbine Meter

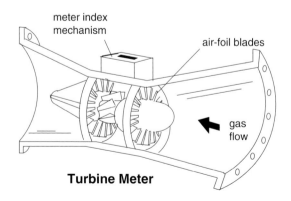

meter index mechanism

air-foil blades

gas flow

Turbine Meter

it can be easily removed for servicing. The drive shaft of the rotor spindle operates a set of internal gears to indicate the gas flow on an index plate at the top of the unit. These meters are much smaller than turbine meters and are ideal for use as commercial secondary meters to monitor local flow of gas. They have an operating range up to 200 m³/h and work in supplies of up to 1.7 bar pressure. *Note:* They must only be installed in horizontal pipework.

Orifice Meter

This is a type of inferential meter, and measures a pressure differential between points each side of an orifice plate that has been positioned in the pipeline. This method of metering is only suitable where the flow is constant. Owing to its simple design and its ability to operate at very high pressures, it is often used for measuring the gas flow through transmission networks. The orifice plate itself is usually made of thin stainless steel or a material that would not be subject to corrosion. It is usually located between two flange joints allowing for regular maintenance to ensure that it does not become dirty, distorted or eroded. This access also allows the orifice plate to be changed for one with a different sized hole, so that different flow rates can be metered. Meter failure does not affect the gas flow as it is completely detached.

Ultrasonic Gas Meter

The first ultrasonic gas meters began to appear in about 1995. They are currently only used in the domestic market with meters operating up to 6 m³/h, the meter being designated the E6. The meter consists of just a tube with a transducer located at each end. A transducer is a device that converts a mechanical signal into an electrical impulse or vice versa; it can also act as a transmitter and receiver.

A 3.6 V battery, with a 10-year life, housed in the base of the meter provides power for the transducer to generate a sound wave. This is sent as an impulse through the moving gas flow, alternately, from each direction. The time of travel each way is measured and the difference between transmission and receiving in each direction is computed to provide an indication of the volume of gas consumed. The flow is displayed on a liquid crystal display (LCD) readout, located on the meter face.

On the right-hand side of the meter is an optical communication port with a standard magnetic coupling from which information can be transmitted into or out of the meter, for purposes such as meter reading. There are several advantages to using this kind of meter including: its compact size (it is smaller than the U series of meters); it is unaffected by air moisture and temperature and it also includes a fraud detection system. Care needs to be taken when removing these meters as some, especially those that use a 'smartcard' credit system, have a tamperproof device fitted that will prevent gas flow until the problem has been investigated by the supplier.

Typical Inferential Meter

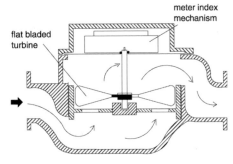

meter index mechanism

flat bladed turbine

Rotary Anemometer

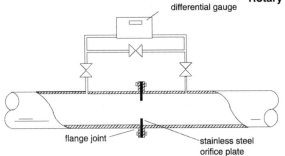

differential gauge

flange joint

stainless steel orifice plate

Orifice Meter

E6 Ultrasonic Gas Meter

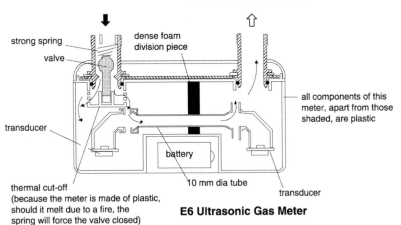

strong spring

valve

transducer

thermal cut-off
(because the meter is made of plastic,
should it melt due to a fire, the
spring will force the valve closed)

dense foam
division piece

all components of this
meter, apart from those
shaded, are plastic

battery

10 mm dia tube

transducer

E6 Ultrasonic Gas Meter

Part 3
Gas Controls

Quarter Turn Gas Control Valves

With this design of control, a change from the no flow condition to full gas flow is achieved by a quarter turn (90°) of the operating handle. Often a groove is cut across the top of the spindle, which when lined up with the pipe, indicates the on position. Any lever fitted should be carefully positioned to align with this groove, so that it will indicate the on/off position. When these valves are installed in vertical pipework it is essential to ensure that the quarter turn movement is restricted to an arrangement that indicates 'on' with the handle, if fitted, looking up and 'off' with the handle looking horizontal. This prevents it opening accidentally should something fall on to the operating lever. There are three designs of quarter turn valve that are used, namely Plug, Ball and Butterfly valves.

Plug Cock or Tap

This is the oldest and possibly the simplest form of gas control. It consists of a tapered plug through which gas flows. 'Cock' refers to a device fitted in the supply line whereas the 'tap' forms the terminal fitting, as will be found, for example, in a school science lab where it is opened to allow gas to flow through to be ignited. The plug tap is often integrated with the appliance (see also the Appliance Plug Tap on page 74). Plug cocks or taps when used with LPG incorporate an additional spring to hold the plug securely in the body of the valve. Plug cocks are made in many sizes, but the largest size selected for gas pipework should not exceed 50 mm unless it is lubricated and in no case should it be more than 100 mm in diameter. Many of the smaller cocks have an incorporated thumb-piece or fan to indicate the on position. When the fan is in line with the pipe it indicates that the supply is on. Sometimes a drop-fan design is used, which allows the fan to fall in the closed position, preventing it from being accidentally turned on.

Ball Valves

These are being fitted quite extensively these days; they incorporate a nylon or PTFE seal to ensure adequate isolation of the gas supply and typically withstand pressures of up to 7 bar.

Butterfly Valves

These are available in all sizes, however, they are only found in commercial situations. These quarter turn valves are different in that they allow for high volumes of flow because of the hinged disc. To ensure complete isolation a soft, neoprene-lined seat is incorporated.

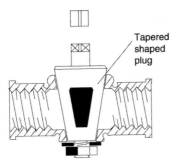

Plug Cock
shown in 'closed' position

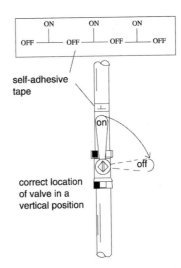

self-adhesive
tape

correct location
of valve in a
vertical position

Typical Quarter Turn Valves
top left butterfly valve; middle drop fan plug cock; top
right lubricated plug cock; bottom middle ball valve

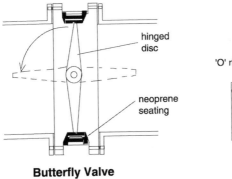

hinged
disc

neoprene
seating

Butterfly Valve

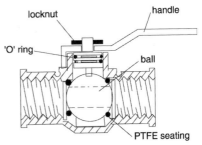

locknut
handle
'O' ring
ball
PTFE seating

Ball Valve
shown in 'open' position

Screw Down Gas Control Valves

This design of valve makes use of the turning motion of a shaft, which slowly opens the gas supply by withdrawing the valve from its seating. This category of valves includes the following specific designs: Disc, Gate, Diaphragm and Needle valves.

Disc Valve

These valves are used for local gas isolation. This valve design, using a disc that closes onto a seating, is more commonly found on automatic controls such as a solenoid or safety shut off device. One design of disc valve that is found in the domestic sector is that referred to as the 'pillar-elbow restrictor', which is associated with the installation of gas fires. It consists of a chrome-plated body with a cover cap that needs to be removed in order to open or close the control.

Gate Valve

This is design of valve not used for the smaller sizes and is restricted to 50 mm upwards; it is therefore only found in commercial situations. In operation, the gate completely lifts out of the pipe run when fully open; therefore it has a minimal loss of flow and pressure.

Diaphragm Valve

This valve consists of a large flexible diaphragm that is raised off the valve seating by the operation of a wheel head. As the wheel turns off the valve it compresses the diaphragm and forces the rubber firmly on to the seating, thus ensuring complete closure even where grit or small particles have lodged under the diaphragm. All moving parts are above the diaphragm, which also provides a real protection against gas escape through the valve head.

Needle Valve

The pressure test nipple on a multi-functional gas valve is one of the commonest forms of needle valve, although, as shown, it can be found as an independent control. It consists of a pointed tapered head, which locates into a similarly tapered seating. As the spindle is turned anticlockwise it draws the two mating surfaces apart, opening the valve and allowing the gas to flow. Needle valves are ideal where limited flow and fine adjustments are required.

Before installation of any of the valves listed on this or the previous page it is essential to consult the manufacturers' data sheets to ensure that the valve is compatible with the gas being supplied.

**Pillar-elbow
Restrictor**

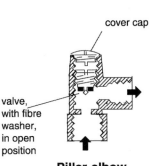

cover cap

valve,
with fibre
washer,
in open
position

**Pillar-elbow
Restrictor**

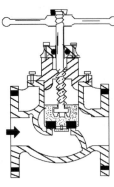

Disc Valve

wheel head

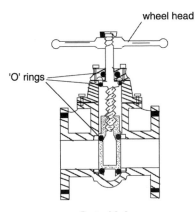

'O' rings

Gate Valve
shown in 'closed' position

Gate Valve

wheel head

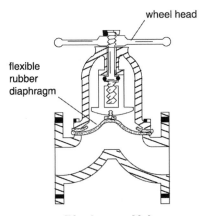

flexible
rubber
diaphragm

Diaphragm Valve
shown in 'open' position

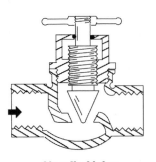

Needle Valve

Spring Loaded Gas Control Valves

There are several designs of spring loaded valve. Any valve that incorporates a spring to assist its operation is technically speaking 'spring loaded'. The valves identified here include the Cooker safety cut-off valve, the Plug-in quick release valve and the Appliance Plug Tap, as found on most cookers and gas fires.

Cooker Safety Cut-off Valve

This type of spring loaded valve is incorporated in devices such as a glass drop-down lid, as found on cookers. There are a number of designs: one, shown opposite, uses a push rod to disengage a shut-off plunger. Another hotplate safety cut-off design for domestic gas cookers is shown on page 311.

Plug-in Quick Release Valves

These are used in conjunction with cooker and tumble dryer flexes, etc. and an assortment of leisure equipment such as barbecues and are designed to facilitate removal and maintenance of the appliance. They act as 'plug-in sockets', requiring a push and turning motion to make or break the connection, which is self-sealing. The two most common valves are the standard bayonet and the micropoint socket. It should be noted that these valves can be regarded as suitable terminal fittings; no additional plugs are needed to seal off the pipe when removing the appliance and flex, although good practice would include an additional isolation valve. The location of these fittings should be such that the length of flex is limited. Micropoint sockets can be obtained as independent fittings or can be supplied to be used in conjunction with a white plastic box.

Appliance Plug Tap

In order to turn on/off the supply of gas to an appliance, such as a gas fire or cooker, a tapered plug is used. Incorporated with the plug will be a spring that is designed to prevent the control tap accidentally being turned on. In order to operate the control, one must first push in the operating knob; this will then allow the tap to be turned. The appliances in which these taps are incorporated are often subject to high temperatures surrounding the valve and, as a consequence, they often dry out and become stiff to use. In this case, the valve will need to be re-greased. This is a simple operation involving removing the tapered plug from its housing. The grease used is usually a high temperature grease such as 'Molykote'. When applying the lubricant, only the smallest smear is needed and particular care should be taken on cooker taps to ensure that the very small simmer port is not blocked with excess grease. Upon re-assembly, after spraying with leak detection fluid, the tap should be checked for correct operation at all settings.

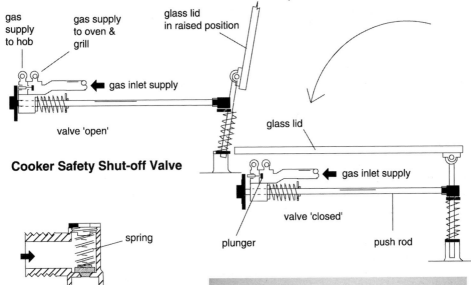

gas supply to hob

gas supply to oven & grill

glass lid in raised position

gas inlet supply

valve 'open'

Cooker Safety Shut-off Valve

glass lid

gas inlet supply

valve 'closed'

plunger

push rod

spring

'O' ring

Bayonet Valve

lug to lock in connector

flexible connection

Bayonet Type Valves

(Micropoint connector shown to the right)

Disconnecting Cooker Plug Tap

valve stripped down in readiness to re-grease

Electrically Operated Gas Control Valves

There are two electrically operated valves that are commonly encountered: the Solenoid valve and the Safety shut-off valve.

Solenoid Valve

This valve, also referred to as the magnetic valve, basically consists of a coil of wire through which electricity is passed. This flow of electricity around the coil creates an electromagnetic field. This draws in an iron plunger that is free to slide inside a brass tube, referred to as the armature. The plunger is connected to a valve, which remains open while electricity is flowing, allowing gas or liquid to flow. When the electricity stops flowing through the coil, the valve closes due to the action of a spring, which pulls the plunger from the coil. Solenoids can operate on both a.c. and d.c. With solenoids operating on a.c., however, the valve sometimes hums because the armature vibrates within the coil as the current changes the magnetic flux back and forth within the coil. Solenoid valves are often incorporated within other more complicated controls, such as multi-functional gas valves and safety shut-off valves.

Safety Shut-off Valve

A safety shut-off valve is a gas valve that maintains a relatively slow opening operation yet has a rapid closure. These valves are often installed on large automatic packaged burners or at the entrance of a large workshop or plant room. These valves are usually classified as classes 1, 2 or 3. The class number simply indicates the quality and the pressures to which the valve can safely operate, class 1 being the better valve. Two such designs are the Electro-hydraulic and the Electro-mechanical.

Electro-hydraulic

This device operates by using hydraulic power to force the valve open against a very strong spring. The oil pressure is obtained by the operation of a small electric motor. When power is supplied, a solenoid operates, closing off a small relief valve within the unit. At the same time, the motor operates pumping the oil from a reservoir through the non-return/check-valve to force the diaphragm up, lifting the valve from its seating. Once fully open, a limit switch trips, breaking the circuit to the motor and the valve is held open as the oil is trapped behind an energised solenoid valve. Switching off the power de-energises the relief solenoid, allowing the oil to flow back into the reservoir, assisted by the spring.

Electro-mechanical

With this device, when a current is supplied it energises a solenoid, pulling in a latch pedestal. This provides the fulcrum for a lever arm that forces the valve open as an electric motor turns a cam, forcing it down. As with the previous device, a limit switch stops the motor when the valve is fully open. Turning off the power de-energises the solenoid, which disengages the latch pedestal.

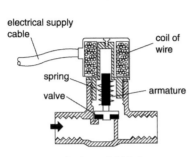

electrical supply cable

coil of wire

spring

valve

armature

Solenoid Valve

valve shown in the 'closed' position
(i.e. no power supply)

when power is supplied to the coil the
valve lifts into the magnetic field created

Solenoid Valve

Electro-Hydraulic Safety Shut-Off Valve

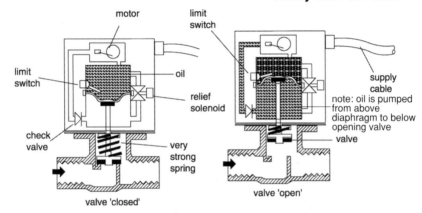

motor

limit switch

limit switch

oil

relief solenoid

check valve

very strong spring

valve 'closed'

supply cable

note: oil is pumped from above diaphragm to below opening valve

valve

valve 'open'

Electro-Hydraulic Safety Shut-Off Valve

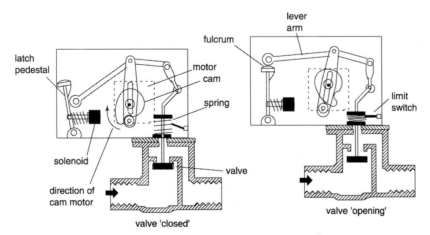

lever arm

fulcrum

latch pedestal

motor cam

spring

solenoid

valve

direction of cam motor

limit switch

valve 'closed'

valve 'opening'

Electro-Mechanical Safety Shut-Off Valve

Heat Operated Gas Control Valves

There are several types of heat operated valves that you will encounter.

Thermal Cut-Off

This is a device designed to close off the gas supply to an installation in the event of a fire. Several designs have been developed over the years; the one shown opposite consists of a powerful spring that has been compressed to keep the valve open. A low melting solder has been used to secure it in this position. If a fire raises the ambient temperature to above 95°C, the solder melts, allowing the spring to force the valve closed. A thermal cut-off is incorporated in the design of an E6 gas meter (see page 67 for an example).

Heat Motor

This gas valve is sometimes found incorporated in a gas cooker or multi-functional control. It consists of a small heater element, which is used to heat up a bi-metallic strip. One can see from the illustration opposite that, as the element coil heats up, the strip bends as a result of the different expansion rates of the two metals, and slowly closes down the supply. The pressure of the gas entering the valve also assists in fully closing the valve diaphragm. When no current is supplied to the element, the bi-metallic strip cools and re-straightens, causing the valve to snap back open. These controls can also be designed to have a normally closed position.

Pressure Stat

This device is designed to close down the flow of gas to appliances, such as pressure and steam boilers, when a pre-determine pressure has been achieved. The diagram opposite shows a connection seen from the top of the control where a steam connection is made. If the steam pressure generated is greater than the tension of the spring, the valve will close. Full closure may be obtained and sometimes a bypass is incorporated to allow for a low fire rate. Other designs of pressure stats allow for pressure adjustment or the stat itself may be incorporated with a control such as a relay valve. Some pressure stats work by operating an electrical connection, which controls a safety shut-off valve.

Thermistor Thermostat

This is a form of temperature control that uses an electrical non-metallic thermal resistor. These thermistors are very sensitive to temperature change. As they heat up, the current flow through them varies. Thermistors are available in many shapes and are suitable for measurement control up to a useful temperature limit of about 300°C. Because of their size (they are no larger than a small bead), they are used to measure temperature in the most inaccessible places. See atmospheric sensing device, page 108, for an example.

See also thermostats on page 100.

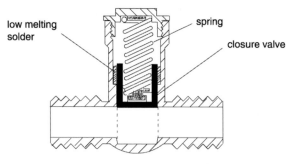

low melting
solder

spring

closure valve

Thermal Cut-Off

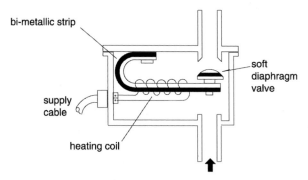

bi-metallic strip

soft
diaphragm
valve

supply
cable

heating coil

Heat Motor

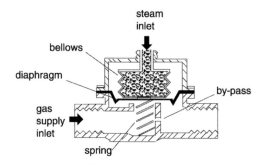

steam
inlet

bellows

diaphragm

by-pass

gas
supply
inlet

spring

Pressure Stat

Pressure Operated Controls Valves

There are several types of controls that detect pressure.

Non-Return Valve

These valves are also known as back-pressure valves or check valves. There are several designs. First, the most basic type, called a roll check valve and sometimes known as a flap valve, consists of a hinged flap, which pivots at a fulcrum to close in a no-flow, or back-flow condition. A second design consists of a metal disc that is held open by slight pressure from a spring. On top of the disc lie two leather or rubber washers that lift as gas flows through. When there is no flow or back flow the washers fall, sealing the outlet. Where excessive back pressure is encountered, the disc itself is forced down to overcome the spring pressure and close completely. The two non-return valves described above both need to be installed in the horizontal position. A different design overcomes this problem and can be fitted in any position. It consists of a pair of rubber diaphragms, housed inside a specially designed frame. Under normal gas flow the diaphragms are pushed apart to allow the gas to flow freely, but under back-flow conditions the back-pressure forces the two diaphragms together, stopping any flow.

Gas Relay Valve

This valve is designed to control the gas flow to the main burner of an appliance. Like the previous valves, it is operated by pressure. The relay valve works in conjunction with an adjoining weep pipe. This allows a small by-pass flow of gas into the appliance, which is ignited by a previously established pilot flame. This bypass or escape of gas through the weep pipe allows the pressure above the diaphragm in the relay to fall below that of the diaphragm. The diaphragm lifts as a result of the negative pressure and, in so doing, pulls open the valve. Installed on the weep pipe will be any collection of controls such as a gas cock, rod type thermostat or possibly a solenoid. In all cases where these should be in the closed position, the valve will prevent the escape of gas through the weep line and thus the pressure will build up above the diaphragm, equalising with that below so that the valve will drop and close under its own weight. It should be noted that the valve needs to be mounted horizontally. Sometimes a valve bypass screw is incorporated to allow the appliance to have a continued high–low control gas rate.

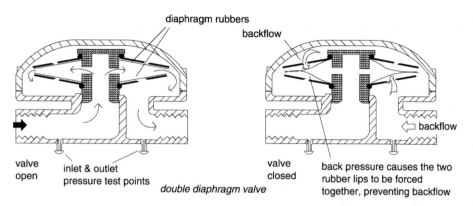

diaphragm rubbers

backflow

valve open

inlet & outlet pressure test points

double diaphragm valve

valve closed

back pressure causes the two rubber lips to be forced together, preventing backflow

backflow

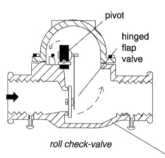

pivot

hinged flap valve

roll check-valve

Non-Return Valves

Note these two non-return valves must be installed in the horizontal position

Roll Check-Valve

Double Diaphragm Valve

Relay Valve

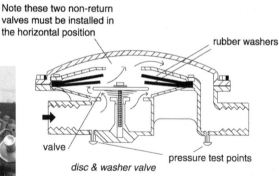

rubber washers

valve

pressure test points

disc & washer valve

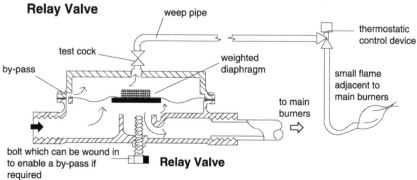

weep pipe

test cock

by-pass

weighted diaphragm

thermostatic control device

small flame adjacent to main burners

to main burners

bolt which can be wound in to enable a by-pass if required

Relay Valve

Low and High Pressure Cut-Off Devices

Low Pressure Cut-Off Device

This is a special valve that prevents gas entering a system of pipework until a pre-determined pressure has been established within the pipework. Its purpose is to prevent gas flowing out of any open ends when the supply is first turned on.

A typical example would be in a commercial kitchen, where the gas supply is electrically isolated when the kitchen is not in use. If an appliance tap was operated with the supply off, releasing the pressure, and a tap was left open, it could pose a potential hazard when the gas supply is restored. The low pressure cut-off, however, prevents this from occurring: all downstream valves would need to be closed and the valve re-set before the gas could flow. The low pressure cut-off device shown operates as follows:

1. Gas enters the system but cannot progress beyond the closed valve.
2. The re-set plunger is held open. This allows gas to flow slowly, by-passing the valve into the installation pipework.
3. If all controls are closed downstream of this valve, the pressure will slowly build up below the main diaphragm to give a surface area pressure greater than that exerted by the spring and the valve will be lifted.
4. Should the supply be isolated upstream, the valve will remain open. If the pressure is lost, however, the valve will drop and need resetting as in (1) above.

Nowadays pressure switches are invariably used in conjunction with a safety shut-off valve to perform the same operation.

Under Pressure Shut-Off Valve (UPSO)

This control is not dissimilar to a low pressure cut-off device and, in effect, performs the same task. Its mode of operation is slightly different, however, and it is invariability incorporated with a second stage regulator, as found on an LPG installation. When the device senses an abnormally low outlet pressure of between 25 and 32 mbar, it causes the diaphragm to drop to a particularly low position, resulting in the UPSO valve closing against its seating. Gas cannot therefore pass into the control valve until the reset rod has been lifted to reset the UPSO valve. This can only be achieved when all appliance controls are closed and pressure can build up in the outlet pipework.

High Pressure or Overpressure Shut-Off Valve (OPSO)

This control operates by detecting high pressure through an impulse duct. This causes the overpressure diaphragm to lift that, in turn, lifts the latch holding the slam shut plunger open. The slam shut is now free to move and a strong spring pushes the plunger, closing off the valve. Some overpressure shut-off valves incorporate a sealed manual reset. The OPSO used for an LPG supply would operate at 75 ± 5 mbar. The OPSO is often incorporated with an UPSO control.

Both high and low pressure shut-off valves would be incorporated within the medium pressure regulator now supplied on a natural gas installation; although the pressure tolerances would differ.

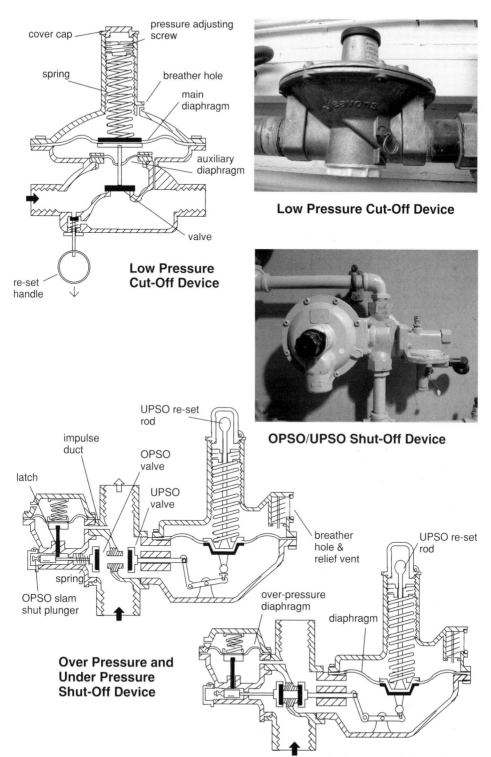

cover cap

pressure adjusting screw

spring

breather hole

main diaphragm

auxiliary diaphragm

valve

re-set handle

**Low Pressure
Cut-Off Device**

Low Pressure Cut-Off Device

OPSO/UPSO Shut-Off Device

impulse duct

OPSO valve

UPSO re-set rod

latch

UPSO valve

breather hole & relief vent

UPSO re-set rod

spring

OPSO slam shut plunger

over-pressure diaphragm

diaphragm

**Over Pressure and
Under Pressure
Shut-Off Device**

valve condition in which both OPSO & UPSO have tripped

Pressure Regulators

There are constant pressure fluctuations within the gas supply pipe caused by various influences on the system, ranging from the operation of appliances, to atmospheric conditions changing. The pressure in the distribution supply main also varies from district to district and, as more and more gas flows due to consumer use, so does the pressure. Therefore it is essential to control the pressure entering a property and appliance to ensure safe and complete combustion of the fuel. This is achieved by the use of a *regulator*, or *governor* as it is also known, which automatically adjusts and opens or closes the gas line.

Constant Pressure Regulator

This regulator is simply constructed, and may be found on the supply to an appliance. It works as follows:

1. With no gas pressure in the pipeline, the spring pushes on the diaphragm, causing the valve to open.
2. When connected to the supply, gas flows past the valve into the appliance or installation. If all the appliance controls are closed, the pressure in the outlet will build up until it equals the pressure of the spring. At this stage the diaphragm will be in a state of equilibrium. Any further pressure rise will simply act on the diaphragm forcing it upwards and, as a result, the valve would equally be pulled up to close the supply. *Note*: When the valve is fully closed it is referred to as 'locked-up' and the pressure within the system is referred to as the standing pressure.
3. Should an appliance be in operation, the 'locked-up' pressure would be released through the appliance injector, which would subsequently allow the spring to push on the diaphragm, so re-opening the valve. Thus, as the gas flows, the diaphragm monitors the pressure: continually rising and falling to meet the demand of the gas flow. The greater the volume of gas being consumed, i.e. the more burners in use, the more the valve will open.

Regulators are sometimes fitted with a weighted diaphragm and as such can only be located in an horizontal plane.

Adjustment of the outlet pressure is made using a manometer when the gas is flowing; it is referred to as the working, or operating pressure. With a manometer connected to the supply downstream of the regulator and with the cover cap removed, the spring pressure adjustor is turned clockwise to increase the pressure on the spring. This subsequently increases the gas pressure within the system, putting pressure on the diaphragm, causing it to close. For pressure readings and adjustments see page 46.

Compensated Constant Pressure Regulators

With the constant pressure regulator described above, the difference between the inlet and outlet pressures is not too great and therefore the upward and downward forces acting on the valve itself are virtually equal. However, as the difference in the

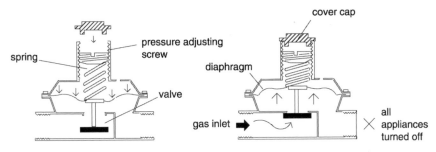

With no gas in the supply, the valve will be fully open, in readiness to receive the gas.

Gas has now been supplied to the valve, yet no appliances are open. Therefore the gas pressure overcomes the spring pressure, forcing the diaphragm up, bringing with it the valve, closing off the supply. Locking the gas within the installation pipe work.

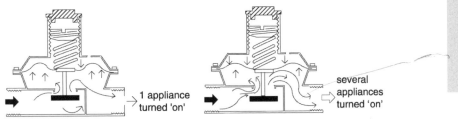

With one appliance on the locked-up pressure is released and the valve re-opens a little to maintain sufficient gas volume.

With several appliances turned on the valve opens further, therefore supplying a greater volume of gas flow.

Operation of a Constant Pressure Regulator

Constant Pressure Regulator Stripped Down, exposing Diaphragm and Spring

pressures become greater, the force on the inlet side of the valve exceeds the force on the outlet side and, as a consequence, the valve tends to close prematurely, reducing the outlet pressure. A typical example would be where gas is taken from the district main with an inlet supply pressure of up to 75 mbar and then reduced to approximately 21 mbar. To overcome this problem there is a need for compensation.

Compensation can be achieved by two basic methods. One is using a design that incorporates two valves, both attached to the same spindle. The diagram opposite shows that the inlet pressure is trying to close one valve but is also trying to open the other, thus resulting in a zero force influence. Smaller regulators tend not to favour this design because of the difficulty in adjusting the valves to close at exactly the same time. Instead they use another method, which uses an auxiliary diaphragm instead of the second valve. In the second diagram opposite, an auxiliary diaphragm is used. It can be seen that the inlet pressure pushes on the valve and the underside of the auxiliary diaphragm. This pressure on the valve is trying to push it open but, simultaneously, the pressure on the auxiliary diaphragm is trying to close it; this results in a zero force influence. As gas passes through the valve seating a small quantity passes up the impulse pipe, where it acts on the spring resistance, closing or reducing the valve opening, an operation previously described in the single diaphragm constant pressure regulator.

High-to-low Pressure Regulator

Nowadays with more and more demand for gas for more houses, the pressure in many districts has been increased above 75 mbar. Occasionally a supply is taken from the high-pressure main prior to the locally regulated low-pressure supply district. In these cases high-to-low pressure regulators become necessary. *Note*: These are sometimes simply referred to as high-pressure service regulators. These regulators are designed to reduce the pressure from around 2 bar down to 21 mbar in one step. An over pressure shut-off device, as previously described, is incorporated with the regulator. A photo of this type of regulator can be seen on page 167.

Zero Governor (Back-loaded Governor)

This design of regulator has a similar design to the compensating regulator above, the difference being that a tension spring is used rather than a compression spring. Thus instead of trying to open the valve, it is continually trying to close it. The valve will open only if the pressure downstream falls to that of atmospheric pressure, caused by suction. This negative pressure is caused by an air stream under pressure sucking gas from the pipeline and mixing it with air. It is used in the combustion process of many forced draught burners.

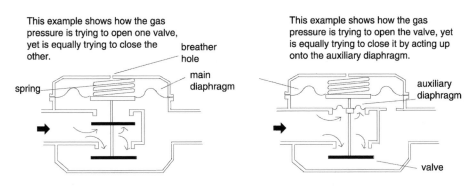

This example shows how the gas pressure is trying to open one valve, yet is equally trying to close the other.

breather hole

spring

main diaphragm

This example shows how the gas pressure is trying to open the valve, yet is equally trying to close it by acting up onto the auxiliary diaphragm.

auxiliary diaphragm

valve

In both examples it can be seen that zero force is being exerted on the valve allowing all influence for valve movement to be generated via the main diaphragm.

Principle of Compensated Pressure Regulation

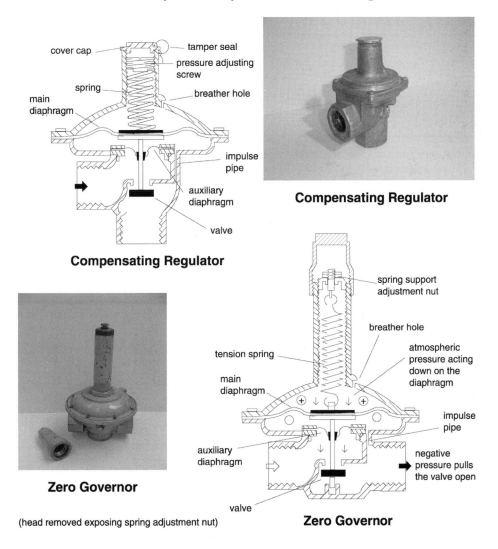

cover cap

tamper seal

pressure adjusting screw

spring

breather hole

main diaphragm

impulse pipe

auxiliary diaphragm

valve

Compensating Regulator

Compensating Regulator

Zero Governor

(head removed exposing spring adjustment nut)

spring support adjustment nut

breather hole

atmospheric pressure acting down on the diaphragm

tension spring

main diaphragm

impulse pipe

auxiliary diaphragm

negative pressure pulls the valve open

valve

Zero Governor

LPG Regulators 1

These regulators work on a similar principle to the natural gas regulators previously described, but because of the higher pressures generally involved they are of a more robust design. They either use two regulators, one reducing the pressure to an intermediate pressure and the other to the final system working pressure; this system is referred to as first and second stage regulation. Alternatively a single stage regulator is used that reduces the pressure to that required in one step.

Gas pressures within the LPG sector depend on the gas being supplied: either butane or propane. The gas is supplied in liquid form and the gas pressure generated is dependant on the ambient atmospheric conditions and temperature. Typical supply pressures would be in the region of 1.5–2 bar for butane and 6–7 bar for propane. The distinguishing difference between the regulators used for these two gases is the gas connection to the high-pressure supply. With butane the connection is either a plug-in bayonet type connection, as found on a cabinet heater or a female gas connection, whereas all propane connections are made via a male thread.

The regulator shown opposite illustrates a typical single stage LPG regulator and works as follows:

1. With the appliance taps closed, the pressure within the system pushes the diaphragm up, drawing the valve with it by means of a connecting lever, to the closed position.
2. Should an appliance valve be opened, the pressure on the underside of the diaphragm would drop, releasing the locked up pressure and the valve would open to allow more gas to flow in through the valve. This operation continues until the pressure builds up, re-closing the valve.

Typical LPG regulators

Make	Colour	Type	Gas	Inlet pressure	Outlet pressure
766P	Red	Single stage	Propane	Cylinder	37 mbar
766B	Blue	Single stage	Butane	Cylinder	28 mbar
315	Light green	First stage	Propane	20 bar	750 mbar
2550	Dark green	Second stage	Propane	750 mbar	37 mbar
Compact 100	Red	Single stage automatic change-over valve	Propane	Cylinder	37 mbar
541	Blue	Single stage bayonet	Butane	6 bar	28 mbar

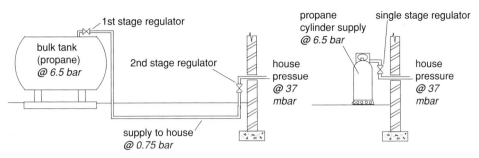

Typical Regulator Locations

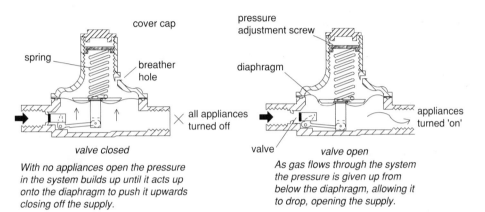

valve closed

With no appliances open the pressure in the system builds up until it acts up onto the diaphragm to push it upwards closing off the supply.

valve open

As gas flows through the system the pressure is given up from below the diaphragm, allowing it to drop, opening the supply.

Single Stage Regulator

Various LPG Regulators

Propane Connection to Gas Cylinder

Butane Connection to Gas Cylinder

LPG Regulators 2

Automatic Change-Over Valves

This is a specially designed regulator that automatically closes the supply from one cylinder or bank of cylinders and opens another in order to ensure continuity of the gas supply, thus preventing any loss of supply when the cylinders have run out of gas. These can be used in conjunction with a second stage regulator to give a high pressure outlet supply or, as is more typically the case, they are used to reduce the pressure down to 37 mbar in a single stage. It will often be found that these controls have an over pressure shut off (described on page 82) incorporated with the valve.

The operation of the automatic change-over is as follows:

1. When the manually operated lever control handle is turned to the on position, the angular base of the actuator applies pressure to the valve it faces. This overcomes the closure spring pressure and opens the valve. The high pressure then enters the valve and operates as previously described in the general operation of regulators.
2. When the supply pressure has finally all gone, the actuator spring pushes down further against the loss of pressure and eventually the valve from the second supply line is made to operate. At this time a red indicator is displayed on the control handle informing the user that the cylinder supply has become exhausted. This indicator was previously white.
3. At this stage the manually operated lever control handle can be turned to face the second gas supply. This causes the indicator to turn back to white but, more importantly, it allows the first exhausted supply valve to close. With the gas still flowing, the empty cylinders can now be exchanged for new full bottles.

Note: If the cylinder capacity was undersized the change-over would occur prematurely with the result that the user would be unable to obtain the full quantity of gas from the cylinders. This is discussed in 'Vapour Off-Take Capacity' on page 144.)

Single Stage Bayonet type Low Pressure Butane Regulator

This type of regulator is used where there is a minimal amount of gas flow required, such as to a small caravan or cabinet heater. The regulator is simply offered to the compact valve located on the top of the butane gas cylinder. The locking device when turned a quarter turn clamps the regulator to the valve and with a second quarter turn allows gas to flow. It works on the same basic principle as the regulator described on the previous page. Note the correct manufactures regulator must be selected as there are slight variances in valve connection sizes.

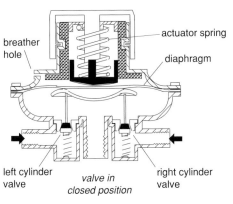

breather hole

actuator spring

diaphragm

left cylinder valve

valve in closed position

right cylinder valve

control handle

actuator

closure spring

valve in 'open position', taking supply from right cylinder

Automatic Change Over Valve

gas supply pressure exhausted

right cylinder supply exhausted & now automatically taking gas from left cylinder

Automatic Change Over Valve

Bayonet Type Butane Regulator

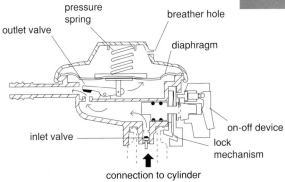

pressure spring

outlet valve

breather hole

diaphragm

inlet valve

on-off device

lock mechanism

connection to cylinder

Bayonet Type Butane Regulator

Flame Supervision Devices

The term flame supervision device (FSD) refers to a gas control fitted into the pipeline. It is designed to turn off the main gas flow to an appliance in the event of flame failure. These devices are sometimes referred to as flame failure devices (FFD). The presence of the flame is detected by one of the following methods:

- using heat generated by the flame;
- allowing electricity to pass through a flame;
- detecting the light emitted from the flame.

Some of the devices identified below are no longer seen as common controls: future developments, aiming for improved efficiencies, will inevitably restrict their long term future.

Bi-metallic Strip FSD

This device is no longer used as a modern gas control, however they are still be found on older appliances, such as water heaters, cookers and warm air units, etc. The device relies on the principle that a strip of two metals, with different expansion rates, and bonded together will bend when exposed to heat. This bending movement forces the gas valve to open, allowing the main gas to flow into the appliance. A pilot flame is used to supply the heat to bend the bi-metallic strip and ignite the main gas flow. Should the pilot flame be extinguished, the bimetallic strip would cool and the main gas valve will close. Problems with this design of FSD include the following:

- If the pilot light goes out, the gas freely discharges gas through the pilot tube. Thus the pilot flame itself is unprotected.
- The continuously burning pilot flame is a waste of fuel.
- Distortion of the strip or valve spindle results in operational failure.

Liquid Expansion FSD

As with the previous control, this device has no control over the pilot flame and, in the event of flame failure, gas continues to enter the appliance through an uncontrolled pilot or by-pass. Few appliances use this kind of FSD but possibly the best example would be that found in a gas oven, as illustrated opposite. A small phial containing a volatile fluid turns into a vapour when heated. This vapour has a much greater volume than the liquid, therefore its expansion causes the bellows to increase in size, forcing the valve to open. *Note*: The initial pilot flame that was used to generate the heat was supplied with gas through a by-pass hole. Should the supply to the unit be interrupted, the flame would go out. The vapour would then condense back into its liquid form and be forced back, by the spring-loaded valve, along the capillary tube to the phial.

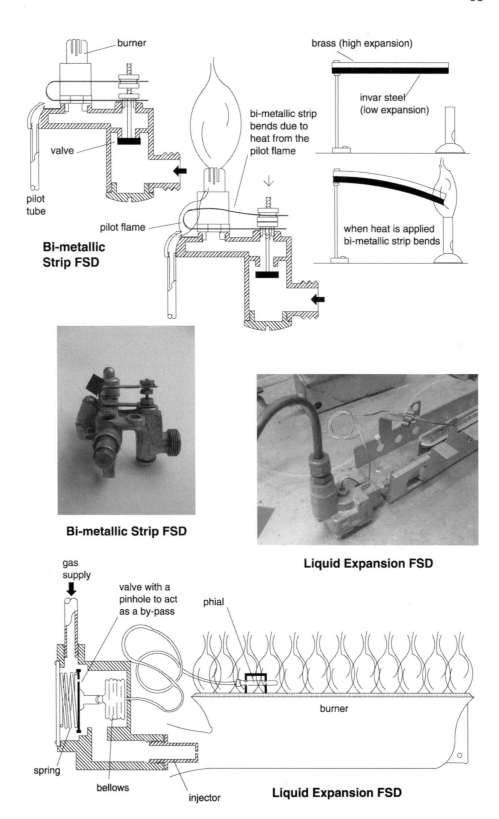

burner

brass (high expansion)

bi-metallic strip bends due to heat from the pilot flame

invar steel (low expansion)

valve

pilot tube

when heat is applied bi-metallic strip bends

pilot flame

Bi-metallic Strip FSD

3 Gas Controls

Bi-metallic Strip FSD

Liquid Expansion FSD

gas supply

valve with a pinhole to act as a by-pass

phial

burner

spring

bellows

injector

Liquid Expansion FSD

Thermo-electric FSD

This device, certainly for the smaller domestic gas appliance, is possibly the most commonly-encountered FSD. However, with greater improved efficiency now being sought from appliances, it may now be replaced by units that rely on electronic ignition. The biggest concern is that the pilot flame burns night and day, wasting fuel gas when the appliance is not in use. It is referred to as a thermo-electric device because it uses heat to generate electricity. The current produced is used to operate an electromagnet (solenoid), which holds the gas valve open. Thermoelectric valves come in all shapes and sizes and can be seen on boilers, cookers, water heaters, fires, etc. The device is often difficult to detect with the untrained eye, and may form part of a larger multi-functional gas control. The give-away is the location of a thermocouple connected between the FSD and pilot flame.

The thermocouple is the device that generates the electrical current. In its simplest form, a thermocouple consists of two different metals connected together at one end. The other ends are connected together by a coil of wire. When heat is applied to the point where the two metals are joined, referred to as the hot junction, a small electromotive force (e.m.f.) of around 15–30 mV is generated. This causes a current to flow along one metal, around the coil and back along the other metal to return to the hot junction, making a circuit. This continues as long as heat is supplied. The current flowing round the coil converts the electrical energy to magnetic energy and, as previously stated, this magnetism is used to keep the gas valve open. Typical metals used for the thermocouple tip are copper and alloys such as constantan, chromel and alumel. The electrical conductor between the thermocouple tip and the electromagnet consists of a small copper tube with an insulated wire running through the centre.

The thermoelectric valve works as follows (see diagram opposite):

1. The reset button is depressed. This closes off the outlet and then allows the inlet valve to open. Gas can now flow through and pass into the pilot tube.
2. Gas discharging from the pilot injector is ignited and the flame burns, playing on to the thermocouple hot junction.
3. After 10 seconds or so, a sufficient e.m.f. has been generated to allow the current to flow round the electromagnet. The pressure can then be taken off the reset button. The magnet continues to hold the inlet valve open and the outlet valve returns to allow the gas to flow to the next stage in readiness to enter the appliance should heat be called for. This main flow of gas would be ignited by the previously established pilot flame. Should the pilot flame go out, the thermocouple hot junction would cool and the valve would fully close, assisted by the inlet valve spring.

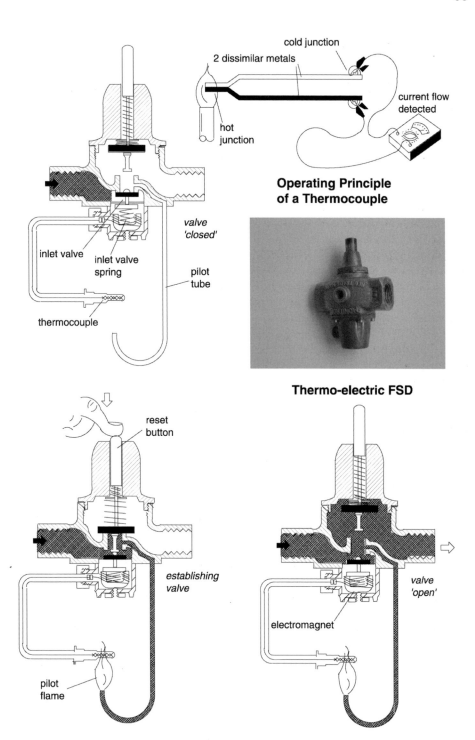

cold junction

2 dissimilar metals

current flow
detected

hot
junction

**Operating Principle
of a Thermocouple**

valve
'closed'

inlet valve

inlet valve
spring

pilot
tube

thermocouple

Thermo-electric FSD

reset
button

establishing
valve

electromagnet

valve
'open'

pilot
flame

Thermo-electric Flame Supervision Device

Flame Rectification FSD

As electricity can flow through a flame, its presence can be detected by positioning a probe in its path. This is located a small distance off the burner head. When a current is supplied, if a flame is present it forms part of the electrical circuit. Should there be no flame, the circuit would be broken and, as a result, the relay coil and the electromagnet would be de-energised and close the gas supply.

Early experiments using this concept used direct current (d.c.), which had a drawback in that, for example, should the probe be dislodged or a conductive object fall on the probe, causing it to touch the burner head, it would complete the circuit, giving the false impression that a flame was present. Therefore in order to avoid this, another method has been developed in which an alternating current (a.c.) is passed round the circuit (a.c. is the flow of the infinitely small electrons first flowing in one direction then, a split second later, in the other direction, continually back and forth throughout the circuit). Because the probe located in the flame is very small compared with the size of the total flame area, many of the electrons miss the probe and so the electron flow or current is not detected at the probe. Current is therefore only detected in one direction, i.e. when the electron flow is towards the burner head. This is referred to as a partially rectified a.c. current or, in simple terms, d.c. Thus a.c. is passed into the probe, but only d.c. is detected back at the control unit.

Should the probe become dislodged and touch the burner head, a.c. current would be detected at the control unit and it would then shut down the gas inlet valve.

Photo-Electric FSD

When a flame burns it gives off ultra-violet (UV) light. Therefore, by careful design, a UV detector would pick up its presence. Effectively the UV detector is an electrical resistor; if the detector senses the presence of a flame it allow electricity to flow in a circuit. Thus, as with the flame rectification FSD, if the circuit is maintained, it allows the electromagnetic gas valve to open. Where no flame is present, the valve will instantly close, shutting off the flow of gas.

The detector consists of a UV transmitting glass, which is filled with inert gas. It has two electrodes positioned parallel to each other with a small gap between them. In the presence of UV light, electricity flows freely across the gap to maintain the circuit.

Flame rectification and photo-electric FSDs have the advantage over the previously used FSDs in that:

1. no permanent pilots are required;
2. complete gas isolation is maintained when the appliance is not in use; and
3. they are very quick to respond to the presence of a flame.

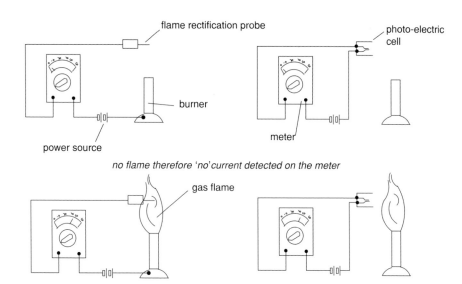

no flame therefore 'no' current detected on the meter

with a flame present current flow is detected at the meter

Operating Principle of Both Flame Rectification and Photoelectric Flame Detection

Photoelectric Cell

Flame Rectification and Spark Electrode
(picture looking onto the head of a force draught burner)

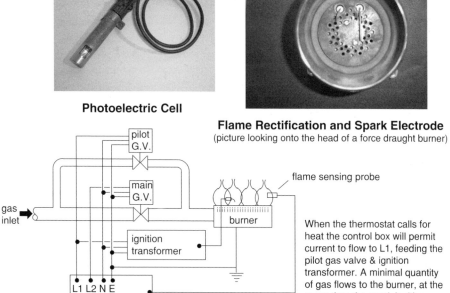

Ignition Sequence using Flame Rectification or Photoelectric cell

When the thermostat calls for heat the control box will permit current to flow to L1, feeding the pilot gas valve & ignition transformer. A minimal quantity of gas flows to the burner, at the same time the spark ignition occurs. Should a flame be detected, the control box will permit a live current to supply L2, the main gas gas valve. If no flame is detected within a set period the control will close the supply to L1.

Testing Flame Supervision Devices

A flame supervision device needs to be checked periodically to ensure that it is still operating effectively within a specified period of time, thereby preventing a dangerous build up of gas within the combustion chamber. The length of this period is determined by the manufacturer. With appliances that use small quantities of gas, such as those in the domestic use, maximum operating times are typically as follows:

Conventional gas fires	180 seconds
DFE gas fires	120 seconds
Cookers	90 seconds
Storage water heaters <8kW	90 seconds
All other appliances <60kW	60 seconds

FSDs such as the bi-metallic strip and the liquid expansion do not shut off fully. However, a check still needs to be made to ensure that they operate, closing the gas down to the by-pass rate. With these control valves the best procedure is to turn off the gas supply for a few seconds, no longer than suggested above, and then turn it back on. The gas flow should have fallen accordingly. For thermo-electric FSDs where the gas is cut off completely it would be appropriate to confirm a gas-tight closure of the valve on to the seating. This can be achieved in one of two ways.

Method 1

With a manometer located on a test point before the FSD, the supply is closed off for a short period, until the sound of the solenoid valve dropping is heard. This should be no longer than the times listed above. The isolation valve is then opened to charge the section of pipe, which can be identified by observing the manometer. This isolation valve is then re-closed. If the FSD has closed fully and there are no leaks on this section of pipework the gas pressure will hold up. Should the pressure be seen to fall, the valve is not closing fully or there is a leak in this section of the installation which must be treated as immediately dangerous until rectified.

Method 2

This method relies on detecting gas at the burner pressure test point. It is not as accurate as the previous method and a small discharge through the pilot flame, etc. may go undetected. As before, with the manometer connected the gas is shut off until the valve drop out is heard. The isolation valve is then opened. Should the valve be letting gas by, a pressure rise will be seen in the manometer.

Checking Flame Rectification and Photo-electric FSDs

The above method can be employed to check appliances that use this method of flame detection. However invariably the appliance will try to re-ignite the fuel gas by initiating the spark ignition process. It is possible to simulate flame failure by turning off the gas supply, then turning off the electricity, thereby enabling a check to confirm that no gas remains flowing to the burner under this fault condition. Alternatively the lead from the flame probe can be disconnected; this would cause the burner to lock out.

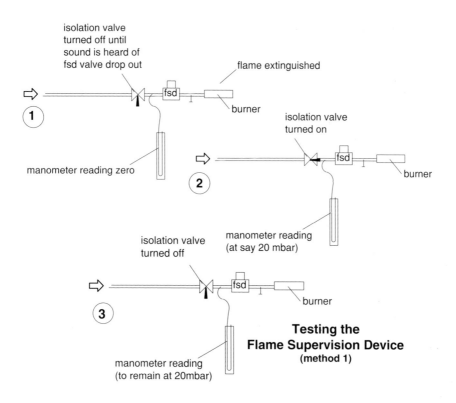

isolation valve
turned off until
sound is heard of
fsd valve drop out

flame extinguished

fsd

burner

1

isolation valve
turned on

fsd

burner

2

manometer reading zero

manometer reading
(at say 20 mbar)

isolation valve
turned off

fsd

burner

3

manometer reading
(to remain at 20mbar)

Testing the
Flame Supervision Device
(method 1)

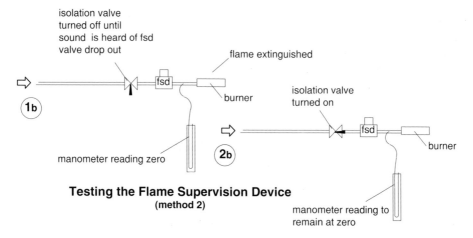

isolation valve
turned off until
sound is heard of fsd
valve drop out

flame extinguished

fsd

burner

1b

manometer reading zero

isolation valve
turned on

fsd

burner

2b

Testing the Flame Supervision Device
(method 2)

manometer reading to
remain at zero

Thermostatic Control

Thermostats are incorporated in many gas appliances to reduce the gas input, thereby reducing the temperature. Some thermostats are designed to provide a gradual or modulating control over the flow of the gas, whereas others are designed to give an on/off control. For example, a gas oven would clearly need to have a variable control, whereas a typical conventional boiler may switch on and off as the thermostat calls for heat. Thermostats work on several principles including: Bi-metallic expansion, Liquid expansion and Electrical resistance, such as the Thermistor as identified on page 78.

Rod Type Bi-metallic Thermostat

Generally these are only found on older gas equipment. The thermostat consists of two dissimilar metals, namely a brass tube, which has a high expansion rate and an Invar steel rod, which has a very low expansion rate. The Invar steel rod is secured at one end inside the brass tube. The brass tube is, in turn, secured to the body of the valve, see diagram. Thus, as the heat of the appliance warms the tube it expands and in turn pulls the Invar rod away from the valve seating allowing it to close, assisted by a spring, turning off the supply. To enable variable operational temperatures, a control knob can be adjusted to alter the distance the rod will need to travel in order for the valve to close fully.

Note: Where these appliances are used in conjunction with a cooker, the valve body incorporates a by-pass hole to allow for the oven to drop to a simmer, and thus prevents the flame going out.

Liquid Expansion Thermostat

These are far more common and are found on most gas appliances. Two designs are found, both working on the same principle, however they differ in that the expansion of the liquid or vapour could either open/close the gas line (typically found in cookers) or could make/break a switch in an electrical circuit (typically found in boilers). Electrical thermostats can also be designed to drop out completely, requiring manual intervention as in the case of high limit thermostats.

The unit consists of a remote phial, which is housed high in the appliance and a capillary tube that joins it to the bellows chamber. Within the phial is a volatile fluid that has a very rapid expansion rate. As the phial heats up the liquid expands, in some cases changing to a vapour, and passes through the small capillary tube to fill the bellows. The bellows in turn increases in size to take up this expansion and, in so doing, breaks the electrical circuit. In the case of a mechanical valve, it closes off the gas supply. On cooling, the fluid contracts and is forced back by the pressure of a spring acting on the bellows, through the capillary tube to the phial. *Note*: As noted earlier with rod thermostats, when a mechanical thermostat is used in conjunction with an oven it requires a by-pass.

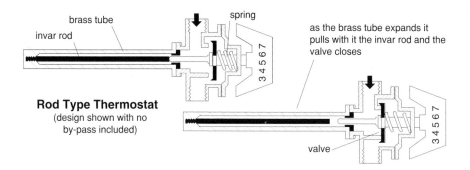

Rod Type Thermostat
(design shown with no
by-pass included)

brass tube
invar rod
spring
as the brass tube expands it
pulls with it the invar rod and the
valve closes
valve

Rod Type Thermostat

Liquid Expansion Thermostats

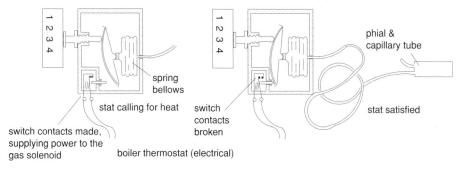

spring
bellows

stat calling for heat

switch contacts made,
supplying power to the
gas solenoid

boiler thermostat (electrical)

switch
contacts
broken

phial &
capillary tube

stat satisfied

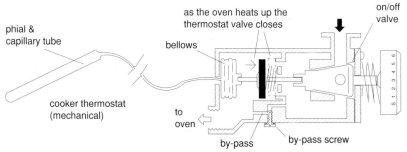

phial &
capillary tube

cooker thermostat
(mechanical)

bellows

as the oven heats up the
thermostat valve closes

on/off
valve

to
oven

by-pass

by-pass screw

Liquid Expansion Thermostats

Multifunctional Gas Valve 1

As the name implies, this gas control has several functions. The valve was developed many years ago, incorporating many controls within one unit, thus saving space and allowing the appliance to be more compact and therefore smaller. The multifunctional control may include any or all of the following components:

- filter;
- safety shut off valve (thermoelectric valve);
- regulator;
- solenoid valve (opens when appliance calls for heat);
- gas inlet pressure test point and burner pressure test point;
- pilot adjustment screw.

There are many different designs of multifunctional control, each offering a slight variation in operation. Two units are described in this book. First, the one shown opposite, that requires a permanent pilot flame and the second, over the page, is a design that works in conjunction with some form of electronic ignition.

Picture 1: Shows the valve off.

Picture 2: Depressing the button opens the pilot valve, allowing gas to flow to the pilot burner, where it can be ignited.

Picture 3: After some 10–15 seconds, the thermoelectric valve will operate, allowing the latch lever to be drawn in, engaging the safety shut-off valve. Thus when the pressure is removed from the button it rises, pulling with it the safety valve lever and so opening the valve. Note: The pilot valve remains open as it is held open by the latch engaging the safety valve lever.

Picture 4: Should the thermostat be calling for heat, current is supplied to the solenoid. This opens the servo regulator valve and closes the weep valve outlet. Gas now flows to the underside of the servo regulator diaphragm and the working pressure diaphragm where it overcomes the valve spring pressure, forcing the main gas valve to open. Gas can now flow to the main burner and be ignited by the previously established pilot flame. Any adjustment to the outlet pressure adjustment screw alters the servo regulator position, which in turn adjusts the pressure to the working pressure diaphragm and so maintains a constant outlet pressure.

When the appliance temperature is sufficient, the solenoid will be de-energised, which will cause the servo regulator valve to close and allow the weep valve to re-open. The pressure within the regulating chamber will drop rapidly to zero, allowing the valve spring to close the main gas valve, as shown in Picture 3.

Picture 5: To shut down, the control the button is simply turned 15°–20° and released. This allows the latch lever to disengage the safety valve lever and closes off the pilot valve. Note: It is impossible to re-establish the control until the magnetic valve is de-energised, re-lifting the latch lever in readiness to grab the safety valve trip.

Principle of Operation of a Multifunctional Gas Valve

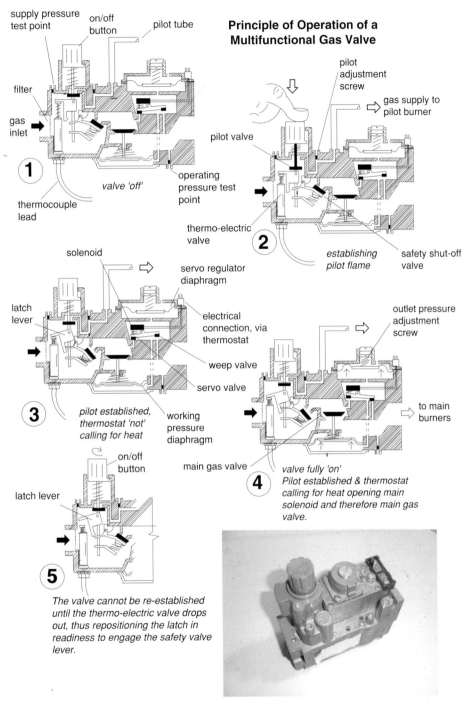

supply pressure test point

on/off button

pilot tube

filter

gas inlet

thermocouple lead

1

valve 'off'

operating pressure test point

pilot adjustment screw

gas supply to pilot burner

pilot valve

thermo-electric valve

2

establishing pilot flame

safety shut-off valve

solenoid

servo regulator diaphragm

latch lever

electrical connection, via thermostat

weep valve

servo valve

3 *pilot established, thermostat 'not' calling for heat*

working pressure diaphragm

outlet pressure adjustment screw

to main burners

main gas valve

4 *valve fully 'on'*
Pilot established & thermostat calling for heat opening main solenoid and therefore main gas valve.

on/off button

latch lever

5

The valve cannot be re-established until the thermo-electric valve drops out, thus repositioning the latch in readiness to engage the safety valve lever.

Multifunctional Gas Valve

Multifunctional Gas Valve 2

The multifunctional valve shown here is typical of a design that is used in conjunction with an appliance that uses electronic ignition and a forced draught burner. The speed of the combustion fan is directly proportional to a variable modulating range of gas rates, thus the greater the fan speed the greater will be the flow of gas and consequently the heat input. As a result, there is a saving in the amount of gas consumed by the appliance because there is no permanent pilot flame burning 24 hours a day, every day. Attached to the side of the valve is the ignition function circuit board, as shown in the photograph.

The valve identified uses a servo pressure regulator that works in conjunction with the fan that supplies air to the appliance. This is shown in the top diagram. The fan causes a negative pressure just in front of the gas injector, pulling the gas from the gas supply. This negative pressure is due to the action of the force draught air supply whose velocity is increased as it passes through the restriction in the air supply line. It could be likened to a similar concept, the zero governor, previously mentioned on page 86. This multifunctional valve eliminates the need for a pressure switch.

The valve shown opposite operates on the following principle:

1. With the thermostat calling for heat, both the safety shut-off and operator solenoids open. This allows gas to flow through the main safety shut-off valve, then into the servo inlet orifice and eventually to the underside of the servo and working pressure diaphragms. This pressure alone would be insufficient to open the main gas valve to the burner.
2. If the fan is operational, the pressure differential each side of the servo diaphragm causes the servo valve to move towards its seating. This causes the gas pressure to build up below the servo and in turn beneath the working pressure diaphragm. Eventually it overcomes the main valve spring, allowing the valve to open and gas to flow to the burner. The greater the pressure on the servo, the more the main gas valve will open.
3. When the thermostat is satisfied, the two solenoids close and the fan ceases to run. With the pressure removed from the servo diaphragm, it lifts as the main valve spring pushes the gas back out from beneath the working pressure diaphragm. Any locked up gas is discharged via the weep line into the appliance.
4. At (2) above, if the fan fails to operate during the initial operation of the appliance, the servo diaphragm will not be pulled down and therefore will not prevent the build up of gas pressure below the working pressure diaphragm. Therefore no gas will flow through to the burner injector. Any gas passing down through the weep line would be insufficient to establish a suitable flame and the appliance would go into lock-out because of the un-detection of a flame by the flame rectification flame failure device, subsequently the solenoids would be made to close the valve.

The top diagram shows the valve in the open position, and the bottom diagram shows the control fully closed. In addition to the components shown, there would be the devices such as pressure test points, filter, etc.

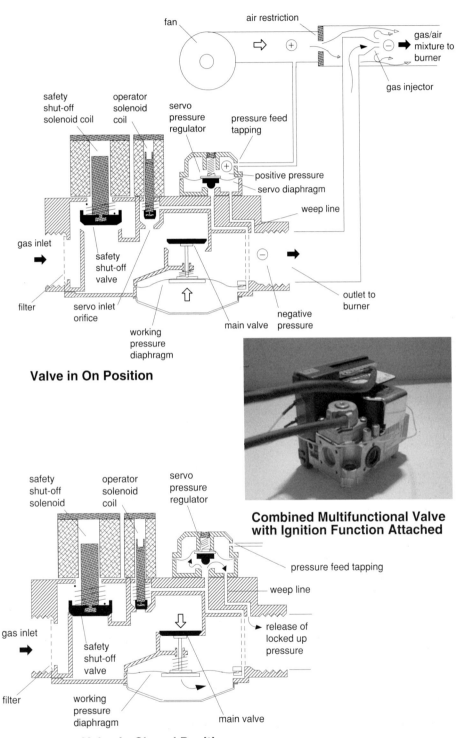

Valve in On Position

**Combined Multifunctional Valve
with Ignition Function Attached**

Valve in Closed Position

Ignition Devices

There are several forms of ignition device as described below.

Filament and Hot Surface Ignition

This design of ignition consists of a filament or coil of high resistance metal, such as platinum, which glows red-hot when a current of electricity is passed through it. As the gas flows past the glowing coil it ignites. The power supply usually comes from a transformer. However, in some of the older filament igniters used on gas fires, a battery was often used. The glow coil igniter is not common, but can still be found in some commercial warm air units.

Piezo-electric Ignition

This type of igniter is commonly found on gas fires, water heaters and existing boilers. This spark ignition device consists of quartz, or similar crystals, which when exposed to pressure or stress produce a voltage of around 6000 V. The one shown uses a press button that, when depressed, allows the plunger head to strike the crystals, causing a spark to be generated. You can see from the diagram how the spark jumps to earth at the point where the electrode is positioned.

Mains Ignition Transformer

This is simply a step-up transformer. It differs from the previous method of spark generation by supplying a continuous rapid spark that jumps or arcs between two electrodes or between one electrode and earth. The voltage generated would be typically 5000–10 000 V. Higher voltages are obtained using two electrodes arcing together, see diagram opposite.

Electronic Pulse Ignition

With this method of spark generation, instead of the continuous arcing across the electrodes as obtained by an ignition transformer, a controlled spark rate, typically of around 4–8 sparks per second is maintained. This is achieved by the use of electronics. The basic operating principle is as follows:

1. When the main switch is made, current flows through the rectifier and charges up the capacitor. [*Note*: A rectifier acts as a non-return, allowing current to flow only in one direction. A capacitor acts like a storage unit, holding a charge or volume of electricity until it can be used or discharged.]
2. After a predetermined period, and at set intervals, the timing switch closes, allowing the current to flow through the transformer. When this switch is closed the capacitor can discharge its contents and, coupled with the normal electron flow, a high impulse flows through the step-up transformer causing a 15–20 000 V spark to jump across the electrode gap. The timing switch then breaks and so the capacitor re-charges in readiness for the next timed cycle a split second later.

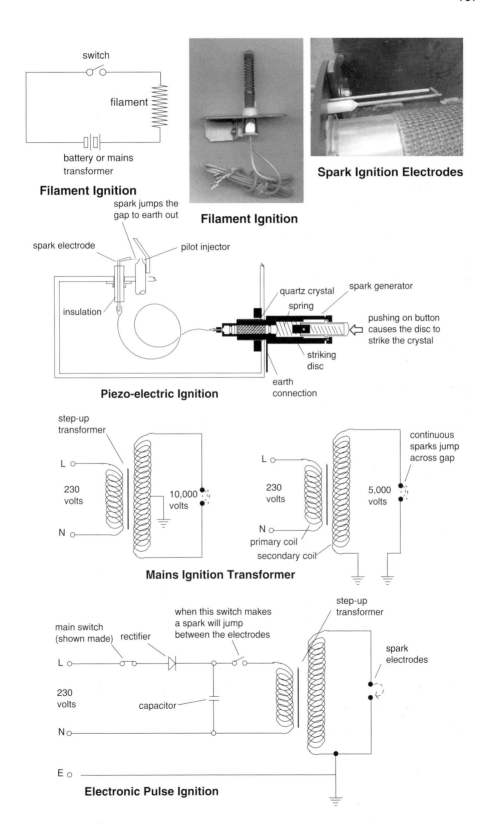

Filament Ignition

switch

filament

battery or mains transformer

Spark Ignition Electrodes

Filament Ignition

spark jumps the gap to earth out

spark electrode

insulation

pilot injector

quartz crystal   spark generator

spring

pushing on button causes the disc to strike the crystal

striking disc

earth connection

Piezo-electric Ignition

step-up transformer

L

230 volts

N

10,000 volts

continuous sparks jump across gap

L

230 volts

N

primary coil

secondary coil

5,000 volts

Mains Ignition Transformer

main switch (shown made)   rectifier

when this switch makes a spark will jump between the electrodes

step-up transformer

spark electrodes

L

230 volts

N

capacitor

E

Electronic Pulse Ignition

Vitiation Sensing Devices

Where an open flued or flueless appliance has been installed, it is possible that the oxygen within the room could become depleted as a result of inadequate ventilation or a poor flue arrangement. The lack of oxygen is called vitiation. Should vitiation occur, incomplete combustion of the fuel will result. A vitiation device is a device that will shut down the appliance, preventing its further use without manual intervention. There are two basic types of vitiation device: those that monitor the oxygen supplying the pilot burner and those that sense heat, caused by spillage. These two devices are described below.

Oxygen Depletion System (OSD)

This method of vitiation detection works by sensing a restriction of the primary air used for combustion of the pilot flame. During normal operation, primary air to supply the pilot flame is initially drawn in through an aeration port, which allows the flame to burn at the point of discharge in the typical stable fashion, characterised in a blue flame playing on the thermocouple. Should the environment lack sufficient oxygen, the amount of primary air would become depleted. This would have the effect of starving the flame of sufficient oxygen and the pilot flame would burn yellow and not be strong enough to play on the thermocouple. As a result, the thermocouple would cool, causing the thermoelectric valve to drop out. With appliances such as gas fires the aeration port is visible and during any service work it should be inspected to ensure that it is not blocked. A blockage would starve the flame, causing the problem identified above. Back boilers, on the other hand, have a sampler tube that passes from the aeration port up to a point above the heat exchanger. So where incomplete combustion occurs, due to spillage, vitiated air is drawn down through the tube and detected by the pilot jet. The back boiler sample tube can also be seen on page 337.

Atmospheric Sensing Device (ASD)

This device is usually fitted inside the down draught diverter and consists of a thermistor or similar heat-sensing device. It has two wires that are connected and join the internal conductor wire of a thermocouple interrupter. Under normal operating conditions, dilution air is drawn over the heat sensor. However, during persistent and continued downdraught and spillage the sensor heats up and breaks the circuit between the thermocouple and thermoelectric device. The valve then drops out and requires manual intervention to reset. Sometimes, instead of interrupting the thermocouple, the ASD is linked to the main burner solenoid or printed circuit, with the same effect.

Many flueless appliances have this type of device fitted within the top third of the unit. It is designed to detect high temperatures that could otherwise damage the appliance, i.e. it senses the atmosphere.

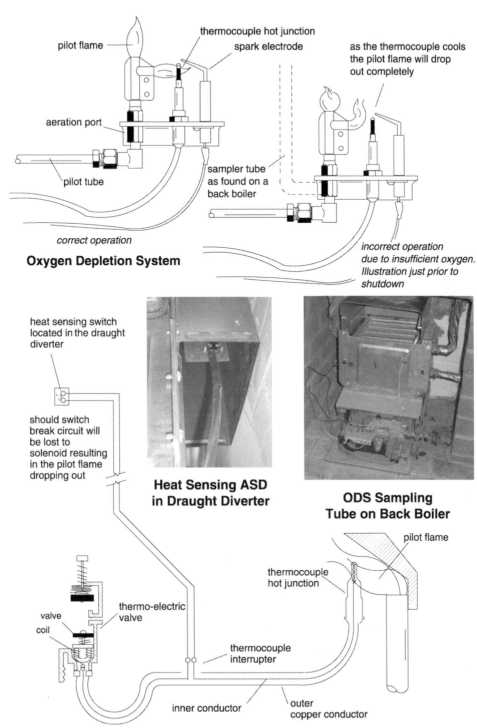

pilot flame

thermocouple hot junction

spark electrode

as the thermocouple cools the pilot flame will drop out completely

aeration port

pilot tube

sampler tube as found on a back boiler

correct operation

Oxygen Depletion System

incorrect operation due to insufficient oxygen. Illustration just prior to shutdown

heat sensing switch located in the draught diverter

should switch break circuit will be lost to solenoid resulting in the pilot flame dropping out

Heat Sensing ASD in Draught Diverter

ODS Sampling Tube on Back Boiler

pilot flame

thermocouple hot junction

valve coil

thermo-electric valve

thermocouple interrupter

inner conductor

outer copper conductor

Atmospheric Sensing Device

Pressure and Flow Proving Devices

Nowadays, to overcome the inherent problems associated with diaphragm type valves, which have a tendency to stick, leak and generally suffer the wear and tear of the large diaphragms, a pressure switch is often used, operating some form of safety shut-off device. Additionally many varieties of modern flued gas appliances now incorporate a fan to ensure positive dispersal of the combustion products. These are situations where the pressure switch and flow switch have been developed and incorporated. They include the following:

Pressure Switch

A device that detects air/gas pressures and uses the movement detected to operate a small electrical switch (micro-switch). A pressure switch may be mounted directly on to pipework or connected via a series of rubber tubes, as in the case of pressure switches fitted to many fan assisted boilers. Inside the pressure switch is a small diaphragm washer. On one side is the air/gas sensing tube and on the other the micro-switch. As the diaphragm moves in response to the pressure applied, so it makes or breaks the electrical contacts.

Flow Switch

This device comprises a vein or paddle that moves in response to air movement. As the paddle moves it pivots at a fulcrum and, in so doing, pushes together the contacts of a micro-switch. As with the previous control switch, this electrical response can then be used to open a gas control valve.

Transducer

This device senses pressure and converts it to an electrical impulse. However, the term 'transducer' relates to any device that converts one energy form to another. Thus, in effect, the above two devices are both transducers, as are piezo-electric ignition devices and thermistor thermostats.

There are several types of pressure/electrical transducers. Some rely on converting the energy of movement to adjust the wiper of a variable resistor, as illustrated. Some use quartz or similar crystals; these have the advantage that they have no moving parts. Transducers are found where printed circuit boards (PCBs) and electronic operation of an appliance is required, as they can be relatively small.

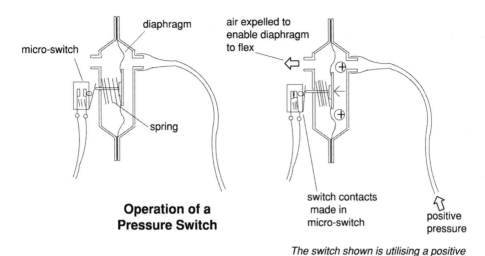

diaphragm

micro-switch

air expelled to
enable diaphragm
to flex

spring

switch contacts
made in
micro-switch

positive
pressure

**Operation of a
Pressure Switch**

*The switch shown is utilising a positive
pressure, however a variation consists
in utilising a negative pressure, sucking
the diaphragm towards the micro-switch.*

Various Pressure Switches

Flow Switch

micro-switch

rubber seal

when a flow is detected
the paddle rocks and
makes the electrical
micro-switch contacts

Flow Switch

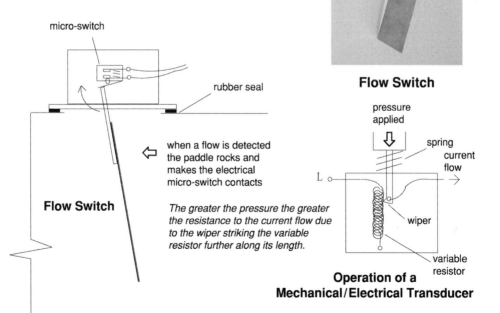

*The greater the pressure the greater
the resistance to the current flow due
to the wiper striking the variable
resistor further along its length.*

pressure
applied

spring
current
flow

L

wiper

variable
resistor

**Operation of a
Mechanical/Electrical Transducer**

Part 4

Installation Practices

Polyethylene (PE) Pipe Jointing

Relevant Industry Documents
BS 6891 and IGE/UP/2

PE pipe can be used for both natural gas and LPG. It is available as either medium or high density. The maximum operating pressures need to be confirmed with the supplier. PE pipe must be fully protected, in most cases by burying below ground. It may rise above ground level, but complete protection is needed to protect it from damage, including that from ultra-violet light from the sun. It would always be preferable to use metallic pipe entries into a building. However, if PE is used then the pipe entering the building must be placed inside a metal fireproof sheath in order to ensure that no gas escape from the plastic pipe could enter into the building. There are three methods of jointing to PE pipe: fusion welded joints, electro-fusion welded joints and compression joints. *Note*: Fusion welded joints may only be undertaken by companies or specialists who are competent and assessed for these jointing methods.

Fusion welded joint

This is a specialist joint that is undertaken using a fine stream of extremely hot air and a filler rod. With the heat directed at the pipe ends and the filler material applied, fusion occurs as the plastics melt together.

Electro-fusion welded joint

An electrical coil is incorporated into each joint. This connects to the outside via two terminals. To make the joint a special transformer that supplies a small voltage of 39.5 V is used for a set time of around 24–90 seconds, depending on the fitting size and manufacturer. A label attached to the fitting will give precise data. First the pipe end should be scraped clean using a special scoring tool, then the pipe is pushed fully into the fitting. At this point a pencil mark to indicate the fully 'in' position is made on the pipe. Failure to do this may lead to a problem as the pipe has a tendency to creep out from the fitting as it is heated. The mark will indicate any lateral movement and so the pipe can be held in place. If the pipe did creep from the fitting, molten plastic would ooze inside and restrict the pipe. With the fitting fully assembled and the connections made, the supply is switched on for the set period. As the fitting melts, markers pop up from the fitting, indicating that sufficient weld temperature has been achieved. *Note*: There are two electrical connections to each fitting, thus with branches for example, the middle connection is a spigot on to which another coupling would be required to complete the connection.

Compression joint

These fittings are usually used where the PE is to join another material, such as copper or steel, and as such they are often referred to as transitional fittings. The fitting consists of a body into which the pipe spigot can enter and be clamped there via a rubber compression ring. In order to prevent the rubber ring twisting out from the fitting as the lock nut is turned, a slip and guide washer are used. This allows the pressure to be spread evenly on to the rubber, forcing it squarely into the fitting. An insert needs to be placed inside the plastic pipe prior to connection; this maintains a solid true bore within the pipe, preventing leakage. The fittings used for an LPG installation may need confirmation with the manufacturer to ensure that the rubber ring is suitable for use.

Electro-Fusion Welding Transformer

Transitional Compression Joint

Label Indicating Heating and Cooling Times

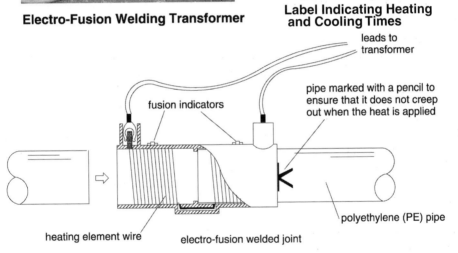

leads to transformer

fusion indicators

pipe marked with a pencil to ensure that it does not creep out when the heat is applied

polyethylene (PE) pipe

heating element wire

electro-fusion welded joint

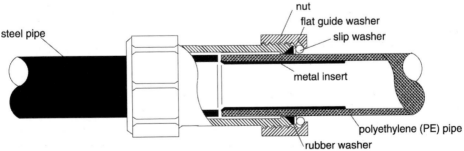

steel pipe

nut

flat guide washer

slip washer

metal insert

polyethylene (PE) pipe

rubber washer

transitional compression fitting

Polyethylene Pipe Jointing

Copper Pipe Jointing

Relevant Industry Documents
BS 6891 and IGE/UP/2

Copper may be used for both natural gas and LPG. However, when used within an LPG installation, soft copper compression rings (olives) instead of those made of brass must be used in compression joints. The copper pipe should be no larger than 67 mm in diameter for systems using natural gas and no larger than 35 mm where LPG in used. Where applicable, copper pipe needs to be suitably protected from mechanical damage and, in the situation where it is to be buried, it must be factory sheathed. Buried pipe must also not be connected to steel pipe and fittings as the surrounding groundwater will act as an electrolyte and cause electrolytic corrosion, eventually resulting in a leak. There are several jointing methods as described below.

Capillary Joints

These joints require the use of solder with a melting temperature below that of the copper to be drawn into the fitting by capillary action. Where the pressures within the gas pipe are to exceed 75 mbar, solders with a melting temperature above 600°C should be used. The type of capillary joint may be either of the end feed or solder ring type. However, whichever method is used it is essential to examine the completed joint visually to confirm that the solder has run fully round the joint.

Method First clean the pipe end and inside the fitting with wire wool and apply an inactive flux. It is possible to use an active (self-cleansing) flux that is active only during the heating process. However, caution needs to be observed to ensure that not too much is applied as the solder will rapidly flow everywhere the flux runs, including inside the pipe. It is therefore recommended that it be applied only to the pipe and not inside the fitting. The purpose of the flux is to maintain a clean oxide free surface on which the solder can readily flow and stick. Heat is now applied to the joint until the solder melts; it will be seen to appear at the mouth of the solder ring fittings or as solder is applied with end feed fittings. Where end feed fittings are used, the amount of solder required is no more than the diameter of the tube. So, for a 15 mm pipe, 15 mm of solder is required for each socket, not 50 mm plus as is often applied. Finally, the joint should be allowed to cool without movement, whereupon any remaining flux residue should be removed with a damp cloth.

Compression joints

Compression joints, where used, must be accessible. Below floors or within voids without removable covers would not be deemed suitable locations. To make a sound joint, first the pipe end needs to be cut square and de-burred, then the nut and compression ring is slid on to the pipe. The tube is now inserted fully into the fitting and the nut hand-tightened on to the tread, with an additional one-and-a-half turns made with a spanner to form a seal.

Threaded connections to brass fittings may be made using a suitable jointing paste or PTFE tape (guidance on this is given in the next section).

Press fitting joints

This is a new method of jointing that has recently been introduced and that can be used where hot works are not permitted. It is essential that the correct design of press fitting is selected, as the seal used to make the joint is different from that used in many other fittings. The fitting is identified by a yellow/tan 'O' ring and a yellow product marking. The joint is made by simply inserting the pipe into the fitting and using a special press fitting tool to compress the fitting tightly on to the pipe. The fittings are available in a range of sizes from 15 to 54 mm.

Various Capillary and Compression Fittings Used for Copper Pipework

Press Fitting Used for Gas Copper Pipework

Note the Gas marking to distinguish the fitting

Making the Press Fitting Joint

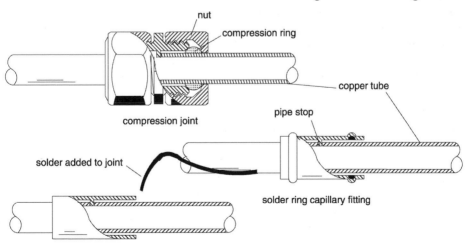

nut

compression ring

copper tube

compression joint

pipe stop

solder added to joint

solder ring capillary fitting

end feed capillary fitting

Types of Joint Used for Copper Pipework

Mild Steel Pipe Jointing

Relevant Industry Documents
BS 6891 and IGE/UP/2

Medium and heavy grade mild steel are very versatile materials that can be used in the installation of gas pipework above ground without fear of damage. In fact, it is the only suitable material where large internal pipe diameters are sought. Pipes greater than 100 mm in diameter always need to be welded. However, depending on the pipe location, welding may be the desired jointing method; this can be seen from the following table.

Pipe size	Below ground				Above ground				High rise or within ducts			
	0–2 bar		2–7 bar		0–2 bar		2–7 bar		0–2 bar		2–7 bar	
	S	W	S	W	S	W	S	W	S	W	S	W
<25	✓	✓	×	✓	✓	✓	✓	✓	✓	✓	×	✓
26–50	✓	✓	×	✓	✓	✓	×	✓	✓	✓	×	✓
51–80	×	✓	×	✓	✓	✓	×	✓	×	✓	×	✓
81–100	×	✓	×	✓	✓	✓	×	✓	×	✓	×	✓
>100	×	✓	×	✓	×	✓	×	✓	×	✓	×	✓

S = screwed joint; W = welded joint   × not acceptable ✓ acceptable

Steel tube is suitable for LPG installations, however galvanised pipe is often to be recommended. Also, with LPG, parallel threads such as with a long screw connector, are not to be used.

Jointing to threaded pipework is made using a suitable jointing paste, as indicated on the side of the tin/tube of compound used or, for the smaller sizes, i.e. up to 50 mm, polytetrafluoroethylene tape (PTFE) may be used. *Note*: PTFE is not to be used with any other jointing medium or compound. Also, the cutting oil used to make the threads needs to be fully wiped from the pipe prior to use. The PTFE tape used must be of the thicker design to BS EN 751 pt 3, and applied with a 50% overlap as shown opposite. Where union connection joints are to be used, they must be located in readily accessible positions. Under floors and within ducts without removable covers are not considered to be accessible areas. The use of hemp is unavoidable in making the grommet for long screw connectors but it should not be used with threaded connections. For large diameter coarse threads over 50 mm, hemp is permitted but its use should not be encouraged if a suitable alternative compound to BS 5292 can be found.

Where welded pipework is used, flange joints may be required to assist in the replacement of components. However, the number of flanged joints should be kept to a minimum, with the flanges welded to the pipes. All welding needs to be completed by a trained operative to ensure that the work meets the required standard.

One can use compression joints when connecting to the smaller sizes, although these are generally restricted to making transitional connections to polyethylene (PE) pipe.

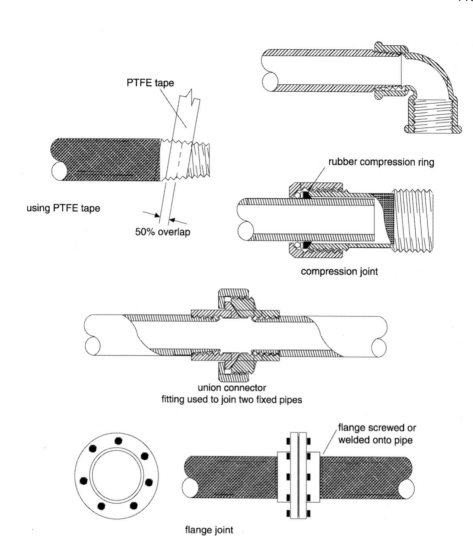

using PTFE tape

PTFE tape

50% overlap

rubber compression ring

compression joint

union connector
fitting used to join two fixed pipes

flange screwed or
welded onto pipe

flange joint

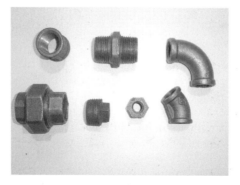

Types of Joint used for Steel Pipework

Semi-Rigid and Flexible Pipe Connections

Relevant Industry Documents
BS 6891 and BS 3212

Flexible Stainless Steel

Small flexible stainless steel connections, known as the anaconda, have been used at the gas meter for many years. However a new range of material in sizes from 10 mm to 50 mm, supplied in coils up to 76 m in length is now available. The pipe is of a corrugated nature and can therefore be easily manoeuvred to suit the design layout of the building, enabling it to overcome obstacles. The pipe is supplied pre-sleeved with a polyethylene membrane coating on the outside and is therefore suitable for use below ground, such as under building slabs and other locations where sleeved piping is required.

Currently, in order to install this piping material, the installer must have successfully completed a manufacturer's installation programme, thus demonstrating competency in installation. No special tools are required as it is simple to cut using conventional tube cutters and the joints are made using two wrenches, clamping a threaded terminal connection to the pipe end and made by flaring the pipe into the fitting forming a metal-to-metal seal. The material may be used for both domestic and commercial installations. It is important, however, to ensure that no flux is allowed to come into contact with this material as it rapidly tends to corrode the metal, leading to pitting.

Flexible Hoses

These are used in the installation of appliances that need to be moved easily, for example cookers and tumble dryers or possibly for the high pressure connections to an LPG cylinder. The hose selected for each application needs to comply with the appropriate current standards, for example natural gas hoses need to comply with BS 699 and LPG BS 3212. The hose should be no longer than is required and it should never be longer than 2 m. Generally a 1 m hose will suffice.

Appliance hoses

These are usually bought as a complete component, with all necessary brass threads and bayonet connections ready for use. LPG appliance hoses are characterised by a red stripe or band running along the tube. Appliances used in commercial situations invariably have a steel outer casing and, where used for catering purposes, are covered with an additional white plastic sheath.

LPG supply hoses

Where these are used to make the connection from the supply cylinders to the regulator they are often referred to as pigtails. They are generally coloured orange. However, where supplied as a complete assembly, with crimped or swaged ferrules, they are sometimes black. The tube should bear the manufacturer's name or identification, type and BS number along with the date of manufacture. Rubber flexible tubing for LPG applications not exceeding 50 mbar working pressure is always supplied as black.

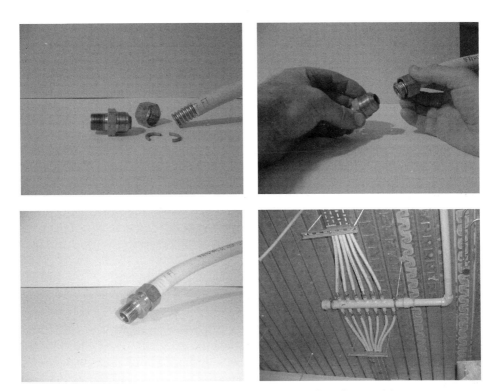

Use of Semi-Rigid Flexible Pipe and Fitting Assembly

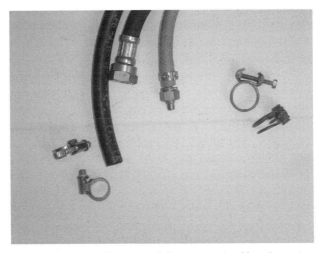

Rubber Hose and Connectors Used
(Note the two hose connectors to the far right
are not of the correct design and therefore not permitted.)

Gas Service Pipework

Relevant Industry Documents
IGE/TD/3 and 4

The service connection from the main in the road to bring gas into the premises is undertaken by the supplier or one of their contractors and not by the gas installer/engineer. In general, all new services today will be PE pipe and connections are made with a saddle tapping tee as shown. First the tee is fused on to the main (see page 114) then the service pipe is run to the desired location, usually just outside the building. Upon completion it can be tested for tightness. Should the tightness test prove successful, a cutter in the top of the fitting is wound down to cut into the pipe. Owing to its design, the plug of PE is retained within the cutter and, when it is screwed up again, gas flows through into the service pipe. Where a steel main has been connected to a PE, a compression joint is made into a service tee which has been cut into the top of the main. To provide additional support to the PE an anti-shear sleeve is incorporated within the first 460 mm. The PE termination at the entry to the building is usually made using a transitional tee or a meter box adaptor, both of which are shown opposite.

The service pipe should not pass through the wall below ground level without adequate protection being provided. Ideally it should rise externally to enter the building above the floor level. Where the pipe is to pass through a solid floor or through a wall it must be suitably sleeved, sealing the internal surface of the sleeve with a flexible fire resistant compound, so that any escaping gas can pass only to the external environment.

The route that the service pipe is to take to the building should avoid unstable structures and ground that is liable to movement and subsidence. Where possible the pipe should travel in a straight line at right angles to the building and service main. If it is necessary to run it alongside a building, it should be kept a minimum distance of 1 m from the wall. Where high pressure mains are encountered, this distance needs to be increased to 3 m.

Protection of Service Pipework Below Ground

The minimum depth of cover for a gas service pipe of up to 50 mm diameter on private property is 375 mm. Where this cannot be achieved, additional protection must be provided. Should the pipe run under public roadways, etc. then this depth would need to be increased to 450 mm. In the case of larger pipes, greater than 50 mm in diameter, the depth below roads and grass verges needs to be increased to 750 mm, this may be reduced to 600 mm below a paved footpath.

Where a possible opening may occur in the protective coating surrounding a steel pipe below ground, the metal would be exposed to direct contact with the soil. Any stray electrical current passing through to earth would undoubtedly flow to earth at this point and cause electrolytic corrosion. To prevent this an anode is sometimes connected to the service pipe, this is destroyed in preference to the service pipe. The anode is simply screwed into a blind tee. The blind tee is a tee fitting with a branch

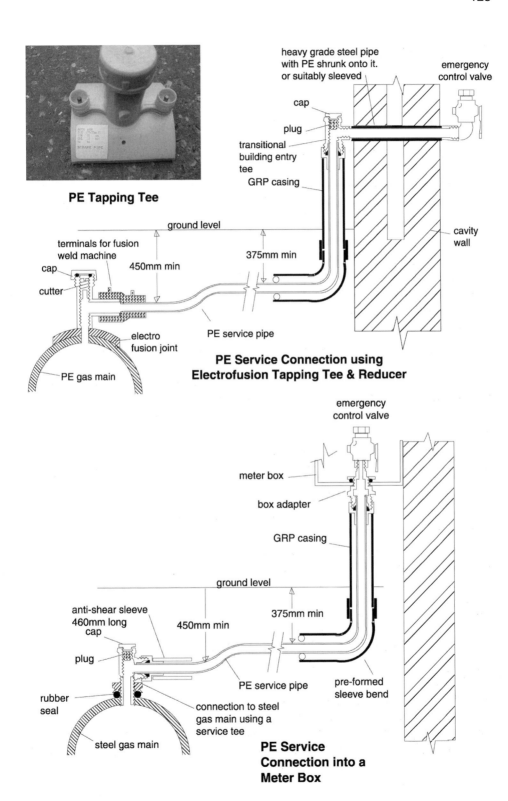

PE Tapping Tee

heavy grade steel pipe
with PE shrunk onto it.
or suitably sleeved

emergency
control valve

cap

plug

transitional
building entry
tee

GRP casing

cavity
wall

ground level

terminals for fusion
weld machine

450mm min

375mm min

cap

cutter

electro
fusion joint

PE service pipe

PE gas main

**PE Service Connection using
Electrofusion Tapping Tee & Reducer**

emergency
control valve

meter box

box adapter

GRP casing

ground level

anti-shear sleeve
460mm long
cap

450mm min

375mm min

plug

PE service pipe

pre-formed
sleeve bend

rubber
seal

connection to steel
gas main using a
service tee

steel gas main

**PE Service
Connection into a
Meter Box**

connection that is blanked off. The tee is fully wrapped with a petroleum-impregnated woven bandage, with a minimum overlap of 50%, leaving the anode exposed. The protective anode is usually positioned just prior to the steel service turning upwards to enter the property. PE pipes should not be laid in ground that may be chemically reactive to the plastic material being laid; this includes ground that contains tar, oils, dry cleaning fluids, etc.

Protection to Service Pipework Above Ground

PE pipework brought to rise above the surface would be subject to ultra-violet light damage from the sun. It can be fully protected by surrounding it with a glass reinforced plastic (GRP) sleeve.

Metal pipework needs to be securely fixed and, in general, a suitable coating of corrosion paint suffices to protect steel pipework. However, wrapping may be carried out in corrosive situations. Copper pipework installed with clips to hold the pipe off the wall does not usually require any additional protection. Internal protection is afforded by the usual paints available.

Service Pipework Insulation

To prevent stray electrical currents passing down metal electrical service pipework one must insulate the service from the internal installation. This is achieved by fitting an insulation joint just inside the building, as shown. The insulation joint may be either of a fire retardant thermoplastic or of a metal design in which the two mating surfaces are prevented from directly touching each other by the use of a mica gasket, and appropriate washers and 'O' ring, thus forming a solid sound fitting. The whole fitting is given a final protective coating to prevent electrical tracking due to moisture across its surface. Where a plastic insulator is used, it needs to be positioned directly on the semi-rigid connector at the meter.

Equipotential Bonding at the Meter

Within a distance of 600 mm from the primary gas meter, or entry to the building, the installation pipe must have a suitable equipotential bonding earthing conductor, no less than 10 mm^2, connected to the main earth bar which is located at the electrical consumer unit (see also page 430).

Emergency Control Valve

An emergency control valve must be located at the point of entry to the building, in an accessible location. The valve should be fitted so that a union connector can be disconnected after the control valve. The valve must have a handle securely fitted along with a suitable indicator label showing the on/off position and instructions for the action to take in the event of an emergency. *Note*: The handle must be positioned so that when it is moved down to its lowest possible position the gas supply is off. Where the supply is to enter individual apartments within a single or multiple installation an emergency control valve needs to be fitted inside each flat or in a suitable communal location. An additional notice must also be displayed giving instructions to the user on the action needed in the event of a gas escape, and providing the telephone number of the emergency services.

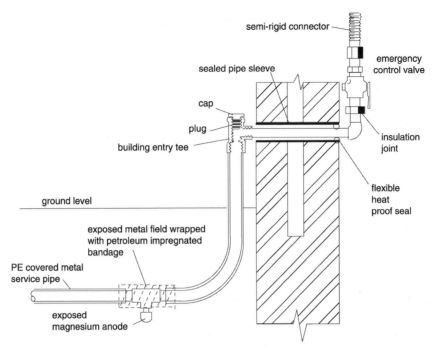

Metal Service with Insulator and Additional Corrosion Protection

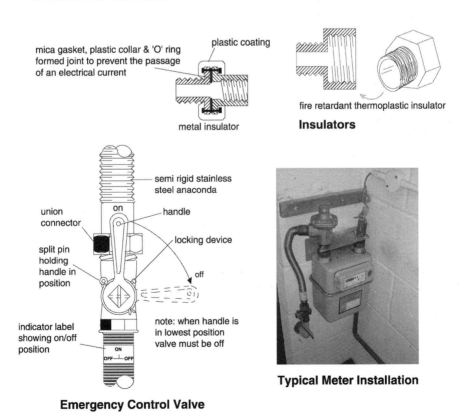

Insulators

Emergency Control Valve

Typical Meter Installation

Connections to High Rise Buildings

Relevant Industry Document
IGE/UP/2

For a multi-storey building it would be necessary to run the service up to the various levels or apartments. To do this an internal fire-resistant shaft could be used. This shaft would, if possible, be located adjacent to an external wall so as to enable ventilation direct to the outside at its highest and lowest position. The required size of the ventilator opening is shown below.

Cross sectional area of shaft (m²)	Minimum 'free air' ventilation size for each opening (m²)
≤ 0.05	Same size as cross-sectional area
0.05–7.5	0.05
> 7.5	150th of the cross-sectional area

Example The service riser shown opposite has a base measuring 0.450 m × 0.3 m, which gives a cross-sectional area of: $0.45 \times 0.3 = 0.135$ m². This clearly falls into the second row in the table above, therefore the ventilation grille fitted at high level and low level would need to be 0.05 m² (or 500 cm²).

Where the riser passes through any wall or floor structures it would need to be suitably fire-stopped to prevent the spread of fire. In order to provide an opening for maintenance purposes a sealed half-hour fire-resistant panel should be provided at each floor level.

The material to be used for a vertical gas riser is heavy gauge mild steel and, if the building is more than four storeys high, the joints would need to be welded. It is essential that the riser is adequately supported at its base; this is usually achieved by means of a concrete plinth. In order to prevent a blockage from particles falling within the pipe, a rust trap should be provided, with a plugged-off 25 mm full bore valve as shown. The riser should maintain a vertical rise throughout its length, without any change in direction and it should be adequately supported at various intervals to allow for thermal movement.

The horizontal pipework connected to the riser at each floor level should incorporate a flexible connector to allow for the differential thermal expansion between the riser and lateral service. This may be of copper with a different expansion rate, which also passes out from the shaft to an environment at a different temperature. To minimise thermal expansion the riser should not be fitted too close to hot water and steam pipes. As it enters the occupied part of the building, the lateral service pipe must include a suitably labelled emergency shut-off valve and suitable means of disconnection.

The gas pressure within the service pipe should be regulated to the low pressure range (21 ± 2 mbar) prior to the pipe entering the building.

In very tall buildings, due to Boyle's law (page 60) it will be found that as the pressure reduces with the loss of atmospheric pressure, the volume of the gas increases. Sometimes special consideration needs to be given to this, possibly in consultation with the manufacturer.

The gas meter/s may be located inside the building as close as possible to the entry point and marked with the dwelling it serves, or it may be located with the apartment it is to serve immediately after the emergency control valve. See also page 131.

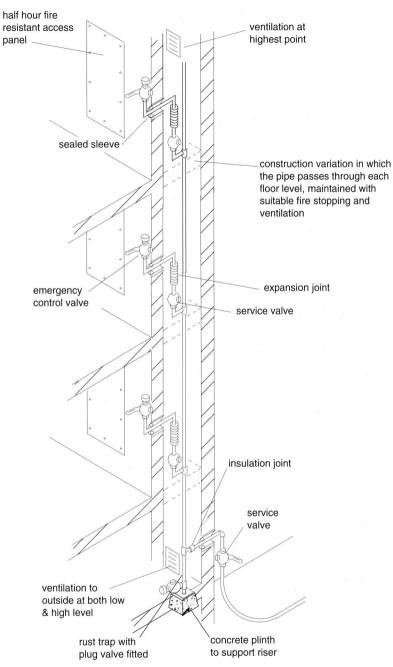

half hour fire resistant access panel

ventilation at highest point

sealed sleeve

construction variation in which the pipe passes through each floor level, maintained with suitable fire stopping and ventilation

emergency control valve

expansion joint

service valve

insulation joint

service valve

ventilation to outside at both low & high level

rust trap with plug valve fitted

concrete plinth to support riser

Gas Service to High Rise Building

Gas Meter Installations

Relevant ACS Qualifications   Relevant Industry Documents
MET1 – MET4   BS 6400 and IGE/GM/1 and 6

Meter Location

The gas meter should be located at a point as close as practical to the point of entry to the building and where the service pipe terminates and may include any of the following locations:

- in a purpose-made meter housing, to include a meter box, located outside the building either adjacent to the property or at a boundary enclosure;
- in a garage or outbuilding;
- inside the building.

The location must provide ease of access for servicing, exchange and reading purposes or, in the case of prepayment meters, to enable easy operation of the coin or token mechanism. The site of any gas meter must be well ventilated and it must be mounted at least 25 mm from the surrounding wall surfaces.

Fitted prior to the gas meter would be the emergency control valve, previously described, with a label attached indicating the 'on' and 'off' positions and the installation regulator. The regulator must not be situated where water from rain or floods could get into the breather hole. Water inside the valve would corrode the mechanism and could freeze during the winter, cracking the valve.

Affixed to the meter, or adjacent to it, must be a completed and dated emergency control notice that tells the occupier what to do in the event of a gas escape, including turning off the supply. It must also contain the telephone number of the emergency service contact.

Meter Boxes

There are various types of meter box, including built-in, surface mounted and semi-concealed designs, all of which are shown opposite. These boxes all have the following basic characteristics:

- They provide adequate ventilation to the external environment.
- They are suitably sized to enable installation, exchange and servicing of all components, e.g. meter, regulator, thermal cut-off, as appropriate.
- They are of non-combustible construction and have adequate fire resistance.
- They are constructed so that access can only be gained by the use of a special key, with which the consumer has been provided.

When installing the meter into these boxes no drill holes should be made except those intended by the manufacturer and all exits should be suitably sealed so that any escape of gas can pass only to the external atmosphere and not into the building. The gas service pipe is run to a point adjacent to the meter bracket, from which the gas meter is suspended. A semi-rigid stainless steel connector (anaconda) connects to the

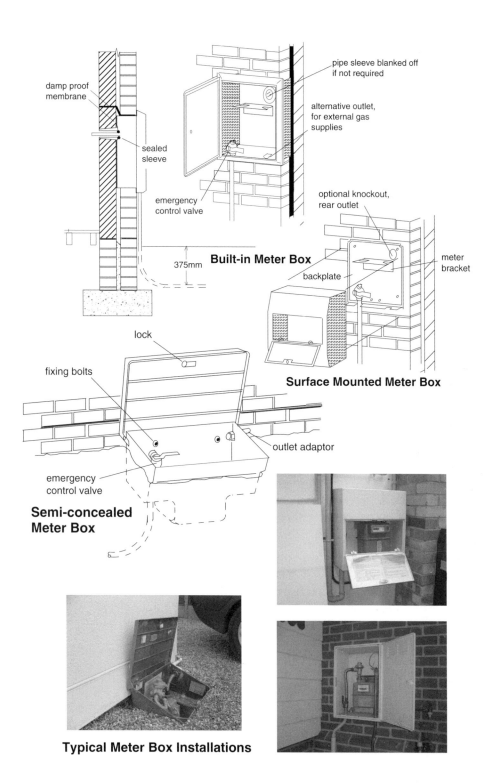

damp proof membrane

sealed sleeve

emergency control valve

375mm

pipe sleeve blanked off if not required

alternative outlet, for external gas supplies

Built-in Meter Box

optional knockout, rear outlet

backplate

meter bracket

Surface Mounted Meter Box

lock

fixing bolts

outlet adaptor

emergency control valve

Semi-concealed Meter Box

Typical Meter Box Installations

emergency control valve and terminates at the regulator, which is, in turn, connected to the inlet of the meter. The outlet from the meter terminates with an elbow facing into the installation serving the building.

Multi-Occupancy Buildings
Should a number of primary meters, grouped together, be required to serve different parts of a building, the following points should be observed:

- The meters need to be sited where access is available at all times.
- The meters should be enclosed in a lockable housing; this may be either as a large single enclosure or as individual meter boxes.
- Each individual meter needs to be clearly marked to indicate the premises it serves, e.g. flat number.
- Each individual premise should have an additional emergency control valve located at the point where the gas pipe enters the property.

Locations where the meter must not be sited
These include:

- locations subject to high temperatures and close to heat sources;
- where the meter could cause an obstruction or be liable to damage;
- in damp or corrosive locations;
- in close proximity to electrical equipment;
- where food is stored;
- under a stairway or in a passageway of building with two or more floors above ground floor (i.e. three storeys).

Although not preferable, if it is unavoidable, for buildings with only one floor above ground floor, a meter may be located along an escape route. However the meter must be of a fire-resistant construction *or* be housed within a fire-resistant compartment with an automatic self-closing door *or* include a thermal cut-off valve, upstream, designed to cut off the gas flow should the temperature exceed 95°C (see page 78).

Removal of the Gas Meter
Where a meter is to be removed for any purpose, a temporary bond needs to be put in to ensure that any stray electrical currents do not cause a spark to jump across to earth. The meter must not be permanently removed without the authority of the meter owner. Where the gas pipe is to be left, plugged off, with a live supply remaining, it must be clearly marked to identify that it contains gas. Where permanent removal of the meter is to take place, the earth bond may also need to be made permanent.

Meter Bypass
Domestic meters are rarely bypassed. However, where continuity of supply is needed, such as in factories or hospitals, a bypass as shown may be required. Two methods can be employed, however, in each case note that it is essential that the supply is regulated. The valve fitted to the bypass is sealed by the supplier and can only be opened in an emergency or during an exchange.

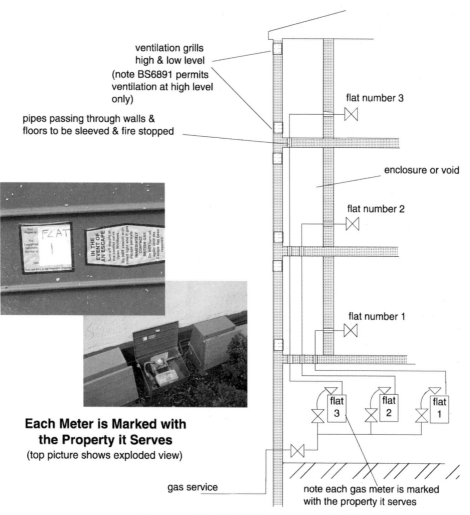

ventilation grills
high & low level
(note BS6891 permits
ventilation at high level
only)

pipes passing through walls &
floors to be sleeved & fire stopped

flat number 3

enclosure or void

flat number 2

flat number 1

flat 3　flat 2　flat 1

**Each Meter is Marked with
the Property it Serves**
(top picture shows exploded view)

gas service

note each gas meter is marked
with the property it serves

Pipework to Multi-Occupancy Buildings

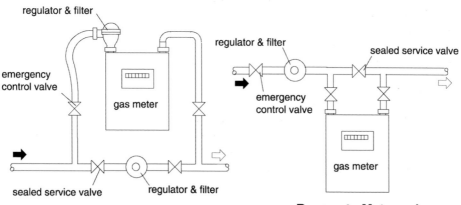

regulator & filter

emergency
control valve

gas meter

sealed service valve

regulator & filter

Bypass to Meter and Regulator

regulator & filter

sealed service valve

emergency
control valve

gas meter

Bypass to Meter only

Commercial Gas Installations

Relevant Industry Documents
BS 6400 and IGE/UP/2

(Specific points in addition to domestic gas installations)

Line Diagram

Where the service pipe is 50 mm or larger a line diagram in permanent form must be attached to the building in a conspicuous and accessible position, such as at the primary gas meter or emergency control valve. In addition, other places could be included such as the security gatehouse or site engineer's office. The diagram is to show the location of all pipes of diameter greater than 25 mm, meters, emergency control valves, pressure test points and electrical bonding points. Its purpose is to inform anyone, especially the emergency services, of the route of the gas supply, thus enabling them to identify and isolate it if necessary. A typical diagram is shown opposite.

Marking of Pipework

All pipework that is accessible for inspection should be permanently marked in such a way that it is easily recognisable as a gas pipe. This does not apply to that pipework which is to be used in living accommodation. The method of identifying the gas pipe is by painting the pipe yellow ochre (BS 4800 reference: 08 C 38). It is possible to simply apply banding at strategic points along the pipe, e.g. each side of valves, entry or exit through walls or at branches. Where other gases, such as LPG, are used on the same site an additional band of primrose yellow (BS 4800 reference: 10 E 53) indicates that it is natural gas. Alternative additional labelling, e.g. 'Natural Gas' could be applied. In addition to marking, sometimes, particularly in confined spaces, a label identifying the direction of gas flow proves most useful. It is the installer's responsibility to ensure that the pipe is initially marked, however the person responsible for the premises must ensure that the pipe remains recognisable as long as it is used to convey gas.

Purge Point

In order to facilitate purging the pipework, a purge point with a plugged off valve should be located at each of the following positions:

- at every point where a section isolation valve is fitted;
- at the end of a pipe run;
- at each side of a gas meter.

Secondary and Check Meters

The location of any secondary gas meters or check meters should be clearly identified on the line diagram. The location of these meters needs to be identified for several reasons: the emergency services needs the information in order to isolate the supply and they also need to be known for purging purposes after disconnection or reconnection. The secondary meter may be for private billing purposes or installed simply to monitor the gas consumption to an appliance.

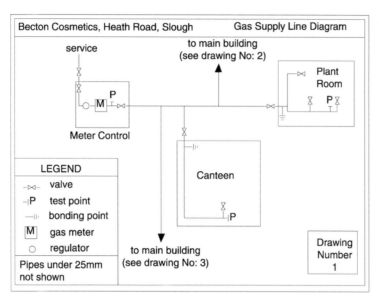

Becton Cosmetics, Heath Road, Slough	Gas Supply Line Diagram

service

to main building
(see drawing No: 2)

Plant
Room

P

Meter Control

Canteen

P

LEGEND

–⋈–	valve
–∣P	test point
–∣∣·	bonding point
M	gas meter
○	regulator

Pipes under 25mm
not shown

to main building
(see drawing No: 3)

Drawing
Number
1

Typical Gas Supply Line Diagram

**Labelling Tape
Showing Direction**

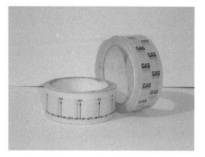

Pipe Labelling Tape

150mm	100mm	150mm
yellow ochre	primrose	yellow ochre

**One Method of Banding to Natural Gas Pipework
in Commercial Premises**

Internal Installation Pipework

The common terminology used to identify the gas pipework to and within a building may vary across the country. However, the following diagram identifies the British Standard terminology and symbols that should be employed for universal recognition.

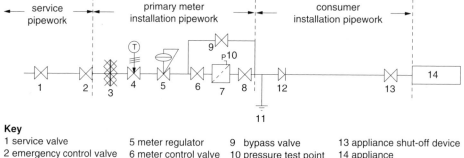

Key

1 service valve	5 meter regulator	9 bypass valve	13 appliance shut-off device
2 emergency control valve	6 meter control valve	10 pressure test point	14 appliance
3 electrical insulator	7 meter	11 equipotential bond	
4 thermal cut-off	8 service valve	12 non-return valve	

[note: not all controls will be fitted to most premises]   *Symbols to BS 1553-1*

The materials that can be used for the pipework were identified earlier in the chapter, and include polyethylene (PE), copper and steel. As the pipework inside the building does not have the protection of being buried in the ground, it is continually subjected to the threat of damage from knocks, alterations or by further unexpected actions and building movement. Therefore the gas installer continually has to look ahead for possible problems. Installing copper pipe runs in a school or college may prove useless if the pipes are soon damaged by students who continually knock against them and stand on them. Slight variations are found in the various industry documents relating to the maximum interval between pipe support brackets, however the table opposite should provide a suitable guide.

When passing a pipe through a wall, movement will occur due to the pipe expanding and contracting, and the building itself will move. If the pipe is not suitably sleeved it will soon suffer the effects of wear as it rubs against the solid wall surface. Placing the pipe in direct contact with a cemented wall would lead to the acids in the mortar corroding the pipework. Again, sleeving will combat the problem. When sleeving a pipe that is to pass through a wall, the material used for the sleeve must be compatible with the material of the pipe, i.e. the pipe material should not react with the sleeve material due to electrolytic corrosion. Therefore a material of the same type is selected, e.g. copper pipe – copper sleeve, alternatively a plastic pipe, such as PE could be used. Where plastic is to be used, the sleeve material itself must be capable of containing gas. Finally, the sleeve must be made good into the wall and one side of the sleeve sealed, between the pipe and the sleeve, with a non-setting heat resistant mastic. The seal is to be on the internal pipework. *Note*: No joints or parts of a joint should be within the sleeve section.

Table identifying maximum interval between pipe supports

screwed mild steel	welded mild steel	corrugated stainless steel	copper	Vertical Distance (m)	Horizontal Distance (m)
—	—	—		0.6	0.5
—	—	—	15	2.0	1.5
—	—	≤ 50	22	2.5	2.0
15	—		28	2.5	2.0
20 - 25	—		35	3.0	2.5
40	≤ 20		42	3.0	2.5
50 - 100	25 - 32	not available	54	3.5	3.0
not recommended	40		not recommended	4.0	3.5
	50			5.0	4.0
	65			5.5	4.5
	80			6.5	5.5
	100			7.5	6.0
	150			8.5	7.0

Material & Nominal bore

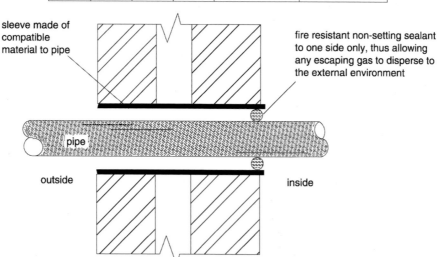

sleeve made of compatible material to pipe

fire resistant non-setting sealant to one side only, thus allowing any escaping gas to disperse to the external environment

pipe

outside

inside

Sleeving of pipes passing through walls

Pipework Laid in Floors

Relevant Industry Document
BS 6891

Pipes Laid in or under Wooden Joisted Floors

Pipes may be run, suspended below timber joists or run parallel between joists. The location of the supports needs to be in accordance with the table on the previous page. Should it become necessary to run the pipe perpendicular to the joist and there is no provision to secure it below the timber it may be possible to drill a hole or notch the timber as necessary. The maximum pipe diameter that could possibly pass through a joist would depend on the depth of the timber and whether a hole was drilled or notch cut. This size is found by making the following calculation:

$$\text{Maximum notch size} = \text{Depth of joist} \div 8$$
$$\text{Maximum drilled hole size} = \text{Depth of joist} \div 4$$

[*Note*: Joists ≤ 100 mm must not be cut or drilled and the maximum depth of joist to be considered is 250 mm, therefore a 300 mm joist is only regarded as being 250 mm.]

So, for example, where the joist is 200 mm deep the largest pipe that could pass would be 25 mm where a notch is made or 50 mm where a hole has been drilled.

The location of the hole or notch also needs to be considered; it must only be positioned within the shaded areas shown in the diagram. Where more than one notch or hole is required, the two must be at least 100 mm apart horizontally. When refitting the floorboards, care needs to be taken to prevent damage. It is also a good policy to mark the floor, where possible, warning of the gas pipe below.

Pipes Laid in Solid Floors

Pipes may be laid within the concrete or floor screed of a floor, providing it is suitably protected against corrosion using one of the following methods:

- Copper: This must be fully annealed and factory sheathed, and is usually laid only into the floor screed. Where it is to be laid within the concrete slab itself, it should be placed inside a plastic sheath, with no joints within the sheath.
- Mild steel: Where this is laid within the floor screed it must be fully protected with a petroleum-impregnated bandage.
- Stainless steel: Only factory sheathed corrugated stainless steel can be laid in floors; no other type is permitted.

Where the pipe is to be laid within the concrete, in addition to the method identified above for the copper, it is possible to lay the pipe in a pre-formed duct, with a protective cover over the pipe. Alternatively another additional soft covering that will allow some movement may be used.

In addition to the points identified, no compression joints should be used below floor level and any joints used should be kept to a minimum. All joints are to be fully protected with a petroleum-impregnated bandage or similar on the completion of a successful tightness test. Where a pipe is simply to pass vertically through a floor it should be suitably sleeved as previously described.

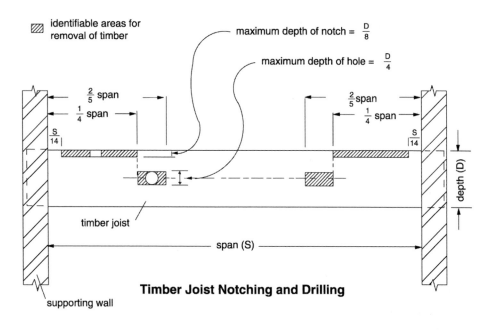

identifiable areas for removal of timber

maximum depth of notch = $\frac{D}{8}$

maximum depth of hole = $\frac{D}{4}$

$\frac{2}{5}$ span

$\frac{1}{4}$ span

$\frac{2}{5}$ span

$\frac{1}{4}$ span

$\frac{S}{14}$

$\frac{S}{14}$

depth (D)

timber joist

span (S)

supporting wall

Timber Joist Notching and Drilling

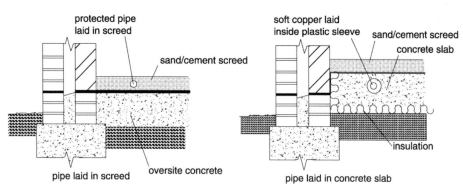

protected pipe laid in screed

sand/cement screed

pipe laid in screed

oversite concrete

pipe laid in screed

soft copper laid inside plastic sleeve

sand/cement screed

concrete slab

insulation

pipe laid in concrete slab

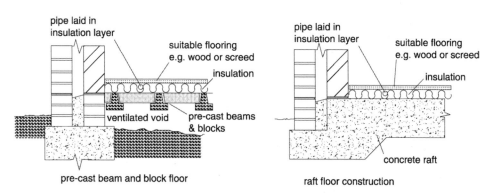

pipe laid in insulation layer

suitable flooring e.g. wood or screed

insulation

ventilated void

pre-cast beams & blocks

pre-cast beam and block floor

pipe laid in insulation layer

suitable flooring e.g. wood or screed

insulation

concrete raft

raft floor construction

Pipes Located in Solid Floors

Pipework in Walls

Whenever possible any pipework placed within the wall structure should run vertically, avoiding horizontal runs. The pipe ideally should be positioned within a duct, however it is possible to cut a pipe chase down the wall into which the pipe, suitably protected against corrosion, can be installed. With the pipe secured in position, a tightness test needs to be completed before applying the final covering material. Joints within this section should be kept to a minimum and in no circumstances are compression joints to be used. Where a pipe is to pass through a wall, it will need to be adequately sleeved as previously shown on page 135. The maximum permitted depth for a chase cut into the brick or block wall should not exceed:

Wall thickness ÷ 6   (*for horizontal pipes*);

Wall thickness ÷ 3   (*for vertical pipes*).

So, for example, for a 100 mm block wall, the maximum depth for a horizontal chase is 16.5 mm; this depth is increased to 33 mm for a vertical chase.

Cavity Walls

It is not permissible to run a pipe down through a wall cavity and, when passing through a cavity wall, the pipe should take the shortest possible route, and will need to be sleeved. The Gas Safety (Installation And Use) Regulations have made provision for 'living flame effect' gas fires, which are installed into the cavity of a building structure. In order to provide a neat gas connection, the gas pipe may be run into the cavity through a suitable sleeve, provided that it takes the shortest possible route and is sealed at the point where the pipe enters the flue box.

Dry Lined Walls

Where dry linings are to be used, one can install the gas pipe behind them. However, the pipe must be totally encased to prevent the void filling up with gas in the event of a gas escape. Two methods could be employed here: either placing two battens, one each side of the gas pipe, along the entire length of the pipe run, or providing a continuous plaster or adhesive dab along its length.

Timber Walls

Where gas pipework is to be contained within a timber stud wall, it should be adequately supported and run within a purpose-designed channel, protected from mechanical damage as appropriate. The number of joints should be kept to a minimum and, as previously shown, no compression joints are to be used in inaccessible places.

Pipework in Ducts and Voids

Where the pipe has been placed within a duct or void, ventilation should be considered to ensure that a minor gas leak would not created a dangerous situation. This has previously been discussed on page 126.

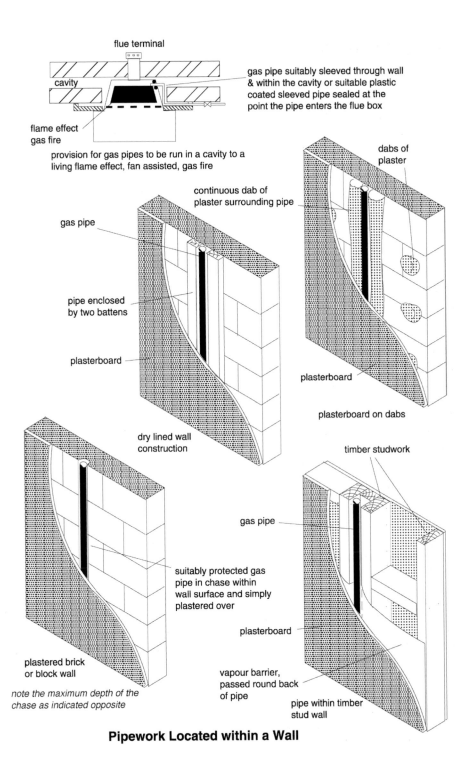

flue terminal

cavity

gas pipe suitably sleeved through wall
& within the cavity or suitable plastic
coated sleeved pipe sealed at the
point the pipe enters the flue box

flame effect
gas fire

provision for gas pipes to be run in a cavity to a
living flame effect, fan assisted, gas fire

dabs of
plaster

continuous dab of
plaster surrounding pipe

gas pipe

pipe enclosed
by two battens

plasterboard

plasterboard

plasterboard on dabs

dry lined wall
construction

timber studwork

gas pipe

suitably protected gas
pipe in chase within
wall surface and simply
plastered over

plasterboard

plastered brick
or block wall

*note the maximum depth of the
chase as indicated opposite*

vapour barrier,
passed round back
of pipe

pipe within timber
stud wall

4 Installation Practices

Pipework Located within a Wall

Pipework Support and Allowance for Movement

Relevant Industry Documents
BS 3974 and IGE/UP/2

There are many designs of pipe support, some hold the pipe firmly in place and others allow for movement due to expansion and contraction or from external forces that move a structure or building.

The diagrams opposite show a few of the many available designs of pipe support, some of which are fabricated on site to suit the needs of a particular situation. It should be noted that where a bracket is fabricated the thickness of the material used is clearly identified in BS 3974 and it is not sufficient to simply use any old piece of metal. For example, the 'U' bolt pipe support shown opposite for a 100 mm diameter pipe requires a 16 mm diameter bar, quite substantial in diameter, the thread length should also be restricted to 55 mm. The two diagrams of the 'U' bolt show (1) a pipe held firmly and (2) a supported pipe that is free to move length-ways. A well designed system requires a mix of pipe supports and fittings that allow for movement, yet maintain a firm sound system.

Expansion due to Temperature Change
As a pipe heats up and cools down due to the changes in temperature of the environment it expands and contracts. The amount of expansion or contraction, is easily worked out using the following calculation:

Pipe length × Temperature change × Coefficient of linear expansion

Example An 80 m long PE gas pipe is installed 370 mm below ground on a very hot day, assume 30°C. During the winter a temperature of –5°C occurs. The amount of contraction expected would be:

$$80\,m \times 35°C \times 0.00018 = \underline{0.504\,m}$$

Typical coefficients of linear expansion

Copper	0.000016
Steel	0.000011
PE	0.00018

Over half a metre! If allowance has not been made for this movement, the pipe would either suffer the effects of being stretched, and therefore become weakened, or a joint could pull out, resulting in a major gas escape.

Provision for Expansion
In the example just illustrated, simply laying the pipe into the trench in a side to side, snake like way, rather than a straight line, allows the bend to straighten out should excessive contraction occur or, conversely, allows the bend to extend if necessary during a rise in temperature. Other methods that allow for movement, particularly above ground, include the use of expansion joints, which are designed to allow the flexing and lengthways movement of the pipe. To allow for deflection, semi-rigid couplings are often used; these incorporate some form of flexible seal. Where lateral displacement is to be expected, two couplings can be used in conjunction as shown.

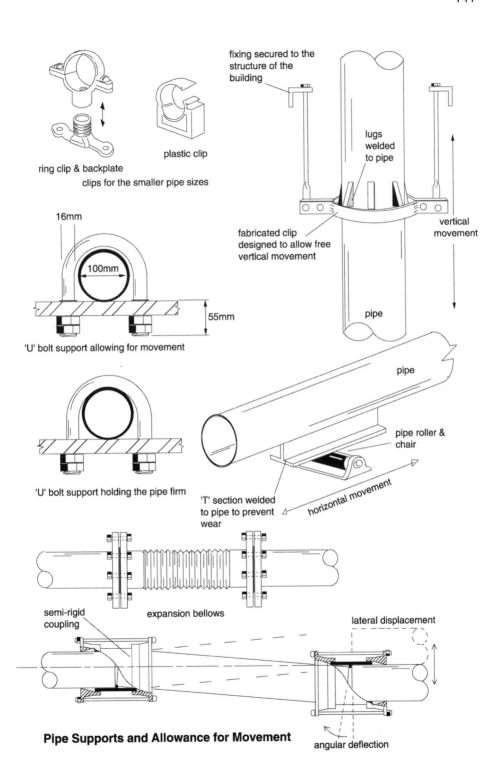

ring clip & backplate

plastic clip

clips for the smaller pipe sizes

16mm

100mm

55mm

'U' bolt support allowing for movement

'U' bolt support holding the pipe firm

fixing secured to the structure of the building

lugs welded to pipe

fabricated clip designed to allow free vertical movement

vertical movement

pipe

pipe

pipe roller & chair

'T' section welded to pipe to prevent wear

horizontal movement

semi-rigid coupling

expansion bellows

lateral displacement

angular deflection

Pipe Supports and Allowance for Movement

Timber Framed Dwellings

Relevant Industry Document
IGE/UP/7

A timber framed building usually consists of a timber inner structure that has been pre-fabricated off site and erected on a suitable foundation, often within a day or two. The outer skin is then positioned; this may be timber cladding or tiles. Alternatively, a more robust material such as brickwork may be built on site, forming a solid outer facade.

Penetrating the Outer Wall

A vapour barrier can be seen immediately behind the plasterboard in the diagram opposite. This is to prevent any internal moist air passing through the wall where it may cool to the dew point and condense. This would lead to rotting in the timber structure. It is therefore imperative that this barrier is not unduly damaged or removed and care needs to be taken when penetrating it to pass pipes or a flue through the structure. In addition moisture could pass from the outer skin, travelling across the cavity by clinging to horizontal pipe runs, including flues. As a consequence the design of flue should include a moisture drip collar or some form of damp-proof membrane. Gas pipe should ideally pass through at a low level, below the level of the timber studwork.

Appliance and Pipework Installation

Plasterboard is technically combustible due to its paper surface, however in practice most appliances can be installed on to or adjacent to combustible walls. However, the manufacturer's instructions have to be consulted to see if any special precautions for fireproofing are required. For new work the builder should cater for the appliance fixing, incorporating a purpose designed frame, timber stud or noggin. The plasterboard itself will have insufficient strength to support any significant load. As a guide, however, for smaller fixings such as pipes and small appliances, one cavity fixing would be required for every 7 kg. Pipework contained within the walls should be run as shown on page 139.

Flue Systems

Vertical Flue Pipes

Routing a vertical flue pipe through a timber-framed building poses the same problems as for a conventional dwelling, use sleeving of a non-combustible material and maintain a 25 mm air gap as shown on page 217. Should the flue penetrate an external wall, the air gap should be packed with a non-combustible thermal insulation material and again care should be observed to ensure that moisture cannot get into the inner timber skin.

Pre-cast concrete flue block systems

These may be incorporated in the timber walls by enclosing the block work in a vertically positioned galvanised channel, maintaining the bond and inserting alternate cut blocks to make up the space. This flue design is further discussed on page 214.

False chimneys

It is possible to include a false chimney breast as shown, enclosing the flue pipe and flue box; this is further discussed on page 294.

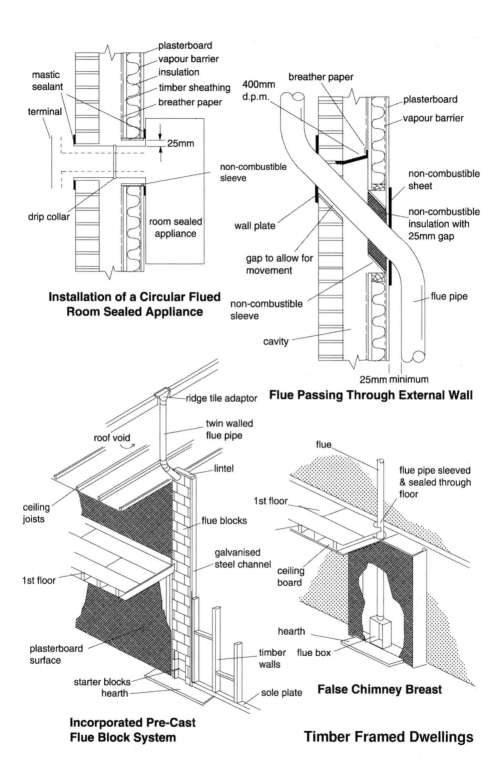

Installation of a Circular Flued Room Sealed Appliance

plasterboard
vapour barrier
insulation
timber sheathing
breather paper
mastic sealant
terminal
25mm
drip collar
room sealed appliance

Flue Passing Through External Wall

breather paper
400mm d.p.m.
plasterboard
vapour barrier
non-combustible sleeve
non-combustible sheet
non-combustible insulation with 25mm gap
wall plate
gap to allow for movement
flue pipe
non-combustible sleeve
cavity
25mm minimum

Incorporated Pre-Cast Flue Block System

ridge tile adaptor
twin walled flue pipe
roof void
lintel
ceiling joists
flue blocks
galvanised steel channel
1st floor
plasterboard surface
starter blocks
hearth

Timber Framed Dwellings

flue
flue pipe sleeved & sealed through floor
1st floor
ceiling board
hearth
timber walls
flue box
sole plate

False Chimney Breast

LPG Storage and Supply

Relevant Industry Documents
BS 5482 and LPGA CP1, 22 and 24

Liquefied Petroleum Gas (LPG) is supplied in liquid form and converted to its gaseous state as previously described on page 26. It is stored either in cylinders, which need to be replaced as and when necessary, or in a large bulk tank on site that is refilled by a delivery tanker as necessary. Usually with a bulk storage vessel the LPG supplier is responsible for its installation and periodic inspection and maintenance.

The same general principles apply to the storage of LPG, whether cylinders or a bulk tank is used. For protection against excessive pressure build up within the storage vessel a pressure relief valve is incorporated. For the bulk tank this is an additional control that needs to be supplied, whereas it forms part of the outlet valve on the smaller cylinders.

Pressure Reduction

In order to reduce the pressure of the high-pressure gas (in the UK this would be typically in the region of 2–7 bar in the vessel), a regulator or series of regulators needs to be fitted. This may consist of a single valve that reduces the pressure in a single one stage operation or it may involve first reducing the pressure to an intermediate pressure of 0.75 bar and then using a second stage regulator to reduce it further to an operating pressure of 37 mbar (for propane) to be used within the building. This design of two stage regulator is used particularly where the stored supply of LPG is some distance from the building as it overcomes the problems arising from too great a pressure drop.

For bulk tank supplies of LPG or where four or more cylinders are used for the supply, an under pressure/over pressure UPSO/OPSO safety device needs to be included in the supply pipeline. This control was described in Part 3, Gas Controls, as was the operation of the regulators.

Vapour Off-take Capacity

The storage vessels used must be suitably large to boil off the gas fast enough to provide the maximum hourly gas rate needed, with everything full on. This rate of gas production is known as the vapour take-off capacity. The take-off capacity is given in cubic metres per hour (m^3/h); a suitable storage capacity can be read from the table opposite.

To calculate the required capacity needed to provide a suitable supply, simply add all the heat input requirements for all the appliances connected to the system and refer to the table.

Example A dwelling has a 28 kW combination boiler, 4.6 kW gas fire and 17 kW cooker. Therefore the minimum storage capacity would need to be:

$$28 + 4.6 + 17 = \underline{49.6\,kW}$$

Therefore either two 47 kg cylinders, three 19 kg cylinders or the smallest size tank available would be needed to supply the installation.

Maximum recommended vapour take-off

Vessel type and base size	m³/h	Approx. equivalent heat input kW
380 litre bulk tank (1.7 m × 0.65 m)	2.3	60
1200 litre bulk tank (2 m × 1m)	5.7	150
2000 litre bulk tank (3.1 m × 1m)	7.1	187
3400 litre bulk tank (3.8 m × 1.2 m)	10.2	269
47 kg propane cylinder	1.27	33
19 kg propane cylinder	0.71	18
13 kg propane cylinder	0.57	15
6 kg propane cylinder	0.42	11
3.9 kg propane cylinder	0.28	7
15 kg butane cylinder	0.28	9
15 kg butane cylinder	0.2	6
15 kg butane cylinder	0.14	4

Note: Butane is not recommended for installations over 8.5 kW.

4 Installation Practices

Bulk Tank Installation

LPG Cylinder Installation

Relevant Industry Documents
BS 5482 and LPGA CP 22 and 24

Where cylinders of propane are to be used, in general an automatic change-over valve would be used, thereby ensuring that the supply is uninterrupted during use. The automatic change-over valve (described on page 90) is a control valve that automatically switches the supply to a new cylinder when the gas runs out. The change-over valve also includes a regulator that reduces the gas pressure to that needed for use in the building. So that a cylinder can be replaced, the supply from the cylinder to the regulator is via a hose, or series of hoses. These hoses are referred to as the pigtails. The hoses are subject to the same high pressure as that found in the cylinder itself and, as a result, need close inspection during maintenance, etc. to ensure that they are not suffering stress from age or damage. UV light from the sun slowly tends to perish these hoses and it is therefore recommended that they be changed every five years.

Location of Cylinders

Propane cylinders must be located in an accessible position outside in the open air. They should be placed on a firm level surface and installed in the upright position, with the valve at the top so that only vapour may be drawn from the vessel when in use. The cylinders should ideally be located against a wall in a stable position where they are unlikely to fall or be knocked over. Where necessary additional protection should be provided to secure the cylinders in place. At no time must any cylinder be used or stored in a cellar, basement or other low-lying area where escaping gas may accumulate.

Cylinders must not be:

- at a distance of less than 1 m horizontally, or 0.3 m vertically, from the valve to any opening into the building, sources of ignition, heat sources or unprotected electrical equipment;
- closer than 2 m horizontally from an un-trapped drain, unsealed gully or opening into a cellar unless a barrier wall is provided;
- within 3 m of any corrosive or toxic substances unless a fire barrier is provided;
- positioned where they obstruct the access to and from the premises.

The associated equipment, e.g. regulators, should be located as close as practical to the cylinders, with the hoses kept as short as possible, allowing just enough room to provide the flexibility needed to exchange the cylinders. In no case should the hose be greater than 2 m in length. For exposed locations some form of protection may be provided, such as a hood for weather protection. The cylinders should not be padlocked, preventing their removal.

When a cylinder needs exchanging or disconnection the cylinder valve and any other valves as applicable must be closed before undoing the supply pipe. This includes situations where the cylinders are empty. All sources of ignition must be extinguished.

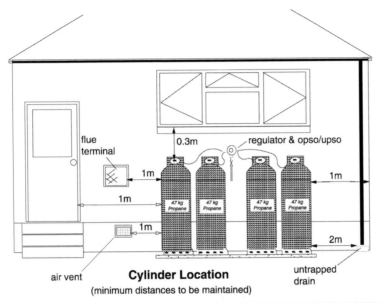

Cylinder Location
(minimum distances to be maintained)

Manual Changeover Valve

**Automatic Changeover Valve
Serving two Banks of two Cylinders**

Wall Block Manifold and Single Stage Regulator

To the left of the photo is a single wall block serving a propane regulator. That on the right is a double wall block manifold serving a butane regulator.

Storage and Transportation of LPG Cylinders

Relevant Industry Documents
LPGA CP 7 and 27

Storage of LPG

Specific guidance on storage can be obtained from the suppliers. However, in general, all LPG cylinders should be stored in the open. A compound to store the cylinders will be required if the storage is greater than 400 kg (e.g. more than eight 47 kg cylinders) with access being restricted to those authorised and notices prohibiting smoking/naked flames should be displayed. All refillable cylinders should be considered as full whatever their state of contents and cylinders containing other gases or hazardous substances should not be stored in the vicinity. Small storage areas should be more than 2 m from a drain; this distance should be increased to 3 m for larger storage compounds. The illustrations opposite gives guidance on storage. Note that the minimum distance to a boundary or building is 1 m. This distance increases as the volume of gas exceeds 400 kg. The maximum height for stacking cylinders when not on pallets should not exceed 2.5 m.

Care with Empty Cylinders

When only a small amount or no more gas can be drawn from an empty cylinder, one could mistakenly think that it could be disconnected and left with the valve in the open position. However the vessel will contain a small quantity of LPG, be it only vapour. During storage the empty cylinders must be treated with great respect and the control valve kept securely closed with the plastic bung reinserted into the cylinder. Failure to do this could result in the volume of gas contraction during the evening when the temperature drops and, as a result, air would be drawn through the valve into the cylinder. There would then be a mixture of gas and air in the cylinder, which may be within the flammability limits for the gas (see page 24). Should a heat source come within close proximity of the vessel, an explosion may result. Whilst the cylinder was filled only with gas any heat applied would cause the pressure relief valve to open, expelling the gas. This could ignite as it issues from the release valve. The cylinder itself would not have expanded as there was no combustion within the vessel. The empty cylinders should be stored with the full cylinders but they should be segregated so that they can be easily distinguished.

Transportation of Cylinders

LPG cylinders should, where possible, be transported in open back vehicles. Small quantities may be transported in closed vans, but the vehicle should carry a suitable label, with the flammable gas diamond sticker. *Note*: The sticker must be removed when no gas is on board. In addition the driver should be trained and made aware of the hazards. There should be a small dry powder fire extinguisher on board. The driver and passengers of the vehicle must refrain from smoking, this being prohibited. When carrying LPG, the cylinders must be stowed and secured upright, with the valves uppermost.

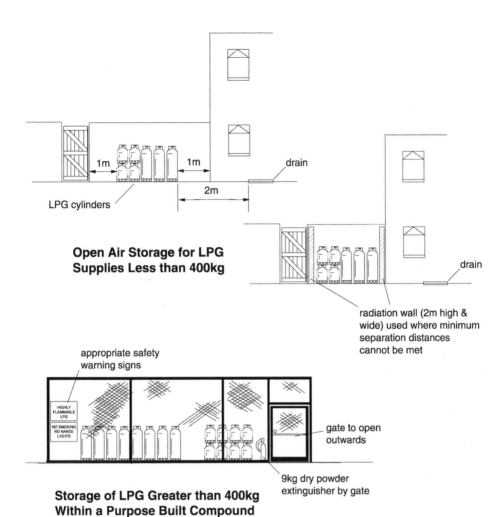

1m   1m

2m

drain

LPG cylinders

**Open Air Storage for LPG
Supplies Less than 400kg**

drain

radiation wall (2m high &
wide) used where minimum
separation distances
cannot be met

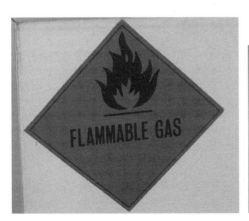

appropriate safety
warning signs

HIGHLY
FLAMMABLE
LPG

NO SMOKING
NO NAKED
LIGHTS

gate to open
outwards

9kg dry powder
extinguisher by gate

**Storage of LPG Greater than 400kg
Within a Purpose Built Compound**

Flammable Gas Diamond Sticker

to be used where LPG is carried within a
closed vehicle

**Plastic Bung Inserted
into Empty Cylinder**

LPG Bulk Tank Installation

Relevant Industry Document
LPGA CP1

Bulk storage of LPG is usually undertaken by the supplier of the gas; it is a specialist operation undertaken by specially trained operatives. The gas stored in a bulk tank is propane and the size of the vessel selected will determine the total volume of gas that can be drawn from the tank in a given time (this was discussed on page 144). The colour of a bulk tank is usually white. This colour is selected to minimise the amount of heat absorbed on a hot day. This absorption would have the effect of expanding the gas and, where suitable expansion has not been allowed for, the pressure relief valve would open. Tanks painted darker colours would need to be much larger in size to hold the same quantity of fuel, thus allowing for a larger ullage space – the void above the level of the liquid LPG.

Storage vessel siting

In general, the storage tank is sited above ground. It is possible to bury the tank, however the vapour take-off would be reduced due to the limited heat that can be taken from below ground level. Where the tank is to be sited above ground it would need to be positioned on a concrete base 150 mm thick. The location of a tank needs to be agreed with the supplier and user, thereby ensuring that it is unobtrusive and easily filled from the road. There are minimal separation distances that need to be observed between the buildings and property boundaries, as indicated in the table opposite. The tank should not be located underneath any part of a building or overhanging structure. Overhanging tree branches and overhead power cables in close proximity also need to be avoided. Where the voltage is less than 1 kV, the tank needs to be sited a minimum distance of 1.5 m to the side of a vertical drop from the cable, this distance increases to 10 m if the voltage is 1 kV or more.

Firewall protection

Where a firewall is constructed, it should be as tall as the vessel itself. It may form part of the boundary wall. For tanks ≤2500 litres, the building may form the fire-wall provided that it is impermeable to the liquid and is of 60 minute fire resistant construction for residential properties and 30 minutes elsewhere.

Warning notices

The markings on a bulk tank must identify the contents and give a clear indication as to their highly flammable nature. There should be a durable, clearly visible suitable sign affixed to an adjoining wall, fence or on to the vessel itself, prohibiting smoking or naked flames and, where the tank is segregated, a warning against unauthorised entry.

Storage vessel fittings

The gas controls fitted to a bulk storage vessel to enable safe filling and use are situated on top of the tank, normally under a hood or cover, which should be locked to prevent unauthorised entry. These are explained in the next page.

Minimum separation distances to be observed for bulk tank installations

Capacity litres	From buildings boundaries and sources of ignition	Distance if firewall is included	Distance between vessels
≤ 500	2.5 m	0.3 m	1 m
>500–2500	3.9 m	1.5 m	1 m
>2500–9000	7.5 m	4 m	1 m
>9000–135 000	15 m	7.5 m	1.5 m
135 000–37 500	22.5 m	11 m	1/4 ϕ of tanks
>337 500	30 m	15 m	1/4 ϕ of tanks

ϕ = diameter

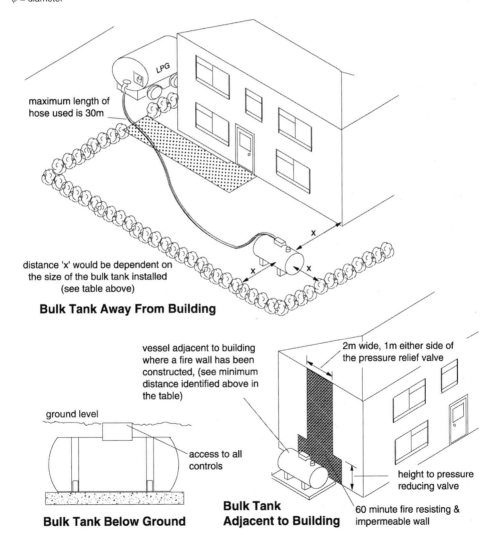

maximum length of hose used is 30m

distance 'x' would be dependent on the size of the bulk tank installed (see table above)

Bulk Tank Away From Building

vessel adjacent to building where a fire wall has been constructed, (see minimum distance identified above in the table)

2m wide, 1m either side of the pressure relief valve

ground level

access to all controls

height to pressure reducing valve

Bulk Tank Below Ground

Bulk Tank Adjacent to Building

60 minute fire resisting & impermeable wall

Gas Supplies from a Bulk Tank Installation

A bulk tank may serve an individual property or it may serve several buildings. The service to a single building may be either low pressure or, where the supply is some distance from the building, a medium pressure installation may be found. Many of the controls listed were discussed in Part 3.

Connected to the outlet from a bulk tank will be a vapour take-off point and service control valve and it is from here that the pipe is run to the building. This pipe is referred to as the service. Along its route will be the regulators and safety controls designed to allow the gas to flow safely and at the correct pressure. The first valve encountered is the first stage regulator. This is a control regulator designed to reduce the gas pressure to an intermediate medium pressure of 0.75 bar. Following this valve may be fitted the second stage regulator, which reduces the gas pressure further to that required inside the building, usually 37 mbar. If the tank is sited some distance away from the building this second stage regulator may not be fitted close by the tank, but just prior to its entry to the building. In this way pressure loss within the pipework is reduced to a minimum. Also along the gas service route to the building will be found the over pressure shut off device (OPSO) and the under pressure shut off device (UPSO). These two controls were described on page 82. They may be installed as separate controls or may be incorporated as a combined unit and also incorporate the second stage regulator. Note that where polyethylene (PE) is used for the gas supply to the building the OPSO will need to be installed close to the tank to ensure that any excessive pressure within the gas line does not cause a problem. Finally, before the entry into the building an emergency control valve needs to be fitted.

Additional Storage Vessel Fittings

A pressure relief valve, designed to operate should the storage pressure become excessive, is fitted to the bulk tank. The pressure relief will discharge the vapour contents into the open air, therefore consideration needs to be given as to a safe method of discharging the gas, which will otherwise flow to lower ground levels and accumulate. A liquid filling connection and its shut-off valve, along with a contents gauge, are also found. Sometimes a combination valve is used, this comprises the filling connection and its shut-off control, the liquid level indicator and the vapour take-off with its appropriate service valve.

Multiple Building Installations

This is where the service is run from the bulk tank, through a distribution network to a number of different locations. Generally the distribution service would be run at a pressure of 0.75 mbar with an OPSO/UPSO, second stage regulator and emergency control fitted at the entry to the premises. A meter can be installed at this point to monitor the volume of gas used.

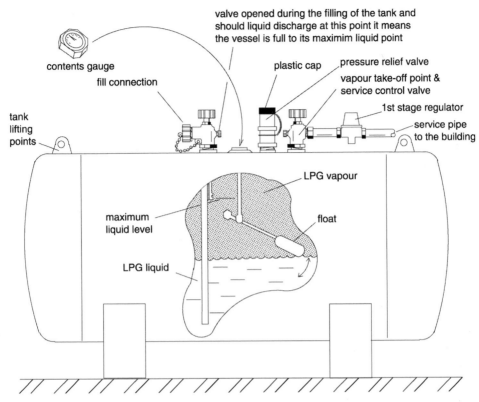

valve opened during the filling of the tank and
should liquid discharge at this point it means
the vessel is full to its maximim liquid point

contents gauge

fill connection

plastic cap

pressure relief valve

vapour take-off point &
service control valve

tank
lifting
points

1st stage regulator

service pipe
to the building

LPG vapour

maximum
liquid level

float

LPG liquid

Typical Bulk Tank Installation

Bulk Tank Combination Valve

Typical LPG Meter Installation

Pipe Sizing LPG Installations

Relevant Industry Documents
BS 5482 pt. 1

LPG installations are allowed a maximum working pressure drop of 2.5 mbar between the outlet of the pressure regulator and the appliance when subjected to full load. In designing a system one of two methods can be employed to calculate the appropriate size.

Method 1

For use where no branch run exceeds 3 m.

- From the table 1, select a pipe size for each branch, depending on the appliance it serves.
- Find the size of the main run, add together the squares of each branch pipe size.
- Take the square root of this total, rounding up the answer to the nearest pipe size.

Example A gas installation has the following appliances connected to individual branches, each less than 3 m in length: a small cooker, multipoint water heater and small gas heater.

Table 1 Branch size

Appliance	Pipe size
Large cooker	12
Small cooker	10
Sink storage water heater	8
Instantaneous water heater	12
Multipoint water heater	15
Refrigerator	6
Small central heating boiler	12
Small space heater	6

From the table, the following branch pipe sizes are selected.

Appliance	Branch size	(Branch size)^2
Cooker	10 mm	$(10^2) = 100$ mm^2
Water heater	15 mm	$(15^2) = 225$ mm^2
Space heater	6 mm	$(6^2) = 36$ mm^2
Total		361 mm^2

The diameter of the main run would be: $\sqrt{361} = 19$, therefore a 22 mm pipe is selected.

Table 2 Flow discharge through copper tube

Pipe size	Max length: Allowing 2.5 mbar pressure differential between each end							
	3	6	9	12	15	18	21	24
6	0.12	0.085	0.071	0.059	0.048	0.048	0.04	0.04
10	0.88	0.57	0.48	0.42	0.38	0.35	0.32	0.29
15	1.49	1.01	0.79	0.7	0.6	0.53	0.5	0.47
22	8.01	5.21	4.19	3.62	3.2	2.86	2.58	2.38
28	15.92	8.86	8.33	7.25	6.51	5.61	5.24	4.87
	Discharge in m^3/h							

Table 3 Flow discharge through steel tube

Pipe size	Max length: Allowing 2.5mbar pressure differential between each end							
	3	6	9	12	15	18	21	24
6	0.5	0.35	0.29	0.25	0.23	0.2	0.19	0.18
15	4.25	2.83	2.26	1.98	1.7	1.53	1.42	1.27
20	8.5	5.95	4.67	3.96	3.4	3.11	2.92	2.72
25	18.7	12.74	9.91	8.5	7.36	6.8	6.23	5.66
	Discharge in m³/h							

Method 2

For use in branch pipes >3 m or where appliances are not listed in Table 1.

This is a much more accurate method and to simplify the calculation it mirrors the method used for natural gas on pp. 48–51 which should be referred to for a rationale as to how the four stages of the calculation are completed.

Example A system, identical to the natural gas example on page 51, is to be designed. The details can be taken from this previous calculation, including the maximum kW rating for each section and the effective pipe length.

Note: To convert the LPG kW rating to m³ the conversion factor has been changed to 0.038. Different flow discharge tables are also used allowing a 2.5 mbar drop, resulting in the following changes to the calculation.

Section	Stage 1 Find the total gas flow in m³. *Max. kW × 0.038*	Stage 2 Determine the effective pipe length. *Actual length + Fittings*	Stage 3 Determine the pipe length for sizing purposes. *Length × No. of sections*	Stage 4 Compare length with required flow.
A–B	(All appliances) 53.8 kW × 0.038 = 2.034 m³	7 m	∴ 7 × 3 = 21 Refer to Table 1.	Suggested pipe size 22 mm
B–C	(Cooker and fire) 21.8 kW × 0.038 = 0.828 m³	2 m	∴ 2 × 3 = 6	Suggested pipe size 15 mm
C–D	(Cooker only) 17.2 kW × 0.038 = 0.654 m³	3 m	∴ 3 × 3 = 9	Suggested pipe size 15 mm
C–F	(Fire only) 4.6 kW × 0.038 = 0.175 m³	3.5 m	∴ 3.5 × 3 = 10.5 Next size up, i.e. 12 m	Suggested pipe size 10 mm
B–E	(Boiler only) 32 kW × 0.038 = 1.216 m³	2 m	∴ 2 × 2 = 4 Next size up, i.e. 6 m	Suggested pipe size 15 mm

Thus, from looking at the completed minimum pipe sizes, it will be seen that for the same kW input, pipe sizes smaller than those for natural gas can be used.

Determining Existing Losses

For an existing installation one can determine whether a pipe size is suitable by following the procedures from page 54, again applying these LPG flow discharge tables and the appropriate conversion factor. One final step would need to be added, however, in order to achieve the actual 'Pressure Loss for each section'. After dividing Stage 2 by Stage 4, the figure found would need to be multiplied by 2.5 to give the actual pressure drop for the section.

4 Installation Practices

Part 5
Tightness Testing

Tightness Testing and Purging

Relevant Industry Document
IGE/UP/1, 1a and 1b and BS 5482

The procedures completed for domestic gas installations, be they for natural gas or LPG, are based on a general concept that the system will be of an average size, containing a limited volume of gas. It has allowed test times and purge volumes to be standard for the gas used. On the other hand, with a commercial installation the size is in effect unlimited and, as a result, more stringent test times and procedures need to be in place to allow for accuracy and safety in all aspects of the work. This includes testing and putting a system into service and the de-commissioning of a supply. The flowchart opposite shows the relevant industry documents that should be complied with in order to undertake the correct procedure. The following pages cover the use of natural gas and LPG at a range of pressures and pipe sizes, however there are many variables to take into account and the relevant industry documents would need to be sought for more in-depth information and advice.

Since the introduction of the latest edition of IGE/UP/1, strength testing is now required for all large gas installations. Strength testing is a new concept that has been introduced to bring testing standards in line with Europe. It involves subjecting the pipework to the worst possible case scenario, in terms of the pressures that the installation may experience. At the time of publication, strength testing to domestic property has yet to be ratified. However, it is something to watch out for as the standards are updated.

The Gas Regulations state that where you perform work on a gas installation that might affect gas tightness, you should test the pipework at least as far upstream and downstream as the nearest valve. With this in mind, it is necessary to understand that you are not required to test the whole system, only that which you are working on. One can see from the following illustration that where an extension has been run from a previously blanked end, the pipework needs only to be tested as far back as the valve marked 'x'.

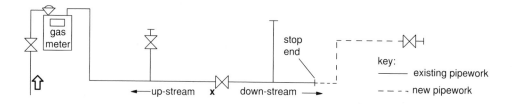

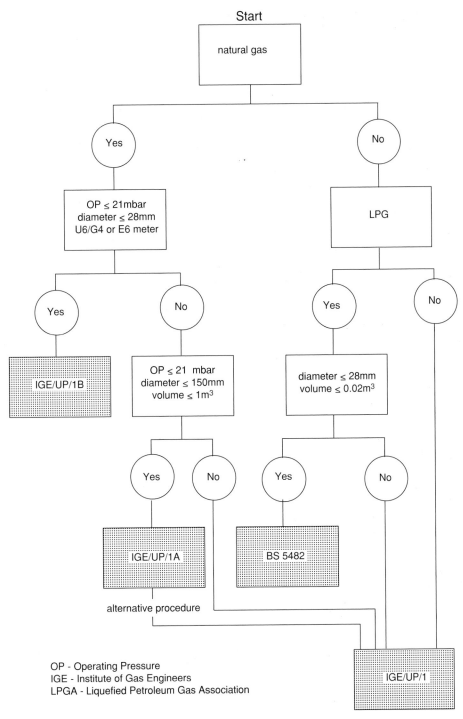

Start

natural gas

Yes — No

OP ≤ 21mbar
diameter ≤ 28mm
U6/G4 or E6 meter

LPG

Yes — No — Yes — No

IGE/UP/1B

OP ≤ 21 mbar
diameter ≤ 150mm
volume ≤ 1m³

diameter ≤ 28mm
volume ≤ 0.02m³

Yes — No — Yes — No

IGE/UP/1A

BS 5482

alternative procedure

IGE/UP/1

OP - Operating Pressure
IGE - Institute of Gas Engineers
LPGA - Liquefied Petroleum Gas Association

Selection of the Appropriate Tightness Testing and Purging Standard

Testing Equipment

Pressure Testing Apparatus

Pneumatic Testing: For testing either a simple hand pump could be used where low pressures are required or it may be necessary to use a compressed air supply for strength testing at high pressures.

Hydraulic testing: This may be undertaken by connection to the water supply main, with water authority approval and protection against back siphonage. Alternatively it can be achieved using a purpose-designed hydraulic test pump.

Pressure Detecting Devices

The manometer: This instrument is used to determine the gas pressure within low-pressure pipework. Several designs will be encountered, the most common being the traditional water filled 'U' gauge, which is available in various sizes. Other manometers include the 'J' gauge, which has the advantage of containing the fluid within an enclosed tube, so if the gauge falls over the liquid will not run out. The J gauge also has the convenience of having only one leg of liquid from which to read the pressure. Battery operated electronic gauges which give a digital readout are now commonplace, however it is essential that where these are used they are regularly checked for calibration. Prior to using a manometer it is essential that the gauge is zeroed to give an accurate reading. *Note*: the water filled gauge is read at the lowest point of the meniscus (see diagram). See also Table 6 on page 177 for the lowest detectible readable movement. When reading a U gauge both columns should always be viewed to ensure that they give the same reading. If they differ, re-set the gauge adjusting the zero adjustor or average the readings by adding them together and dividing the total by 2.

High specific gravity gauge: This type of gauge is almost the same as the U gauge just described but the liquid used has twice the specific gravity of water, i.e. it is twice as heavy. Therefore, in effect, the high specific gravity gauge shows half the height of that of a water-filled manometer.

The Bourdon gauge: This type of gauge can be used to read both high and low pressures. However, for testing purposes it is generally restricted to checking higher pressures. The gauge consists of a flattened tube, formed in a ring. Gas flows into the tube and in so doing tries to straighten it out. This causes a series of cogs to turn the needle of a dial indicating the pressure within.

Metering Devices

Volumetric metering: Any gas meter that records in m^3/h may be used for this purpose, provided it is of a suitable size, to give volume/h.

Flow metering: There are several designs of flowmeter, including those that operate by the turning of a series of vanes, and others that have a weight floating on a cushion of flowing gas.

Gas Sampling Equipment and Detectors

There are many types and designs of gas detectors, including some that sample for all gas types and give a warning bleep at any sign of 'impurity', including natural gas, LPG, CO and spillage and some that detect only specific fuels. One such detector is the 'Gascoseeker'. This particular device can be adjusted to read:

- 100% gas, as used when purging to fuel gas or N_2;
- 100% Lower Explosive Limits (LEL), also called Lower Flammable Limits (LFL), used to monitor an environment for levels of gas concentration that may lead to an explosion. This scale is also used when purging to air;
- 10% LEL, used to locate gas leaks around pipes and fittings.

It should be noted that the Gascoseeker is designed to detect one specific fuel, e.g. methane, propane, etc. and the correct unit therefore needs to be selected.

Electronic Equipment and Calibration Certificates

All test equipment used in conjunction with strength testing, tightness testing or purging operations must be intrinsically safe, i.e. it must not cause a spark to be generated where fuel gas is concerned and above all, it must be maintained and tested prior to use and, where applicable, annually certified as correctly calibrated.

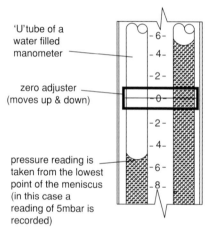

'U'tube of a water filled manometer

zero adjuster (moves up & down)

pressure reading is taken from the lowest point of the meniscus (in this case a reading of 5mbar is recorded)

Section from Water Filled Manometer

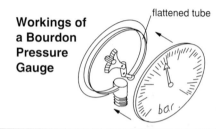

Workings of a Bourdon Pressure Gauge

flattened tube

bar

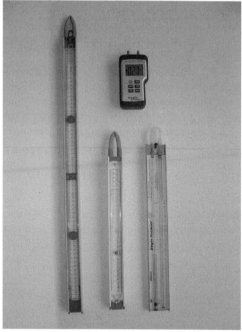

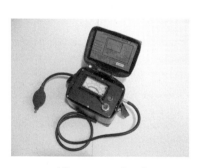

'Gascoseeker'

Various Manometers
far left 60mbar U gauge; centre 30mbar U gauge; far right J gauge; top middle electronic U gauge

Tightness Testing with Air (Domestic)

Relevant Industry Documents
IGE/UP/1B and LPGA TM-62 and BS 5482

Natural Gas and LPG Pipework

When a new gas carcass or installation has been completed and before gas is distributed throughout the system it is always good practice to check for leaks with air, thereby avoiding problems with uncontrolled discharging gas entering the property. This procedure must be undertaken prior to painting the pipework. The test is undertaken using either a water filled manometer or an electronic gauge.

The methods employed for tightness testing with air for natural gas and LPG systems are virtually the same, with the exception that different test pressures are used and different times allowed for temperature stabilisation. Temperature stabilisation is a period of time allowed for the gas to expand or contract due to the temperature environment in which the gas is contained. For example, the air supply may be coming from a cold outside location, but the installation pipe itself is in a warm room. Following a successful air test, the system needs to be tested with the fuel gas that is to be used. The test described below is for installation pipework operating at the normal supply pressures that are encountered within domestic premises, i.e. 21 mbar for natural gas, and 28 or 37 mbar where LPG is the supply fuel.

Test Procedure

1. Survey the system to ensure it meets the required standards. Make sure that all open ends are securely capped off, except one, to which is secured a test tee with associated pump, valve and manometer, as shown opposite The manometer should be adjusted to zero.
2. Operate the pump to increase the pressure, turning the valve off at the following pressures:
   - 20 mbar for natural gas installations;
   - 45 mbar for LPG installations, except on a boat when the test pressure needs to be 70 mbar.
1. Allow a temperature stabilisation period of:
   - 1 minute for natural gas installations;
   - 5 minutes for LPG installations.
   During stabilisation the pressure reading may rise or drop slightly.
2. Then observe the gauge over the next 2 minutes. If there is no pressure drop, the system is deemed to be gas tight.

Because LPG is supplied locally via a bulk tank or cylinders on site, the service pipe to the dwelling also needs to be checked. Because this operates at an elevated pressure a variation to the above test needs to be undertaken. This is achieved as follows:

1. Connect an electronic pressure gauge or Bourdon pressure gauge (with a dial with minimum diameter 150 mm) into the pipeline, via a test tee.
2. Pressurise the pipework to a pressure of 1.5 times its operating pressure. For example, where the pressure is 0.75 bar, the test pressure would be: $0.75 \times 1.5 = 1.125$ bar. The pressure is then isolated.
3. Leave 5 minutes for the system to stabilise.
4. Then observe the gauge over the next 15 minutes. If there is no pressure drop, the system is deemed to be gas tight.

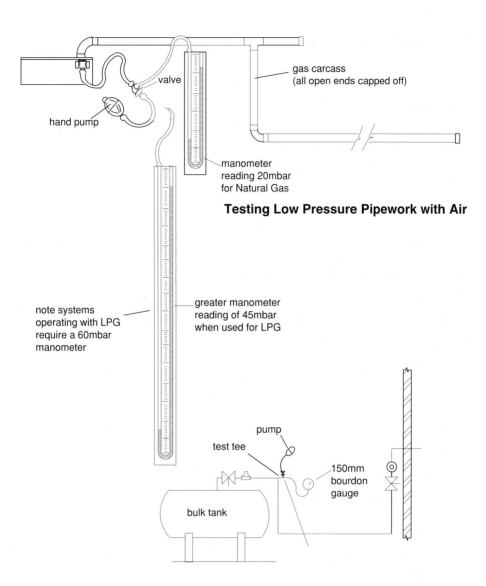

valve

hand pump

gas carcass
(all open ends capped off)

manometer
reading 20mbar
for Natural Gas

Testing Low Pressure Pipework with Air

note systems
operating with LPG
require a 60mbar
manometer

greater manometer
reading of 45mbar
when used for LPG

pump

test tee

150mm
bourdon
gauge

bulk tank

Testing Intermediate
Stage with Air

temporary tee inserted and
pressure raised to 1.5 times
intermediate pressure

5 Tightness Testing

Tightness Testing and Purging with Natural Gas (Domestic)

Relevant Industry Document
IGE/UP/1B

Owing to the increased demand on the national gas distribution network, it has become necessary to increase the supply pressure in certain districts. This has led to the installation of medium pressure regulators prior to the gas meter. Therefore the method of testing is dependant on the supply regulator that is installed, see photographs opposite. The test procedure for systems subjected to medium pressure is described on the following page.

Testing systems that do not have a medium pressure regulator fitted, and systems with medium pressure regulators and an additional test valve fitted between this regulator and the gas meter

1. Survey the system to ensure that it meets the required standards. Make sure that all open ends are securely capped off with the appropriate fitting.
2. Ensure that any appliance isolation valves are in the open position, with cooker lids up and any appliance control tap or pilot flames turned off.
3. Turn off the supply with the appropriate valve, this would be the emergency control valve or the test valve for systems with a medium pressure regulator.
4. Connect a manometer to the outlet of the gas meter and zero the gauge.
5. Open the supply valve to increase the pressure to approximately 10 mbar, then re-close the valve. Observe the gauge reading over the next 1 minute to make sure the pressure reading does not increase by more than 0.5 mbar for a water gauge, or 0.3 mbar where a digital gauge is used. Should this not be the case it will mean that the supply valve is not fully closing off the supply and is letting-by. The valve would therefore need further inspection by disconnecting and spraying leak detection fluid into the valve to confirm let-by. The test would need suspending.
6. Following a successful let-by test, slowly open the supply valve, increase the pressure and re-close the valve at a reading of 20 mbar. Then allow 1 minute for temperature stabilisation, in which the pressure reading may rise or drop slightly.
7. Now observe the gauge over the next 2 minutes. There must be no pressure drop greater than 4 mbar where a diaphragm meter (U6 or G4) is installed or 8 mbar where an ultrasonic meter (E6) is installed. Where a drop is evident, the appliance isolation valves need to be closed to confirm that the drop is not on the system, but may be caused by an appliance control valve letting-by. No drop is permitted on the system pipework with the isolation valves closed. A pressure drop no greater than 4 mbar where a diaphragm meter (U6 or G4) is installed or 8 mbar where an ultrasonic meter (E6) is installed is permitted where the loss is suspected to be through an appliance i.e. with the isolation valves open.
8. Upon satisfactory test, the manometer is removed and the test point re-sealed. The inlet supply valve is now fully re-opened and leak detection spray applied to the test point and all pipework preceding the gas meter.
9. Finally there must be no detectable smell of gas for the system to be deemed gas tight.

**E6 Meter with
Low Pressure Regulator**

**U6 meter with Medium Pressure
Regulator and Test Valve**

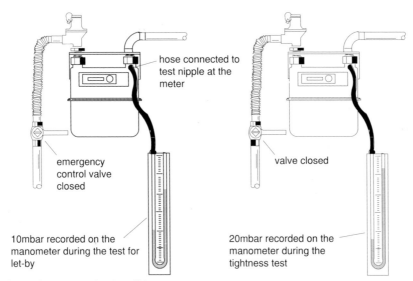

hose connected to
test nipple at the
meter

emergency
control valve
closed

valve closed

10mbar recorded on the
manometer during the test for
let-by

20mbar recorded on the
manometer during the
tightness test

if the valve was not shutting off the
flow of gas fully the pressure reading
on the manometer would rise

the pressure reading on the manometer
should not drop during the test period

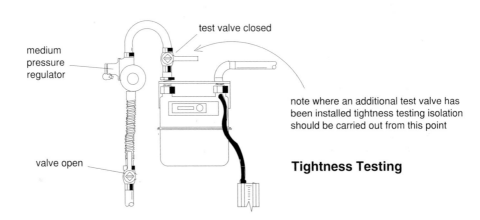

test valve closed

medium
pressure
regulator

note where an additional test valve has
been installed tightness testing isolation
should be carried out from this point

valve open

Tightness Testing

**Testing systems that have a medium pressure regulator fitted
and no additional test valve**

1. Survey the system to ensure that it meets the required standards. Make sure that all open ends are securely capped off with the appropriate fitting.
2. Ensure any appliance isolation valves are in the open position, with cooker lids up and any appliance control tap or pilot flames turned off.
3. Turn off the emergency control valve (ECV). Remove the test nipple at the meter outlet and release all the gas pressure. This would require holding open the release mechanism on the regulator (see photograph).
4. Connect the manometer to the test point and zero the gauge.
5. Hold open the reset mechanism for a period of 1 minute. Observe the gauge reading over this period to make sure the pressure does not increase by more than 0.5 mbar, thus confirming no emergency control valve let-by.
6. The reset mechanism is allowed to return to its rest position, the ECV is opened and the gauge is observed for another 1 minute. Again there should be no increase in pressure. This confirms that the regulator mechanism itself is not letting-by.
7. Following successful let-by tests, re-close the ECV and re-open the reset mechanism. This will allow the regulator to operate and allow the locked up pressure, between the ECV and regulator, to enter the system.
8. The ECV is now slowly opened to increase the pressure to 19 mbar and re-closed. Now allow 1 minute for temperature stabilisation, in which the pressure reading may rise or drop slightly.
9. Now observe the gauge over the next 2 minutes in which there must be no pressure drop greater than 4 mbar where a diaphragm meter (U6 or G4) is installed or over 8 mbar where an ultrasonic meter (E6) is installed. If there is, the appliance isolation valves will need to be closed to confirm that the drop is not on the system, but may be due to an appliance control valve letting-by. No drop is permitted on the system pipework with the isolation valves closed.
10. Upon satisfactory test, the manometer is removed and the test point re-sealed. The inlet supply valve is now fully re-opened and the release mechanism operated to charge the system. Leak detection spray is now applied to the test point and all pipework preceding the gas meter.
11. Finally there must be no detectable smell of gas for the system to be deemed gas tight.

Note: Only 19 mbar is used for the test pressure, thus ensuring the regulator is in the open position and that no high pressure gas within the pipe between the emergency control valve and regulator could make up for any pressure loss.

**Close up View of
Pressure Regulator**
(Reset Lever is in Rest Position)

**Meter Installation with G4 Gas Meter
Medium Pressure Regulator and No Test Valve**
(Picture shows test at stage 6, testing for let-by.)

Domestic Purging Natural Gas Systems

Purging relates to the discharge of air or gas from the pipework in order to commission or de-commission a system. The purge volume of gas to be discharged from meters with a metric index should be 0.01 m^3 and for meters with an imperial index this should be 0.35 ft^3. The procedure to observe when purging is as follows:

1. Inform all appropriate personnel of the purge and prevent any potential sources of ignition. Also open any windows and doors.
2. With all appliances turned off, open the emergency control valve and note the meter test dial reading.
3. Open the appliance or control tap furthest from the meter for a short period and turn off again. Where an appliance such as a boiler has a pilot flame it may be necessary to purge through a disconnected union, spraying any joints with leak detection spray on final completion.
4. At the meter note the volume that has passed and, where necessary, allow more gas to flow until the correct purge volume has been achieved. Where possible try to detect the smell of gas at the appliance end.
5. Finally establish a stable flame at each appliance.

Note: For an installation where gas has been supplied to an appliance that has not been commissioned, the appliance must either be commissioned or disconnected from the gas supply and labelled.

Tightness Testing and Purging
with LPG (Domestic)

Relevant Industry Documents
LPGA TM-62 and BS5482

Before tightness testing with gas, the system needs to be fully purged. This is because the molecular structure of LPG is such that it would find the smallest of holes within a pipeline, including those that air testing would not detect.

Purging

The process of purging LPG is one in which the initial gas flow through into the system is achieved by allowing the gas to be discharged to an appliance, such as a cooker, holding a permanent flame close by or ignition spark operated to ignite the gas immediately as it begins to issue from the burner head. For appliances such as boilers with pilot flames, a temporary burner may be required, such as a Bunsen burner connected to a test point. Gas should not be allowed to flow freely from an open end as the issuing gas would fall to a low level and, where undetected, may lead to an explosion. Prior to tightness testing with gas, it is essential that all branches are fully purged.

Tightness Test Procedure

When tightness testing, first establish that the supply valve feeding the system is operating correctly. In other words, is it shutting off the supply fully? Failure to check this control valve may mean that gas, or let-by gas, can pass into the supply, thereby making up any losses during the test period. The valve is tested for let-by as follows:

1. Survey the system to ensure it meets the required standards. Make sure that all open ends are securely capped off with the appropriate fitting.
2. Ensure any appliance isolation valves are in the open position with cooker lids up and that any appliance control taps or pilot flames are turned off.
3. Turn off the supply with the appropriate valve; this would be the emergency control valve or the cylinder/bulk tank valve.
4. Connect a manometer to the outlet side of this valve and zero the gauge.
5. Slowly open the supply valve and allow the pressure to rise to that where the regulator locks up. This would normally be 45–50 mbar for a system operating at 37 mbar. Failure to lock up would indicate regulator adjustment or that replacement is required.
6. Close the supply valve and then turn on one appliance to reduce the pressure to 5 mbar. Gas discharge from the burner should be burnt off. Where an UPSO is fitted this would need to be operated to release any locked up pressure in the service pipe. Thus the pressure may rise when the reset rod is operated, and would subsequently need to be lowered to 5 mbar.
7. Wait for a temperature stabilisation period of 5 minutes.
8. After the stabilisation period, observe the gauge for a further 2 minutes. There should be no rise in pressure, thus indicating no let-by.

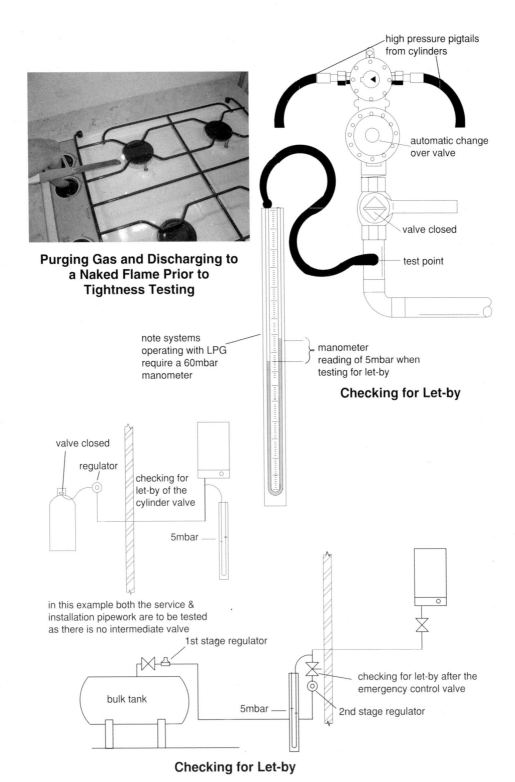

high pressure pigtails
from cylinders

automatic change
over valve

valve closed

test point

**Purging Gas and Discharging to
a Naked Flame Prior to
Tightness Testing**

note systems
operating with LPG
require a 60mbar
manometer

manometer
reading of 5mbar when
testing for let-by

Checking for Let-by

valve closed

regulator

checking for
let-by of the
cylinder valve

5mbar

in this example both the service &
installation pipework are to be tested
as there is no intermediate valve

1st stage regulator

bulk tank

5mbar

checking for let-by after the
emergency control valve

2nd stage regulator

Checking for Let-by

With no let-by confirmed, the test can proceed as follows:

1. Slowly re-open the supply valve and allow the gas to rise until regulator lock-up. Then re-close this valve.
2. Wait a temperature stabilisation period of 5 minutes.
3. Open an appliance and burn off a small quantity of gas, reducing the pressure to either:
   - 37 mbar propane or 28 mbar butane, if an inline valve downstream of the supply regulator is being used to isolate the supply. (e.g. (1) opposite), or
   - 30 mbar propane or 20 mbar butane, if the supply valve is upstream of the supply regulator. This test with a reduced pressure ensures that the regulator is open, releasing any locked up pressure from the high pressure stage (e.g. (2) in the diagram opposite).
4. Now observe the pressure gauge over the next 2 minutes during which time there must be no drop in pressure. Where a drop is evident, the appliance isolation valves need to be closed to confirm that the drop is not on the system but may be due to an appliance control valve letting-by. No drop is permitted on the system pipework with the isolation valves closed. If it is identified that the drop is due to an appliance, providing the drop does not exceed that indicated in the following table and no smell of gas is evident, the system may be deemed gas tight.

Permissible drop with appliances connected

Dwelling type	Test pressure			
	37	30	28	20
Permanent dwelling with a meter	0.5	0.4	0.4	0.3
Permanent dwelling without a meter	1.5	1.0	1.0	0.5
Large holiday caravan home (single supply)	0.5	0.4	0.4	0.25
Coach built motor caravan or tourer	5.0	4.0	4.0	2.5
Residential park home	2.0	1.5	1.5	1.0

5. Upon satisfactory test, the manometer is removed and the test point re-sealed. The inlet supply valve is now slowly fully re-opened and leak detection spray applied to the test point.

The high pressure pipework preceding the supply valve or regulator, where applicable, would also need to be confirmed as sound. For small cylinder fed systems this is achieved using a leak detection fluid, however, for larger systems, such as bulk fed supplies it is achieved by the use of a electronic gauge or Bourdon gauge (150 mm dia.) fitted to the pipe preceding the emergency control valve at the building or into the intermediate pipeline, between the first and second stage regulators. First the valve serving this line is checked for let-by, then the pipe is pressurised to a pressure of 30 mbar, where it is a low-pressure service or to 80% of the normal operating pressure for intermediate pressures (e.g. if 0.75 bar the test pressure would be 0.6 bar).

This ensures that the locked-up tank pressure prior to the first stage regulator is released into the pipework. Therefore, with the pressure set, 5 minutes is waited for temperature stabilisation, then no drop should be recorded over the next 2 minutes if a low pressure service or 15 minutes if intermediate pressure. The test tee should be removed and made good. With the gas re-established, all remaining joints and high-pressure connections should be tested with leak detection fluid.

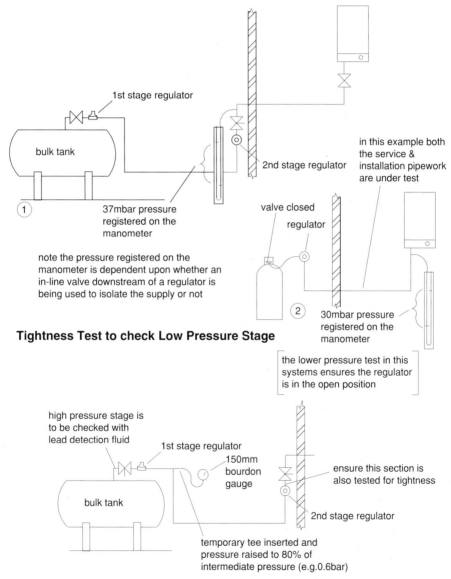

Tightness Test to check Low Pressure Stage

Tightness Test to check Intermediate Stage

Strength Testing for Commercial Pipework

Relevant ACS Qualification
TCPC1

Relevant Industry Document
IGE/UP/1

Where a new system or extension has been installed, the pipe needs to be tested not only for tightness but also to ensure that it can withstand the stresses that may occur during its life without leakage. This is referred to as a strength test. Existing pipework should not be subjected to strength testing. There are two types of strength test: those that use air or nitrogen, called pneumatic tests, and those that use water, called hydrostatic tests. The method selected depends on location, system size and design. For example, a pneumatic test may be more appropriate where water damage of the internal pressure equipment may occur or where the weight of the water in the system on test may cause damage.

Strength Test Pressures, Test Criteria and Method

The actual pressure to be applied to the system under test depends on the material used and the maximum operating pressure of the system. Table 1 opposite gives the test criteria to be observed. Strength testing must be considered at the design stage. It should be noted that applying high pressures to pipework needs considerable care and therefore it is essential that a suitable risk assessment is completed and that the pipework is suitably anchored to withstand the strength test pressure. All components that could be damaged, should be removed or disconnected, inserting bridging or stool pieces, or the section should be blanked off. Do not rely on valves to isolate. These sub-assemblies need to be tested separately to an appropriate standard.

The strength test itself is carried out as follows, ensuring that all valves are open within the section to be tested. As far as possible all pipework should be exposed. The pressure in the system is 'slowly' raised to the required strength test pressure (as the pressure is raised to above 1 bar it is increased in 10% stages) and then the required stabilisation period waited. After this period, the test pressure is recorded; it is recorded again at the end of the strength test duration to identify any drop. Finally a calculation is made as to the drop in pressure, if any, and it is checked against Table 1 to ensure that it does not exceed the permitted drop. For a worked example see page 174.

Specific Notes to Hydrostatic Testing

1. Make sure there is provision to remove the water from the low points at the end of the test and to remove the air at the start. Warm air can be circulated through the pipework to help dry it out.
2. Take pressure readings at the highest point. Allowance needs to be taken for the effects of the weight of water at the lowest points.

Specific Notes to Pneumatic Testing

1. Where test pressures are to exceed 1 bar, an exclusion zone where no operative is permitted to enter must be maintained, as shown in Table 2.

Table 1 Test criteria for strength testing up to 150 mm diameter

Maximum operating pressure Metallic pipes	Diameter (mm)	Strength test pressure	Stabilisation period (minutes)	Strength test duration (minutes)
≤ 100 mbar	All	MOP × 2.5 or *	5	5
> 100 mbar ≤ 2 bar	All	MOP × 1.75 or *	10	5
> 2 bar ≤ 16 bar	≤ 25	MOP × 1.5 or *	15	30
> 2 bar ≤ 7 bar	> 25	MOP × 1.5 or *	30	30
PE pipes				
≤ 100 mbar	All	MOP × 2.5 or *	5	5
> 100 mbar ≤ 200 mbar	All	MOP × 1.75 or *	10	15
> 200 mbar ≤ 1 bar	All	MOP × 1.5 or *	15	15
> 1 bar ≤ 3 bar	All	MOP × 1.5 or 3 bar or *		

Note 1 *MIP × 1.1.
Note 2 The maximum permitted loss of pressure during the test period should not exceed 5% when hydrostatic testing and 20% when pneumatic testing.
Note 3 If the calculated strength test pressure exceeds 10.5 bar or 150 mm in diameter pneumatic testing is not allowed.
Note 4 After strength testing, PE pipe should be allowed to relax at the proposed operating pressure for at least 3 hours.

MOP = maximum operating pressure MIP = maximum incidental pressure (e.g. supply pressure)
• See IGE/UP/1 for pipes/pressures outside those listed above.
The MOP is the pressure at which the system is designed to operate. The MIP is the pressure under fault conditions, such as when a regulator fails or a gas booster is incorporated, etc. These terms are further defined on page 44.

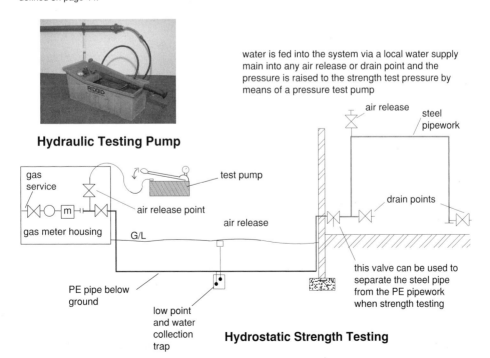

Hydraulic Testing Pump

water is fed into the system via a local water supply main into any air release or drain point and the pressure is raised to the strength test pressure by means of a pressure test pump

air release
steel pipework

gas service
test pump
air release point
gas meter housing
G/L
air release
drain points

PE pipe below ground
low point and water collection trap

this valve can be used to separate the steel pipe from the PE pipework when strength testing

Hydrostatic Strength Testing

5 Tightness Testing

Table 2 Exclusion zone for pneumatic tests exceeding 1 bar pressure

Test pressure (bar)	Installation volume (m³)									
	1	2	3	4	5	6	7	8	9	10
>1 ≤ 2	1.0	1.3	1.6	1.8	2.0	2.2	2.4	2.7	2.7	2.9
>2 ≤ 3	1.5	1.9	2.4	2.7	3.0	3.3	3.6	3.8	4.1	4.3
>3 ≤ 4	1.75	2.4	3.0	3.5	3.9	4.2	4.6	4.9	5.2	5.5
>4 ≤ 5	2.0	2.9	3.7	4.1	4.6	5.0	5.4	5.8	6.2	6.5
>5 ≤ 6	2.4	3.3	4.1	4.7	5.2	5.7	6.2	6.6	7.0	7.4
>6 ≤ 7	2.6	3.7	4.5	5.2	5.8	6.4	6.9	7.4	7.8	8.3
	Distance to be maintained between operatives and centre-line of pipework									

The installation volume is estimated and described in 'Commercial Tightness Testing'.
For volumes in between, round up to the next highest figure.

This is to protect against fittings, etc. shooting off from the pipe with explosive force, causing injury. When hydrostatic testing, this exclusion zone is not required because, unlike air, water cannot be compressed. If the pipework is overhead, this exclusion zone will extend to each side of a vertical line down to ground level as if the pipe were at this elevated position.

2. Leaking pipework is traced using leak detection fluid, with the pressure reduced to 1 bar.
3. A pressure relief valve, suitably adjusted, to open just above the correct strength test pressure, is included in the test apparatus.

Worked Example of Strength Testing

A gas installation of 4 m³ has been installed. The maximum operating pressure (MOP) is to be 30 mbar, and the system is to be supplied from a medium pressure supply with a maximum incidental pressure (MIP) of 2 bar.

The strength test pressure to be applied would be whichever is the greater:

$$\text{MOP @ 30 mbar} \times 2.5 = 75 \text{ mbar or MIP @ 2 bar} \times 1.1 = 2.2 \text{ bar}$$

$$\therefore \ \textit{Strength test pressure} = \underline{2.2 \text{ bar}}$$

The stabilisation time, based on the MOP, is taken from Table 1 and would be 5 minutes; similarly, the test duration would also be 5 minutes.

When hydrostatic testing, the system should be slowly filled with water, observing the guidelines previously given for hydrostatic testing, and allowing a maximum pressure loss of 0.11 bar (5% of 2.2 bar) over the test period. However, where pneumatic testing is undertaken, when the air pressure exceeds 1 bar all operatives should be excluded from the work area to a distance of 2.7 metres, as identified in Table 2 above. The maximum pressure loss during the test is 0.44 bar (20% of 2.2 bar).

Where a pneumatic test is satisfactory, the operating pressure can be reduced and a tightness test may be carried out immediately, subtracting the time used so far in strength testing from the tightness test stabilisation period.

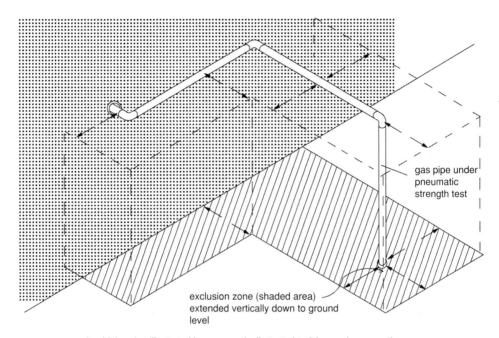

gas pipe under
pneumatic
strength test

exclusion zone (shaded area)
extended vertically down to ground
level

should the pipe illustrated be pneumatically tested to 3 bar and assume the
installation volume was 1.2m³, the exclusion zone would need to be 1.9m

Exclusion Zones when Pneumatic Pressure Testing

Set up for Pneumatic Pressure Testing

Note the pressure reducing valve, pressure gauge and pressure
relief valve fitted to hose prior to entering the pipework

Commercial Tightness Testing

Relevant Industry Document
IGE/UP/1

The method described here may be used on all commercial systems, using various gases up to 16 bar pressure. It is possible to use another test method as identified in IGE/UP/1a, however, this is restricted to 21 mbar natural gas supplies ≤ 1 m^3 volume and up to 150 mm in diameter.

In order to tightness test an installation, one needs access to several tables from IGE/UP/1. These have been reproduced opposite, with their kind permission of the Institute of Gas Engineers and Managers (IGEM), but to my own design. Note the table numbers correspond to the table numbers in the IGE document to assist further cross reference.

Tightness Test Procedures

1. A calculation is made to find the tightness test duration, see over the page.
2. All components that may trap the gas pressure, e.g. non-return valves and regulators, must be temporarily by-passed, and all valves on the system must be in the open position, thus ensuring all necessary pipework is subjected to the test.
3. Open ends need to be blanked off and all valves to and from the section under test spaded or capped off, to include the final connection to the appliance.
4. A review of the ambient temperature and barometric conditions needs to be made to ensure that they will remain stable throughout the test.
5. With the test apparatus connected, slowly raise the pressure within the system to the required test pressure, indicated in Table 5, which is generally the system operating pressure. *Note*: For existing installations, or where the gas supply is being used to pressurise the system, a test for 'let-by' will be required. This is achieved by slowly pressurising the system to 50% of its test pressure and then closing the supply valve. A period is waited, equal to the length of time that will be used for the duration of the test, during which there must be no rise in pressure, thus confirming that the valve is shutting off correctly. After completing a satisfactory let-by test, the supply is raised to the operating 'test' pressure.
6. With the supply valve left open the temperature is now allowed to stabilise for a minimum period of 15 minutes or the length of the tightness test duration, whichever is longer. The source of the pressure is then isolated.
7. The tightness test duration is then waited, during which time no pressure drop should be shown on the gauge for the system to be deemed gas tight. For an existing installation a small pressure drop is permitted, however the leak rate must not exceed the maximum permitted leak rate calculated from (see page 180; F3 is from Table 11): *F3 × Pressure loss × Installation volume ÷ Test duration.*
8. After proving tightness, the pipework, where applicable, should be purged and fuel gas introduced. Where possible, all joints in inadequately ventilated areas should be checked with a suitable intrinsically safe gas detector.
9. Finally, the tightness test should be documented on a certificate, which should be given to the person responsible for the property.

Table 3 Volume of gas meters

Meter type	Total volume (m^3)	Badge rating (m^3/h)	Capacity/ Revolution (m^3)
G4/U6	0.008	6	0.002
U16	0.025	16	0.006
U25	0.037	25	0.01
U40	0.067	40	0.02
U65	0.100	65	0.025
U100	0.182	100	0.057
U160	0.304	160	0.071
E6	0.0024	N/a	N/a
Turbine	0.79d^2l*	N/a	N/a
Rotary	0.79d^2l*	N/a	N/a

d = diameter, l = length
* = or equivalent pipe length

Table 4 Volume of 1 m of pipe

Material and nominal Size (mm)		(in)	Volume of 1 m pipe (1 m^3)
Steel:	15	$^1/_2$	0.000264
	20	$^3/_4$	0.000506
	25	1	0.000704
	32	$1^1/_4$	0.00121
	40	$^1/_2$	0.00165
	50	2	0.00264
	65	$2^1/_2$	0.00418
	80	3	0.00594
	100	4	0.0099
	125	5	0.0154
	150	6	0.022
	200	8	0.0385
	250	10	0.0583
Copper:	15		0.000154
	22		0.000352
	28		0.000594
	35		0.000924
	42		0.00132

Note: The above includes 10% for fittings.
For PE and larger sizes see IGE/UP/1.

Table 5 Tightness Test pressures

Pipe Type	System Pressure	Tightness Test Pressure
Metallic	All	@ Operating pressure
PE Pipe	≤1 bar	@ Operating pressure
	>1 bar	See IGE/UP/1

Table 6 Selection of pressure gauge

Gauge type	Range (mbar)	Readable movement (mbar)	Maximum test duration (minutes)
Water (SG = 1.0)	0–120	0.5	30
High SG (SG = 1.99)	0–200	1.0	45
Electronic to 1 decimal place	0–2000	0.5	30
Electronic to 2 decimal places	0–200	0.1	15
Electronic to 0 decimal places	0–20000	5.0	60

SG = specific gravity
For other gauges and longer test durations see IGE/UP/1.

Table 7/8 Maximum permitted leak rate (MPLR)

Gas Type	New installations and existing installations 'inadequately vented'	Existing installations ≤60 m^3, 'adequately vented' (rate per m^3 of ssv)	Existing installations 'adequately vented' Volume >60 m^3 or externally exposed or buried
Natural	0.0014	0.0005	0.03
Propane	0.00044	0.00016	0.0098
Butane	0.00057	0.0002	0.0123

ssv = smallest space volume
For other gases see IGE/UP/1.

Table 10 Factor (F1) to apply when calculating tightness test duration

Gas type	F1 if using fuel gas to test	F1 if using air or nitrogen to test
Natural	42	67
Propane	102	221
Butane	128	305

Table 11 Factor (F3) to apply when calculating leak rate

Gas type	F3 if using fuel gas @ operating pressure	F3 if using air or nitrogen @ operating pressure
Natural	0.059	0.094
Propane	0.059	0.126
Butane	0.059	0.134

5 Tightness Testing

Tightness Test Duration and the Completion of the Calculation Table

Column 1: Make an accurate survey of the system and divide it into sections, i.e. at branch connections and changes in diameter. Test different materials separately, e.g. steel and PE.

Column 2: Enter the pipe diameter for each section.

Column 3: Enter the actual length of the section.

Column 4: Enter the volume per cubic metre taken from Table 4, depending on the pipe size. *Note:* The table given in this book includes 10% for fittings IGE/UP/1 assumes this to be added to the installation volume.

Column 5: Multiply column 3 by column 4 to give the total volume of the section. Add the meter volume, taken from Table 3, then total up all the sections to give a total installation volume and record it in the box at the base.

Column 6: Enter the gauge readable movement, taken from Table 6, which depends on the gauge used.

Column 7: From Table 10, enter a factor number based on the gas type to be used and the test medium.

Column 8: Complete the calculation along the bottom row to give the test time to be selected for all 'new installation work' and existing situations where the environment is inadequately ventilated and it is not possible to test all joints with leak detection fluid.

For existing installations that are adequately ventilated a further calculation needs to be made. This will depend on the degree of ventilation:

- for internal rooms less than 60 m^3 in volume, multiply the test time above by 2.8 and divide by the room volume, columns 9–10;
- for external environments or internal rooms greater than 60 m^3, the test time is multiplied by 0.047.

Tightness Test Duration Calculation Table

Note: 1. Round up to the nearest minute & never test for less than 2 minutes
Note: 2. Ensure test time does not exceed test duration for gauge selected from table 6
Note: 3. For test durations outside those listed in table 6 and extended test periods see IGE/UP/1

Worked Example:

Below is a schematic pipe layout of a welded steel Natural Gas Installation.

Specific Details:
Section A-B = 150mm dia & runs through an outside area
Section B-C = 100mm dia & runs through an occupied room of 30m³
Section B-D = 100mm dia & runs through an occupied room of 23m³
Section D-E = 80mm dia & runs through a large workshop area
Section D-F = 50mm dia & runs through a large workshop area
Section E-G = 50mm dia & runs to an outside area

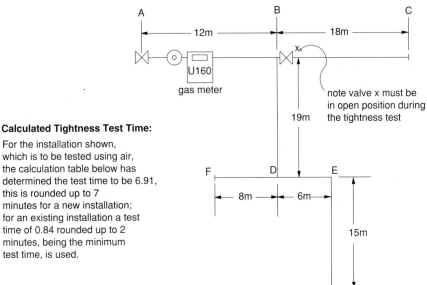

Calculated Tightness Test Time:

For the installation shown, which is to be tested using air, the calculation table below has determined the test time to be 6.91, this is rounded up to 7 minutes for a new installation; for an existing installation a test time of 0.84 rounded up to 2 minutes, being the minimum test time, is used.

Tightness Test Duration Calculation Table

1 section	2 diameter (mm)	3 length (m)	4 volume per metre (m³/m)	5 section volume (m³)	6 gauge readable movement *(electronic gauge to 2 decimal places selected)*	7 factor 1 *(based upon air being the test medium)*	8 test time in minutes *(new installations & inadequately ventilated existing pipework)*	Adequately Ventilated Existing Installations		
									9 room volume	10 test time in minutes
A-B	150	12	.022	0.264						
B-C	100	18	.0099	0.1782				internal & ≤ 60m³		
B-D	100	19	.0099	0.1881					(m³)	
D-E	80	6	.00594	0.0357				→ x 2.8	÷ 23	= 0.84
D-F	50	8	.00264	0.021						
E-G	50	15	.00264	0.0396						
meter volume:	(U 160)			0.304	(mbar)			internal > 60m³ external or buried		
total installation volume:				1.031 x	0.1 x	67 =	6.91	→ x 0.047	=	

Note: 1. Round up to the nearest minute & never test for less than 2 minutes
Note: 2. Ensure test time does not exceed test duration for gauge selected from table 6
Note: 3. For test durations outside those listed in table 6 and extended test periods see IGE/UP/1

Maximum Permitted Leakage Rate

With the tightness test period calculated, the test is performed as described on page 176. If a leak was evident on a new installation it must be found. However, for an existing installation, it is possible to leave an untraceable leak, providing there is no apparent smell of gas and the maximum permitted leak rate determined from the following calculation is not exceeded:

$$F3 \times Pressure\ loss \times Installation\ volume \div Test\ duration$$

where F3 is a factor taken from Table 11.

From the previous example in an existing gas installation, if a drop in pressure of 0.26 mbar is experienced during the test time, the above calculation needs to be made to determine whether the system could be left in service. Hence:

$$0.094 \times 0.26 \times 1.031 \div 2 = \underline{0.0126\ m^3/h}$$

This quantity (0.0126 m^3/h) is now compared with the maximum permitted leak rate given in Table 8. In this example, the middle column of the table would need to be selected because our system is <60 m^3, which suggests that the leak should not exceed 0.0005 × *Smallest space volume*. Our smallest space was 23 m^3, therefore 0.0005 × 23 = 0.0115 m^3/h would be the maximum leak allowed. Our value of 0.0126 exceeds this figure and, as a consequence, the leak would need to be found.

Successful Installation Tightness Test and Completion

Immediately following the tightness test and when the pressure is removed from the system, any tools such as spades should be removed and disturbed joints checked with leak detection fluid. The system should be purged to fuel gas. If this is not done immediately, a further tightness test will be required prior to purging. Once fuel gas is contained within the pipework, all accessible joints in inadequately ventilated areas will need to be checked with a suitable intrinsically safe gas detector set on the 0–10% LFL scale which should show no movement of the needle. Finally the tightness test results should be recorded on a formal certificate, such that shown opposite, and a copy given to the owner or person who is responsible for the property.

Appliance Connection

It would have been noted that the appliance connection was isolated during the tightness test. With a successful test this section of pipework now needs to be checked for tightness prior to commissioning or re-commissioning the appliance. This may be achieved by undertaking a tightness test in accordance with IGE/UP/1B, previously described under domestic tightness testing, or where the pipe exceeds 28 mm in diameter or 0.18 m^3, IGE/UP/1 or 1A may be applied.

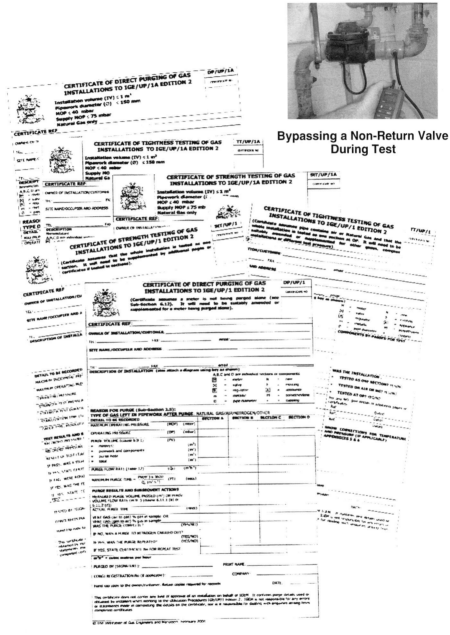

Bypassing a Non-Return Valve During Test

Sample Strength Testing, Tightness Testing and Purging Certificates
The above certificates available in pad form can be purchased from IGEM
(address will be found at the front of this book).

Direct Purging Commercial Pipework

Relevant Industry Document
IGE/UP/1

Direct purging refers to the exchanging of air in a pipeline for fuel gas or, conversely, removing the gas by replacing it with air. It differs from 'indirect purging' in which nitrogen (N_2) is introduced into the pipe between the full gas or air condition. Both direct and indirect purging processes have a certain degree of danger attached, for example, the danger of an explosion in the case of a direct purge and asphyxiation when nitrogen is used; therefore many safety checks or risk assessments are needed before you start. The risk assessment checklist shown opposite encompasses many of the general considerations listed below.

General Considerations when Purging

1. The system to be purged should be surveyed and a tightness test carried out immediately prior to purging. Where a 'ring main' or looping system exists, the system needs to be divided to enable a complete purge.
2. The impact of releasing fuel gas into the air should be considered and, where practical, the gas should be flared or burnt off. Gases such as LPG should be flared in any case, preventing their accumulation in low-lying areas.
3. The pressure within the pipe during the purge should not exceed the maximum operating pressure. Conversely, where the fuel gas is being drawn from a connecting supply, this too should be prevented from dropping to a pressure affecting its operation, such as below 50% of its operating pressure.
4. Purging of meters should only be undertaken with the agreement of the owner.
5. All pipework, not forming part of the purge, should be capped as appropriate, or if impractical the section should be locked off to prevent unauthorised use.
6. Sufficient gas operatives need to be available, each with their specific duty.
7. In all but the simplest purge operations, prepare a written Risk Assessment.
8. In the event of not achieving the required purge rate (example over) the purge should be aborted and restarted if the cause can be determined. Alternatively, the system should be indirectly purged with nitrogen. Provision for it should therefore be included at the planning stage.
9. Purging needs to be done progressively and if there are branches these should be done in order of volume.

Purging Internally

It is possible to purge small volumes of gas directly into a well ventilated area, providing the following criteria are met:

- The operating pressure must not exceed 21 mbar.
- The volume of the ventilated internal space is not less than 30 m^3.
- The installation volume of the pipework does not exceed 0.02 m^3.
- There is no sources of ignition within a distance of 3 m.
- Gas concentrations of the room are monitored and where the level reaches 10% LFL the purge should be stopped immediately.

5 Tightness Testing

Risk Assessment
Activity: Direct Purging Operation

Location:

Risk Assessment Dated:

Completed By: ....   **Signature:**

Who might be affected:
- Operatives within controlled environment • Other personnel with the workplace

Hazard	Control Measures	A × B = *	Risk
Release of gas into the environment leading to possible fire risk	• Gas operatives trained in the purge operation • Warning signs posted, to include no naked flames and no smoking. Area of purge stack cordoned off • Fire fighting apparatus readily available and staff trained in its use • Monitoring of atmosphere around purge stack and supply shut done if 20% LEL recorded	4 × 2 = 8	Tolerable
Air/gas mixture causing risk of explosion or fire	• All equipment to be used is intrinsically safe • Calculations undertaken in accordance with IGE/UP/1 to determine minimum velocity to be maintained (see calculation sheet N° 1, attached) • Safety operative stationed at gas inlet to shut down supply if required in aborted purge. • Nitrogen (N_2) and associated equipment available for indirect purge if required • All control measures as above maintained	4 × 2 = 8	Tolerable
Purge aborted requiring the use of N_2	• Calculations re-checked (see calculation sheet N° 2) • All valves within installation suitably locked off and labelled • Pipework spaded off as appropriate • Installation, including purge equipment tightness tested immediately prior to purge operation • Test equipment checked as working correctly immediately prior to purge operation and where appropriate calibration certificates available	2 × 1 = 2	Acceptable
Existing gas supply being subject to reduced pressures	• Safety operative stationed at inlet point with manometer attached to pipe recording the pressure within and instructed to turn off valve should pressure drop to < 50%	4 × 2 = 8	Tolerable
Existing gas supply being subject to increased pressures	• Supply valve tested for let by prior to purge and safety operative stationed at this point with manometer attached, instructed to turn off N_2 should pressure rise to above system operating pressure	4 × 2 = 8	Tolerable
Affixation where N_2 leakage occurs	• N2 stored in well ventilated area • Appropriate COSHH information sheets available	5 × 1 = 5	Tolerable
Injury due to general practical work activities	• Appropriate COSHH information sheets available • Staff trained in safe working practices • All tools maintained in safe working condition	1 × 1 = 1	Acceptable
Equipment stored or transported containing fuel gas	• All equipment to be purged with air immediately following completion of purge process	2 × 1 = 2	Acceptable

A: Severity	B: Likelihood	*Risk Factor: (A × B)
5 = Death 4 = Major injury 3 = Off work more than 3 days 2 = First aid only 1 = Minor injury	5 = 50% chance (likely) 4 = 25% chance 3 = 10% chance (possible) 2 = 5% chance 1 = 2 % chance (unlikely)	1–2 = Acceptable 3–10 Tolerable Over 10 Danger – Do not proceed

Purge Velocity

The rate at which the gas or air is dispelled from the pipe determines whether the purge would be successful or not. If the velocity is insufficient, the two gases may stratify and mix together to form a combustible mixture within the pipe. Two methods can be employed to verify that the purge velocity is maintained including:

1. using a suitable flow meter, which can directly read the volume of the gas/air flowing through the pipe (e.g. in m^3/h), *or*
2. using a suitably sized 'volume' meter, either already fitted in the section to be purged, or incorporated for the test. The volume meter is used in conjunction with a timer so that the flow rate can be calculated.

In addition to maintaining the velocity, the discharging gas/air needs to be tested to confirm that the percentage of gas is greater than 90%, where inserting the fuel gas; or less than 1.8% gas when purging to air. Note LPG is indirectly purged with N_2.

One method of finding the purge velocity is to complete a table as shown opposite, carrying out each step as described below.

Purge Velocity and the Completion of the Calculation Sheet

(A worked example is shown over the page.)

Column A Enter the installation volume for the various sections to be purged, taken from the tightness test table previously completed (column 5). Also enter the meter volume for the E6, rotary or turbine meter. Also enter the volume of the purge equipment. This is calculated in the same way as the installation pipe was determined for tightness testing. For the appropriate installation volume of the hose use Table 4, which will be acceptable for this calculation.

Sub-Section A Where a diaphragm meter is installed, enter the capacity/revolution (cyclic volume) taken from Table 3 into the appropriate box.

Column B Multiply the figures in column 'A' by 1.5 and any diaphragm meter by 5 to give the required purge volume for each part of the supply.

Column C Referring to the schematic drawing made in the survey, total up the purge volumes of each individual run and determine which has the highest total purge volume. This is usually the section to purge first. List the pipe sections in sequential order on the table.

Column D Enter the total purge volume for each of the sections shown at C. Do not forget that the purge equipment is added to each individual section.

Column E Multiply Column 'D' by 3600.

Column F Using Table 12, enter the minimum purge flow rate, based on the largest pipe within the section to be purged.

Column G Divide column E by column F to give the maximum purge time in seconds.

Column H Enter the minimum purge velocity, taken from Table 12 opposite.

With the table completed, sufficient information is now available for either of the purge methods, mentioned above, to be adopted.

Table 12 Minimum flow rates and associated purge stack dimensions

Nominal pipe diameter (mm)	Minimum purge velocity (m/s)	Minimum purge flow rate (m³/h)	Maximum purge length (m)	Purge point nominal bore (mm)	Purge hose and vent stack size (mm)	Flame arrestor nominal size (mm)
20	0.6	0.7	N/a	20	20	20
25	0.6	1.0	N/a	20	20	20
32	0.6	1.7	N/a	20	20	20
40	0.6	2.5	N/a	20	20	20
50	0.6	4.5	N/a	25	40	50
80	0.6	11	N/a	25	40	50
100	0.6	20	N/a	25	40	50
125	0.6	30	N/a	40	50	50
150	0.6	38	N/a	40	50	50
200	0.7	79	500	50	100	200
250	0.8	141	500	80	100	200
300	0.9	216	500	80	150	200
400	1.0	473	500	100	150	200

For larger pipe diameters see IGE/UP/1.

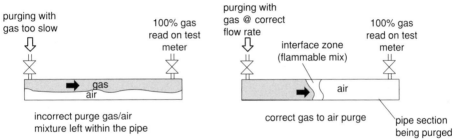

purging with gas too slow

100% gas read on test meter

gas
air

incorrect purge gas/air mixture left within the pipe

purging with gas @ correct flow rate

interface zone (flammable mix)

100% gas read on test meter

air

correct gas to air purge

pipe section being purged

Purge Velocity and Timed Flow Calculation Table

A		B	C	D	E	F	G	H
	Volume (m³)	purge volume (m³)	sequential purge order	total purge volume (m³)	column 'D' x 3600	minimum purge rate (m³/h)	time in seconds (sec's)	minimum purge velocity (m/s)
section		multiply by 1.5						
purge equipment								
diaphragm meter capacity/rev	☐ x 5 =							

5 Tightness Testing

Worked Example of Calculating the Purge Sequence and Velocity

We refer back to the previous tightness test, as completed on page 179 and assume that this system has now to be purged. It will be seen that valve 'x' within the pipework could be isolated; this would therefore allow the initial purge operation to be completed, from A through to points F or G. Either route could be selected for the initial purge, however, ideally the route with the largest installation volume should be chosen. Totalling up each individual section it is found that the greater volume passes through to G. Thus the first purge would be from A to G.

When gas is detected at G a purge from point F should begin immediately. Therefore ideally two purge stacks would be required. With most of the installation purged all that remains to be completed now is the section B–C, which was previously isolated. For this, a stack from the first purge could be re-used.

Installation volume of the purge stack

The size/diameter of the purge stack and hose to be used is based on the largest diameter of pipework in the section being purged and is taken from Table 12. With the size selected, in this case 50 mm and a measurement of the total length of hose and stack (assume the hose to be 10 m in length), the installation volume can be determined by simply following the same format as that used to find the volume of the pipework. *Note*: It is acceptable to use Table 4 for this purpose. Thus, following the same format from page 178:

1	2 (size)	3 (length)	4 (volume/m)	5 (s/volume)
Purge equipment	50 mm	10 m	0.00264	0.0264

*If a meter is included with the purge equipment it would be to be added to this volume.

Completion of table opposite

The table is completed by following the steps laid down on the previous page. The information is simply added to the table and, where necessary, the calculations carried out.

To assist in understanding how the total purge (row D) was determined, the following separate calculations needed to be made, totalling up the volumes through which the gas flow would pass.

Purge volume for A–G:		Purge volume for D–F:	
Gas meter	0.355	D–F	0.0315 +
A–B	0.396	Purge equipment	0.0396
B–D	0.2822 +		0.0711
D–E	0.0536		
E–G	0.0594	**Purge volume for B–C:**	
Purge equipment	0.0396	B–C	0.2673 +
	1.1858	Purge equipment	0.0396
			0.3069

Worked Example of Purging Calculations

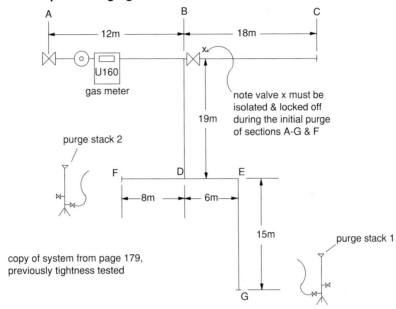

copy of system from page 179,
previously tightness tested

Purge Velocity and Timed Flow Calculation Table

A		B	C	D	E	F	G	H
section	Volume (m³)	purge volume (m³)	sequential purge order	total purge volume (m³)	column 'D' x 3600	minimum purge rate (m³ /h)	time in seconds (sec's)	minimum purge velocity (m/s)
A-B	0.264	0.396						
B-C	0.1782	0.2673						
B-D	0.1881	0.2822						
D-E	0.0357	0.0536	A-G	1.19	4284	38	113	0.6
D-F	0.021	0.0315						
E-G	0.0396	0.0594	D-F	0.072	259.2	4.5	58	0.6
			B-C	0.31	1116	20	56	0.6
purge equipment	0.0264	0.0396						
diaphragm meter capacity/rev	0.071 x 5 =	0.355						

(multiply by 1.5)

With the calculation table completed, all risk assessments undertaken and purge equipment erected, the purging operation can be undertaken. This consists of allowing the gas/air to flow through into the system and simultaneously starting the timer, measuring the flow of purge gas by the chosen method, i.e.

- using a suitably sized 'flow' meter that passes a minimum gas rate (in column F), or
- using a 'volume' meter, passing a total volume (in column D).

For the full purge procedure and method, see over.

Commercial Purging Procedure

Purge Equipment and Vent Stack

For any purge to succeed, the purge points, hoses and vent stack need to be of the correct size as previously listed in Table 12, columns 5–7 (page 185). In addition any valves used in the purging operation would need to be of the full-bore type. If these sizes cannot be achieved it is possible to use multiple vent stacks simultaneously, providing each point is adequately supervised. The hose used to convey the gas to the vent stack needs to be suitable for the gas used and be secured firmly to the pipework and vent. In order to avoid sparking where an externally armoured hose is used, the hose needs to be suitably earthed. Hoses of polyethylene should not be used as they may generate static electricity.

The vent stack itself should include a flame arrestor; it may be in-line where the gas is to be burnt off or as a termination fitting where the gas is not flared. In addition, a sampling point is needed to check for the presence of gas. Where flaring or burning off the gas, the inline arrestor should be fitted at least 2 m from the gas discharge point. A source of ignition is also required at the burner, and care needs to be taken with the issuing flame. The outlet should be in the open air and terminate at a distance of at least 2.5 metres above ground. Where venting, the outlet needs to be located at least 5 m downwind of any sources of ignition and any drifting into buildings should be prevented.

Purging Procedure

Before beginning the purge, ensure that all the general considerations for purging have been met including the completion of the appropriate Risk Assessment (see page 183). Ensure safety barriers and signs are in place and that sufficient fire fighting apparatus is available. All radios used should be intrinsically safe and their operation tested. The communication procedure should be practised by all operatives involved in the purge operation to ensure that they know their specific duties.

Method

1. A manometer should be located at the inlet to the system and an operative stationed at this point. Their job is to keep the supervisor informed should the pressure rise dramatically when supplying air, via an external source, or fall to a point nearing 50% of the operating pressure when supplying fuel gas from an existing pipeline. In addition, they should turn off the supply if instructed to do so by the supervisor.
2. At the purge point/vent stack the valve should be opened to allow the gas to flow and at the same time the timer should be started. At about half way through the purge time, samples of gas/air should be taken to test the gas concentration using a gas detector looking for:
   - 90% fuel gas when purging to gas, or
   - <1.8% fuel gas when purging to air or N_2.

   Note: It is also possible to select the 100% LFL scale when purging to air looking for a reading <40%.

   The purge should be aborted if the gas/air concentration is not achieved within the purge time. It may be re-attempted if the reason can be determined (e.g. blocked hose), alternatively an indirect purge using nitrogen should be undertaken.

3. Upon completion of a satisfactory purge, the equipment should be removed and any disturbed joints checked with leak detection fluid. Inadequately ventilated areas also need to have all accessible joints checked using a suitable gas detector. All connected appliances must be commissioned as necessary, or disconnected from the supply. The purge apparatus should also be purged with air before storage. Finally, an appropriate purging certificate, as shown on page 181 should be completed and given to the owner or person responsible for the property.

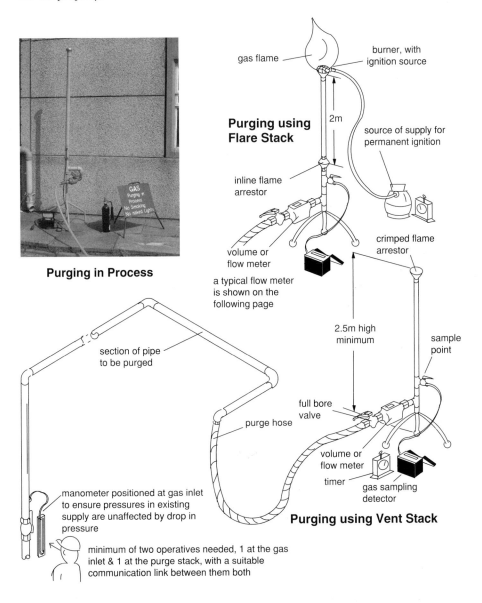

Purging in Process

Purging using Flare Stack

gas flame

burner, with ignition source

2m

source of supply for permanent ignition

inline flame arrestor

volume or flow meter

a typical flow meter is shown on the following page

crimped flame arrestor

2.5m high minimum

sample point

section of pipe to be purged

full bore valve

purge hose

volume or flow meter

timer

gas sampling detector

Purging using Vent Stack

manometer positioned at gas inlet to ensure pressures in existing supply are unaffected by drop in pressure

minimum of two operatives needed, 1 at the gas inlet & 1 at the purge stack, with a suitable communication link between them both

5 Tightness Testing

Indirect Purging Using Nitrogen (N_2)

Indirect purging would only be carried out if:

- the risk assessment deemed a direct purge to be unsafe, such as when purging LPG;
- insufficient purge points are available;
- a direct purge had been unsuccessful and aborted.

Note: It is possible to re-attempt a direct purge if the cause for its abandonment can be determined, such as an error in the calculation or a restriction in the purge line. However, when such repeated purges are undertaken particular care needs to be observed due to the nature of a gas/air mixture being expelled from the pipe.

When purging to N_2 particular attention needs to be paid to avoid asphyxiation, especially in basements or confined places. It is also important to monitor the situation and pressure to ensure that no purge gas enters the distribution network supplying the installation. When purging with N_2 gas, the aim is to provide a complete displacement based on installation volume and sampling at the vent stack and, where possible, maintain the purge velocity as for direct purging. It is important to ensure that any dead legs are adequately purged. N_2 is an inert gas and will therefore not readily mix with the fuel gas; where any mixing does occur no combustible mixture will be formed. Thus with the installation volume worked out (as previously described in the section on direct purging and tightness testing), gas is released into the system to the volume calculated. Sample gas readings are taken at the various vent points to give the following safe purge quantities/gas concentrations:

Natural gas
- <7.5% gas (on the 100% gas scale) if purging to N_2, or
- >90% gas (on the 100% gas scale) if purging to fuel from N_2.

Propane gas
- <3.5% gas (on the 100% gas scale) if purging to N_2, or
- >90% gas (on the 100% gas scale) if purging to fuel from N_2.

Butane gas
- <3.0% gas (on the 100% gas scale) if purging to N_2, or
- >90% gas (on the 100% gas scale) if purging to fuel from N_2.

As a guide to the calculation of the total number of N_2 cylinders required for adequate inert purging, it should be noted that a standard 1.5 m high bottle contains approximately 6.5 m³ of N_2. This could be discharged at a rate of approximately 60 m³/h through a standard single stage regulator. Where inert N_2 purging has been completed to allow for hot work to be undertaken, it is essential to take special care. Possibly consider localised monitoring for traces of fuel gas, as small pockets may remain as a result of the effects of stratification, particularly in dead legs.

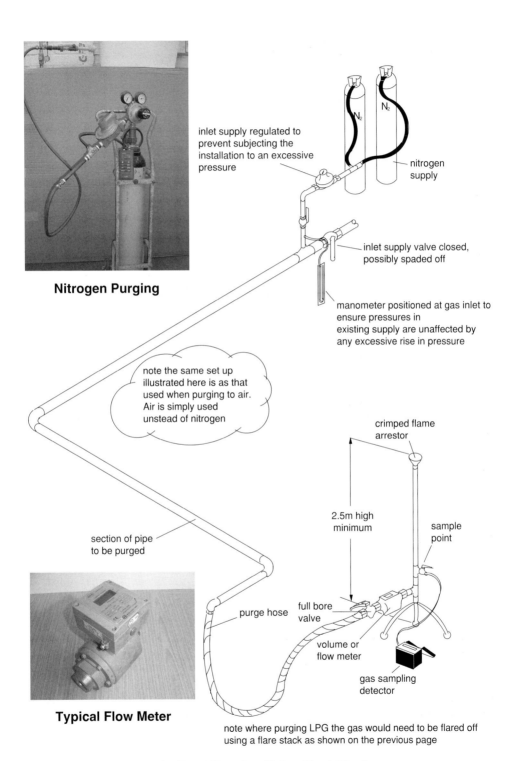

Nitrogen Purging

inlet supply regulated to prevent subjecting the installation to an excessive pressure

N₂   N₂

nitrogen supply

inlet supply valve closed, possibly spaded off

manometer positioned at gas inlet to ensure pressures in existing supply are unaffected by any excessive rise in pressure

note the same set up illustrated here is as that used when purging to air. Air is simply used unstead of nitrogen

crimped flame arrestor

2.5m high minimum

sample point

section of pipe to be purged

Typical Flow Meter

purge hose

full bore valve

volume or flow meter

gas sampling detector

note where purging LPG the gas would need to be flared off using a flare stack as shown on the previous page

Indirect Purging Using Vent Stack

Part 6
Flues

Flue Classification

Relevant Industry Documents
BS 5440, BS 6644, IGE/UP/10 and BS EN 1443

Appliances generally fall into one of the following classifications: **Type 'A'**: Flueless; **Type 'B'**: Open Flued and **Type 'C'**: Room Sealed. The letter A, B or C is immediately followed by a number that further defines the type of flue system design as shown below.

Flue type		Flue design	Natural draught	Induced draught*	Forced draught
Flueless		Not applicable	A1	A2	A3
Open flued	B1	With draught diverter	B11	B12 or B14	B13
	B2	Without draught diverter	B21	B22	B23
Room sealed	C1	Horizontal and 'in balance' to outside	C11	C12	C13
	C2	Inlet and outlet duct connections to multi-stack system (SE duct)	C21	C22	C23
	C3	Vertical and 'in balance' to outside	C31	C32	C33
	C4	Inlet and outlet connections to leg of 'U' duct system	C41	C42	C43
	C5	Ducted flue and air supply system 'out of balance'	C51	C52	C53
	C6	Appliance purchased without flue or air supply inlet ducts	C61	C62	C63
	C7	Vertical flue, with draught break in roof void above ducted air intake to appliance, also in roof (e.g. vertex)	C71	C72	C73
	C8	Flue connected to common duct with air supply ducted from outside and therefore 'out of balance'	C81	C82	C83

*Fan downstream of heat exchanger but prior to draught diverter, except B14 where the fan is also downstream of this assembly. See types B12 and B14.

The flueless appliance takes its air for combustion from within the room in which the appliance is situated and discharges the products of combustion into the same environment.

Open flued appliances also take the combustion air from the room, but remove the combustion products to the outside air. It should be noted that the appliance formally referred to as a 'closed flue' has the same design concept but had no draught diverter and it is now also called an open flue (namely B2 above).

Room sealed appliances do not take any air from the room in which the appliance is situated, but may take the air from a void, such as a roof space. The term 'balanced flue' refers to a system where the air intake is taken from a point adjacent to the flue gas extract and at the same pressure zone, i.e. in-balance.

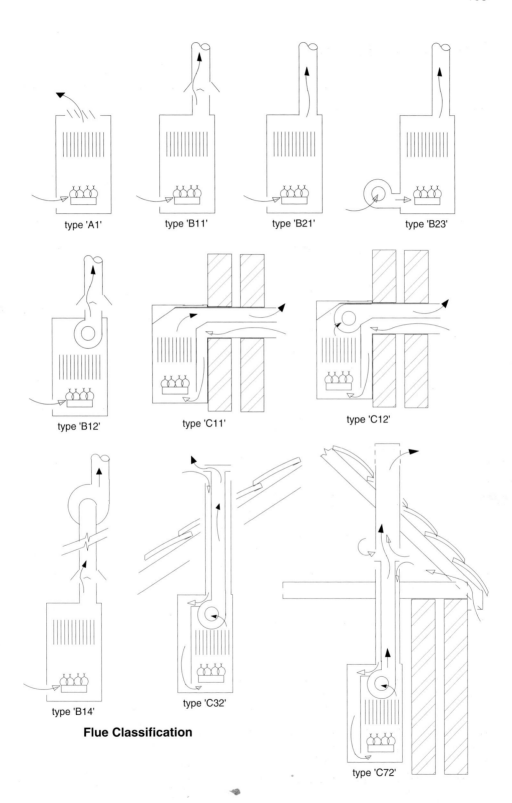

type 'A1'

type 'B11'

type 'B21'

type 'B23'

type 'B12'

type 'C11'

type 'C12'

type 'B14'

type 'C32'

type 'C72'

Flue Classification

Flue Material and Specification

Relevant Industry Documents
BS 5440, BS 6644, IGE/UP/10 and BS EN 1443

The flueway is generally rectangular or round in shape, depending on the material used. Its cross-sectional area is dependant on the manufacturer's design and the heat input.

The Building Regulations (Part J) give the minimum size of any flue system as the appliance outlet cross-sectional area. However, for a gas fire a cross-sectional area of 120 cm^2 (12 000 mm^2) is needed, which in effect needs to be 125 mm in diameter if the flue is round or 165 cm^2 (16 500 mm^2) if the flue is rectangular, with a minimum dimension of 90 mm in any direction.

For many years, operatives have often referred to a flue system as being class 1 or class 2. This was an old British Standard definition relating to the design temperature for which the flue system was made. It was effectively dropped many years ago in favour of specifying the temperature in which a flue system could safely work. The term class 1 refers to those used for solid fuel. Today, however, a new system of flue specification is used and materials are classified according to the prescribed format of BS EN 1443, which uses a system of numbers and letters to identify what the flue is suitable for.

Defining flue specification details

Defining characteristics		Example
Temperature class (max. working temperature)	T250	
Pressure class: N (negative) P (positive) H (high positive)	P1	
Sootfire resistance class S (with) O (without)	O	
Condensate resistance D (dry) W (wet)	W	T250 P1 O W 1 R22 C50
Corrosion resistance (class 1 gas 2 and 3 oils and solid fuel)	1	
Thermal resistance (in units of m^2 K/W $\times$100)	R22	
Minimum distance to combustible material	C50	

So, for the example given, the flue material would be suitable for a system with a maximum flue temperature of 250°C, operating under positive pressure. It would have no resistance to sootfire, but would be resistant to condensate with a corrosion resistance of 1, i.e. for gas. It would have a thermal resistance of 2200 W/m^2 K temperature change and, finally, it should be installed no nearer than 50 mm to any combustible surface.

Several materials may be used for the passage of flue gases and it is essential to choose a material that is suitable in accordance with the Building Regulations. The following table lists some of the available options.

Typical flue materials

Material	Temperature range
Acid resistant brick and insulated brick	$\leq 200°C$
Pre-cast concrete	$\leq 300°C$
Glazed clayware	$\leq 200°C$
Enamelled mild steel	$\leq 450°C$
Stainless steel	$\leq 500°C$
Aluminised steel	$\leq 600°C$
Cast iron	$\leq 500°C$
Aluminium	$\leq 300°C$
Plastic materials (e.g. GRP)	Variable generally $\leq 150°C$

Notice Plates

Where a flue, chimney, hearth or fireplace opening is included in a new building or as part of a refurbishment the information applicable to the flue system should be indelibly marked on a durable robust indicator plate fixed at an unobtrusive, but obvious location. This might be positioned:

- next to the chimney opening or
- next to the water, gas or electrical supply inlet.

The notice plate should convey the following information:

- the location of the beginning of the flue system, e.g. fireplace;
- the category of flue and types of appliances that may be fitted to the system;
- the manufacturer's name of the flue material and the type and size;
- the installation date.

Typical notice plate

IMPORTANT SAFETY INFORMATION
Property Address:
164 Beachcroft Road, Anytown, Essex.
The hearth and chimney installed in the front lounge is suitable for:
Any type/design of gas fire
Flue: 175 mm id Clay liner supplied by 'All-Flue Ltd'
Installed: 10.12.2004
Flue Designation to BS EN 1443: T250 N1 S D 1 R22 C50
Installer: Baytree Construction, Anywhere, Essex.
This label must not be removed or covered.

Note: The flue designation and installer details are optional.

6 Flues

Natural Draught Open Flue Systems

This design of flue system is the oldest form of flue design and for many years was referred to as a conventional flue system. It consists of a flueway through which the products of combustion can pass out to the external environment. The air supply for combustion is drawn into the appliance from the room in which the appliance is installed. Hot air rises up through the flue way by convection.

Convection

As the air or flue gas molecules are heated they expand and in so doing occupy less space per volume. From the illustration opposite it can be seen that there are only four large hot air molecules in the left-hand basket of the scales, whereas there are eight cooler, more denser air molecules on the right-hand side. The weight of each molecule has not increased, only its size. Therefore the basket with the greater number of molecules bears down with a greater force and causes the lighter volume to lift. By applying this principle to the heated gases in a flueway it will be seen that the cooler air surrounding the burner forces the lighter gases upwards. To achieve good convection it is essential that air is free to enter the room in which the appliance is situated, otherwise the supply of heavy dense cold air would soon be depleted and convection would slow down, and the appliance would spill flue gases into the room. For appliances with a heat input of less than 7 kW, adventitious ventilation (air that comes in through openings in windows and doors) is usually all that is needed, but as the heat input increases so does the need for a fixed air supply.

Natural draught open flued appliances incorporate a draught diverter, which must be installed within the same room as the appliance, to assist in the movement of the combustion products. It does this by maintaining a lower pressure within the flueway. In addition the draught diverter:

- assists in diluting the flue gases and thereby reducing the CO_2 content;
- breaks the pull of the secondary flue;
- prevents downdraught blowing back down into the appliance.

When wind blows it creates positive and negative pressure zones along and around the surfaces of the building. The side on which the wind blows has positive pressure. In designing a good flue system the aim would be to position the flue terminal just outside the influence of these pressure zones. Unfortunately wind speeds are not constant and as the wind speed increases so does the size of the pressure zone. Wind gusts continually vary and so the terminal is occasionally contained within the pressure zone and a negative or positive draft will blow through the flue route. Wind from a down-draft will hit the baffle and blow into the room; conversely where an excessive up-draught pulls on the products, air is drawn in at the draught diverter, rather than through the restricted route of the heat exchanger. Both cases help prevent the loss of flame stability.

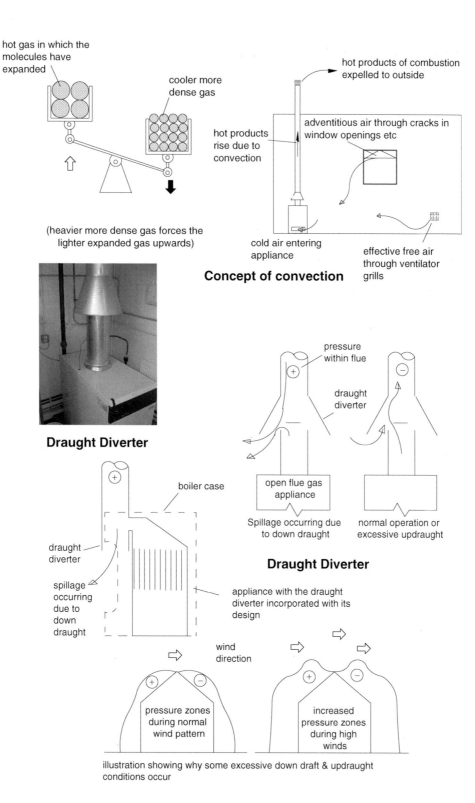

hot gas in which the molecules have expanded

cooler more dense gas

(heavier more dense gas forces the lighter expanded gas upwards)

hot products of combustion expelled to outside

adventitious air through cracks in window openings etc

hot products rise due to convection

cold air entering appliance

effective free air through ventilator grills

Concept of convection

Draught Diverter

pressure within flue

draught diverter

open flue gas appliance

Spillage occurring due to down draught

normal operation or excessive updraught

Draught Diverter

boiler case

draught diverter

spillage occurring due to down draught

appliance with the draught diverter incorporated with its design

wind direction

pressure zones during normal wind pattern

increased pressure zones during high winds

illustration showing why some excessive down draft & updraught conditions occur

Installation of an Open Flue

Restricted Locations

As an open flue appliance takes its air for combustion from the room in which it is situated there is always the possibility that the air may become vitiated (lack oxygen). As a consequence, the Gas Regulations restrict the use of such appliances, stating that they must not be installed in, or take their air supply from, a bathroom/shower room. In addition to this, in rooms that are used for sleeping, i.e. bedrooms and bed-sits, an open flued appliance installed there must not have a gross heat input greater than 14 kW. It is possible to install an appliance with an input rating less than this, however, it must have some form of safety control to shut down the appliance before there is any build up of combustion products in the room such as a vitiation sensing device.

Component parts of an Open Flue

An open flue system consists of four component parts, namely: primary flue, secondary flue, draught break or diverter and terminal.

Primary and Secondary Flue

The primary flue is the section of pipe from an appliance to a draught break, where fitted, and the pressure within this section is that experienced in the appliance itself. From this point the flue is run to the external environment by what is known as a secondary flue. In order to facilitate the disconnection of an appliance from a flue system a disconnecting joint will often be found just above the appliance, as shown in the diagram.

Draught break

This is a component that allows pressure fluctuations to be alleviated within the flue system. Two different types of draught break can be found: the draught diverter (described on the previous page) and the draught stabiliser. The draught stabiliser, shown opposite, is generally restricted to commercial appliances and consists of a hinged door that opens if the pressure within the flue is different from the pressure surrounding the flue.

Terminal

In general, appliances with a flue size less than 170 mm in diameter require a terminal; those that are larger may not. See page 206 for terminal specification.

Factors affecting open flue performance

Many factors influence the operation and performance of an open flue system including the following, which are individually described on the previous and following pages:

- materials: shape and cross-sectional area;
- flue height;
- flue route;
- terminal position;
- heat loss and temperature.

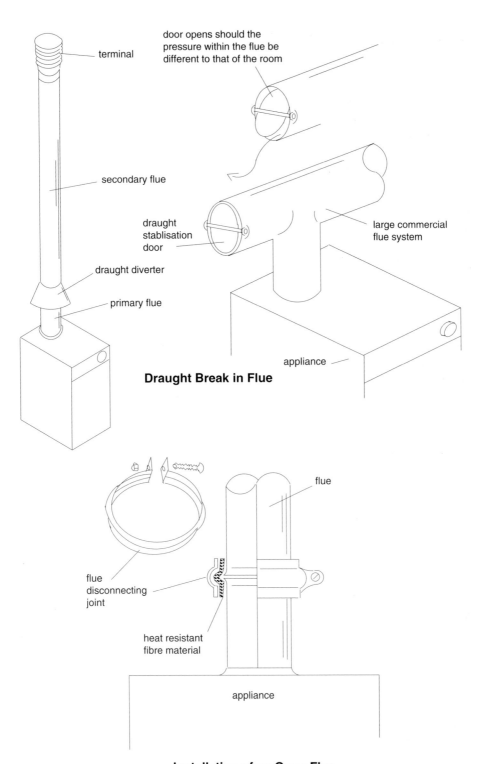

terminal

door opens should the
pressure within the flue be
different to that of the room

secondary flue

draught
stablisation
door

large commercial
flue system

draught diverter

primary flue

appliance

Draught Break in Flue

flue

flue
disconnecting
joint

heat resistant
fibre material

appliance

Installation of an Open Flue

6 Flues

Minimum Open Flue Heights 1

Relevant Industry Document
BS 5440 and IGE/UP/10

Height is required for convection currents to work. The flue needs to be full of heated expanded gases that weigh less than the cool air surrounding the appliance. However this height cannot be infinite, as sooner or later the hot gases will cool to a temperature where they will begin to fall back down the flue. Also, as the flue gases cool to the dew point of water, around 55°C, condensation will begin to form on and run down the internal surface of the flue, causing all sorts of problems. The effective flue height depends on the appliance and installation type and can be read from the table opposite.

Flue Route

The route that the flue takes as it passes up through the building should ideally be as straight as possible, travelling vertical throughout its entire length. A certain flue height is required, as discussed above, to create the draught, but not a particular length – every time there is a change in direction, there is frictional resistance, which decreases the velocity the flue gas. Bends should be avoided and, where the system is to rely on natural draughts caused by convection, they should be restricted to angles of not less than 135°. The distance from the appliance to the first bend needs to be at least 600 mm in order to get a good start to the up-draught, unless the manufacturer's instructions state otherwise.

Effective Height

The effective flue height for a given flue system is not the same as the measured height, because resistance is experienced as the products of combustion flow through the flue system. Instead an equivalent height is calculated using the following formula. However, this is only applicable for flues up to and including 150 mm diameter, for larger flues see IGE/UP/10:

$$He = Ha \times (Ki + Ko)/[(Ki + Ko) - KeHa + \Sigma K]$$

where:

He = equivalent height;
Ha = vertical height;
Ki = inlet resistance from appliance;
Ko = outlet resistance from flue;
Ke = resistance factor per m run of selected flue;
ΣK = sum total of resistance due to total pipe length (not height) and fittings.

To complete the above calculation, see Tables 2 and 3 on the following pages for the resistance factors (i.e. Ki, Ko and Ke).

To help in understanding the above calculation, the example on the following page should be completed.

Table 1 Minimum equivalent flue height and size

Appliance	Minimum height (m)	Minimum diameter (mm)
Gas fire connected to pre-cast block system	2	125
Other gas fires and gas fire/back boiler unit	2.4	
All other appliances <70 kW net input (excluding DFE and ILFE gas fires)	1	Same diameter as appliance flue spigot
Appliances of 54 kW–3.5 MW net input	3	

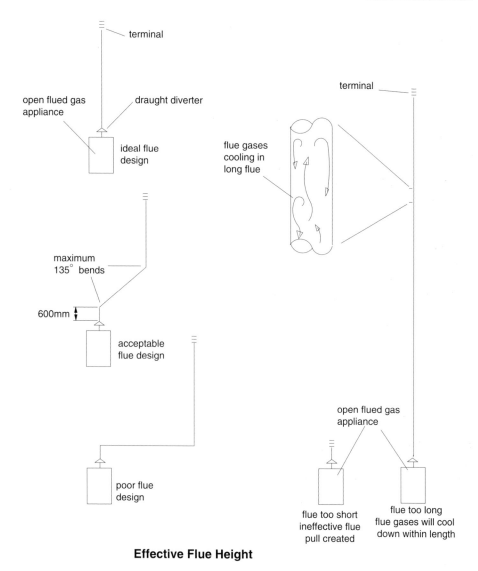

Effective Flue Height

Minimum Open Flue Heights 2

Worked Example

The formula for calculating the effective flue height is given on the previous page. However, for ease of calculation, the following table can be used, entering the data as explained below. *Note*: The tables referred to are given opposite and on the previous page.

Column 1 [*Ki*] Enter the internal resistance for the appliance (taken from Table 2).
Column 2 [*Ko*] Enter the external resistance for the flue (taken from Table 2).
Column 3 Enter the sum total of columns 1 + 2.
Column 4 [*Ke*] Enter the resistance factor m/run of flue used (taken from Table 3).
Column 5 [*Ha*]Enter the total measured vertical height, in metres, above the draught diverter.
Column 6 Enter the total for column 4 multiplied by column 5.
Column 7 Enter the total for column 3 minus column 6.
Column 8 [ΣK] Enter the sum total of resistance factors due to pipe/fittings (Table 3).
Column 9 Enter the sum total of columns 7 + 8.
Column 10 Enter column 3 divided by column 9.
Column 11 Enter the final equivalent height: column 5 multiplied by column 10. This should exceed the minimum height from Table 1.

Table to calculate equivalent flue height

1	2	3	4	5	6	7	8*	9	10	11
(Ki)	(Ko)	1 + 2	(Ke)	(Ha)	4 × 5	3 − 6	(ΣK)	7 + 8	3 ÷ 9	5 × 10

Example 1 Using the following diagram, find the equivalent flue height and decide if the flue system is adequate in length.

Completed table:

1	2	3	4	5	6	7	8*	9	10	11
(Ki)	(Ko)	1 + 2	(Ke)	(Ha)	4 × 5	3 − 6	(ΣK)	7 + 8	3 ÷ 9	5 × 10
2.5	2.5	5.0	0.78	1.1	0.858	4.142	2.756	6.898	0.725	0.798

To calculate ΣK:

Pipe length = 1.2 m × 0.78 = 0.936
Two 135° bends @ 0.61 = 1.22
Terminal @ 0.6 = 0.6
Total = 2.756 metres

Thus the effective flue height is found to be 0.798 m. Therefore, referring to Table 1 on the previous page, the flue pipe is too short.

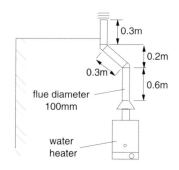

Table 2 Inlet and outlet resistance factors

Appliance	Internal resistance (*Ki*)	Flue diameter (mm)	External resistance (*Ko*)
Appliances excluding gas fires			
100 mm spigot	2.5	100	2.5
125 mm spigot	1.0	125	1.0
150 mm spigot	0.48	150	0.48
Gas fire	3.0		
Gas fire and back boiler unit	2.0		

Table 3 Resistance factors for flue components

Component	Internal size (mm)	Resistance factor (*Ke*)	Component	Internal size (mm)	Resistance factor (*Ke*)
Pipe	100	0.78		100	0.61
(per metre run)	125	0.25		125	0.25
	150	0.12		150	0.12
			135° bend	197 × 67	0.3
Brick chimney	213 × 213	0.02		231 × 65	0.22
(per metre run)				317 × 63	0.13
	317 × 63	0.35	**Terminal @ ridge**	100	2.5
Pre-cast	231 × 65	0.65		125	1.0
blocks	197 × 67	0.85		150	0.48
(per metre run)	200 × 75	0.6	**Terminal not @**	100	0.6
	183 × 90	0.45	**ridge**	125	0.25
	140 × 102	0.6		150	0.12
	100	1.22	**Raking block**	Any	0.3
90° bend	125	0.5	**(per block)**		
	150	0.24	**Transfer block**	Any	0.5

6 Flues

Example 2 Using the following diagram, find the equivalent flue height and decide if the flue system is adequate.

Completed table:

1	2	3	4	5	6	7	8*	9	10	11
(Ki)	(Ko)	1 + 2	(Ke)	(Ha)	4 × 5	3 − 6	(ΣK)	7 + 8	3 ÷ 9	5 × 10
2.0	1.0	3.0	0.65	4.5	2.925	0.075	4.275	4.35	0.7	3.15

To calculate ΣK:

Pipe length 3.6m × 0.25 = 0.9
Flue blocks 2.5 × 0.64 = 1.625
Transfer block @ 0.5 = 0.5
One 135° bend @ 0.25 = 0.25
Terminal @ 1.0 = 1.0
Total = 4.275 metres

The effective flue height is 3.15 m Therefore, referring to Table 7, the flue pipe is of adequate length.

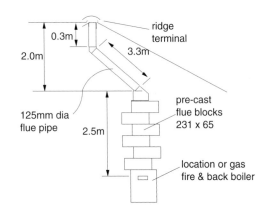

Open Flue Terminal Design and Location

Relevant Industry Documents
BS 5440, BS 6644 and IGE/UP/10

Flues with a diameter less than 170 mm require the use of a terminal. The terminal serves many functions, including:

- preventing the entry of rain, snow or debris and even small birds, etc.;
- assisting in the discharge of flue gases;
- minimising up and down-draughts.

Where a terminal is incorporated, its outlet opening needs to be such that it maintains an area twice that of the cross-sectional area of the flueway it serves. This outlet should also be of a design that will admit a 6 mm diameter ball, but will not allow the entry of a 16 mm diameter ball.

It is essential that the terminal design does not compromise the efflux velocity or the speed of the flue gases passing up through the flue pipe, or plume gas may be seen discharging from the terminal, clinging to the pipe as it falls. For natural draught boilers up to 2.2 MW the target efflux velocity should be 6 m/s.

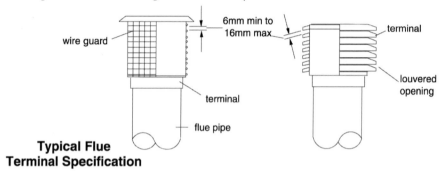

Typical Flue Terminal Specification

Flue discharge location

The flue discharge or terminal, where fitted, should not be located where it is likely to cause a nuisance. It also needs to be outside any pressure zone that may affect its performance. The following table gives an indication of the minimum height to be maintained. *Note*: Unless the outlets are more than 300 mm apart they should terminate at the same height.

Flue discharge position

	Pitched roof		At ridge, or 600 mm above, or at least 1.5 m measured horizontally to the roof line*
≤54 kW Net input	Flat roof	With parapet or external flue route	600 mm above the roof line*
		Without parapet and providing internal flue route	250 mm above the roof line*
>54 kW Net input	Pitched or flat		1000 mm above the roof line‡

*If within 1.5 m of a nearby structure the flue needs to be raised to 600 mm above that point.
‡ If within 2.5 m of a nearby structure the flue needs to be raised to 1000 mm above that point.
For 'fan draught' open flue terminal position, see page 229.

Where the flue discharge is adjacent to a location of general access to the public, it will need to be raised to a height of 3 m above the level of access.

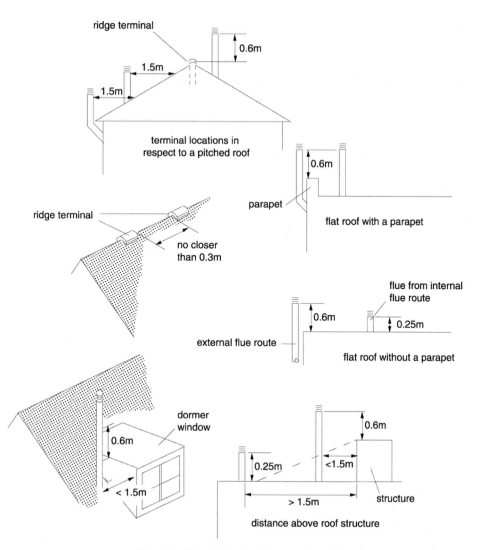

ridge terminal

0.6m

1.5m

1.5m

terminal locations in respect to a pitched roof

0.6m

parapet

flat roof with a parapet

ridge terminal

no closer than 0.3m

flue from internal flue route

0.6m

0.25m

external flue route

flat roof without a parapet

dormer window

0.6m

0.6m

0.25m

<1.5m

< 1.5m

> 1.5m

structure

distance above roof structure

Termination Height Above a Roof for Appliances Less than 70kW Net Heat Input

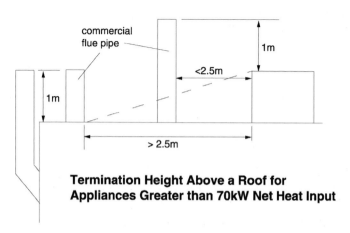

commercial flue pipe

1m

<2.5m

1m

> 2.5m

Termination Height Above a Roof for Appliances Greater than 70kW Net Heat Input

Condensation Within Open Flue Systems

Prevention of condensation within any flue system needs to be a major consideration at the design stage. Under full load condition a correctly designed flue system should not give rise to any condensation problems.

Heat loss and temperature

When an appliance is first lit, the products have to struggle to make their way up through the cold flue and, in so doing, rapidly cool. This causes much condensation and spillage at the draught diverter. Therefore it is essential that within a short period the flue warms up to assist in the transportation of the flue gases to the outside environment. However, as the flue gas temperature increases, so does the velocity. This increased flow rate creates a greater frictional resistance and increased turbulence, which again slows the flue flow to an ultimate maximum flow velocity. Heat loss from the flue must be prevented to minimise any cooling effects and the associated problems. Pipes run externally and through roof voids, etc., where they are subject to cooling, must be of a twin wall design. The high flue temperatures needed to make natural draught systems work effectively reduce the efficiency of an appliance and, as a result, fan draught systems, which are more positive in the extraction of flue gases, are superseding natural draught flueing. With their increased efficiencies and lower flue gas temperatures, flue systems are invariably more liable to condensation.

Condensation Problems

In a large commercial application, especially where several appliances are connected to a modular flue system, the flue is often basically oversized for much of the time. This can result in continuous condensation problems. To prevent the condensation from flowing back down into the appliance, it is sometimes possible to fit a 22–25 mm condensate pipe at the lowest point, which can drain the liquid away to a suitable drain or soakaway. Because the Building Regulations do not permit any openings into the flue system, apart from a draught diverter or stabiliser, a trap should be fitted to this pipe, suitably insulated, thus ensuring that no flue products can exit via this route. The material used for any condense pipe should be non-corrodible. Copper and copper alloys should not be selected due to the acidic nature of the water produced. With unavoidable sustained condensation, the flue wall or lining and all jointing materials used must be non-permeable, with joints sealed to prevent the condensation running out from them. Where condensation occurs in an existing masonry chimney that was previously used for an oil or solid fuel appliance it can mix with the soot deposits and sometimes cause staining to the inner walls of the brickwork.

When designing any flue system, consideration needs to be given to the formation of condensation and the following tables give the approximate known maximum condensate-free length of flue installed. No tables are available for flues over 70 kW net for which IGE/UP/10 should be used.

6 Flues

Approximate maximum condensate free length for individual gas fire

Flue material	Internal	External
225 × 225 brickwork and standard pre-cast flue blocks	12 m	10 m
125 mm single walled flue pipe (not insulated)	20 m	N/A
125 mm twin walled flue pipe	33 m	28 m

Approximate Maximum Condensate Free Length for Other Appliances up to 80% Efficient

Net Heat Input (kW)	225 × 225 brickwork & standard pre-cast flue blocks (m)		125 mm twin walled flue pipe (m)	
	Internal	External	Internal	External
5	4	2	17.5	5
10	6	4	22.5	15
15	10	6	26	19
20	13	9	29	23
25	15	11	32	26
30	17	14	34.5	29
35	18	15.5	37	32
40	18.5	16.5	39	34
45	19	17.5	41.5	37.5
50	20	18	43	40
55	21	19	45	42.5

Condensate Free Flue Length

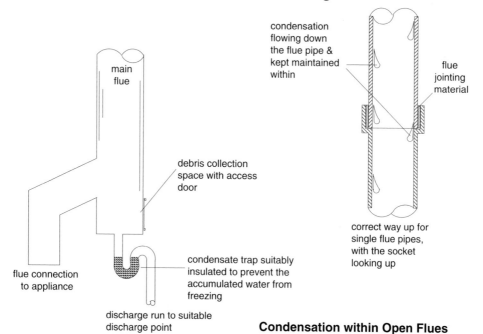

main flue

condensation flowing down the flue pipe & kept maintained within

flue jointing material

debris collection space with access door

correct way up for single flue pipes, with the socket looking up

flue connection to appliance

condensate trap suitably insulated to prevent the accumulated water from freezing

discharge run to suitable discharge point

Condensation within Open Flues

6 Flues

Brick Chimneys

Brick has been the traditional material for flues since the earliest times. However, in most instances it is quite porous as is the jointing mortar used, and it is therefore generally unsuitable for use. Since 1966 brick chimneys have been lined with one of the following:

- clay liners to BS EN 1457;
- rebated and socketed clay pipes;
- rebated liners made from kiln-burnt aggregate and high-alumina cement;
- metallic liners to BS715.

It is sometimes possible to line brick chimneys with poured/pumped concrete, certificated by an accredited test house. Lining the flue allows gas appliances to be connected without fear of the internal fabrication deteriorating and being subjected to condensation problems. A brick chimney offers a very stable and effective flue system into which most appliances can happily discharge their products, more major problems are encountered where a flue is oversized for the installation in question. All new chimneys, including those for the installation of a gas fire, must have a minimum cross-sectional area of 120 cm^2 (12 000 mm^2).

Connection to a Brick Chimney

The traditional construction of a chimney in a domestic dwelling consisted of a 225 mm × 225 mm cross-sectional void through which products could pass. Commonly-used clay flue liners used have an internal diameter of 175 mm. Both these sizes meet the minimum dimensions required. Where it is necessary to connect to the chimney, the first consideration is to ascertain its age and whether it has been lined. If it has been used for other fuels it will certainly first require sweeping. Any dampers or restrictor plates will need to be removed. It is possible to secure a damper in the open position where removal is difficult. For a simple gas fire where the heat input is relatively small it is permissible to install the appliance to an unlined chimney but, in general, for all other appliances a flue lining is essential. Connections to clay flue liners can be achieved as shown, putting the pipe into the liner a minimum distance of 150 mm. Alternatively, a flexible flue liner could be considered as shown on the following page.

Debris Collection Space

At the base of open flued chimneys and large flues where several appliances may join a main flue, a debris collection space should be provided with an access door or similar, and there should be regular maintenance to inspect for potential blockage of the flue. The minimum height of this void generally needs to be 250 mm and it must be of sufficient volume to accommodate any falling debris, e.g. birds, falling mortar, etc. BS5440 suggests that this void may need to be a minimum volume of 0.012 m^3, which is equivalent to, say, 250 mm high × 300 mm × 160 mm.

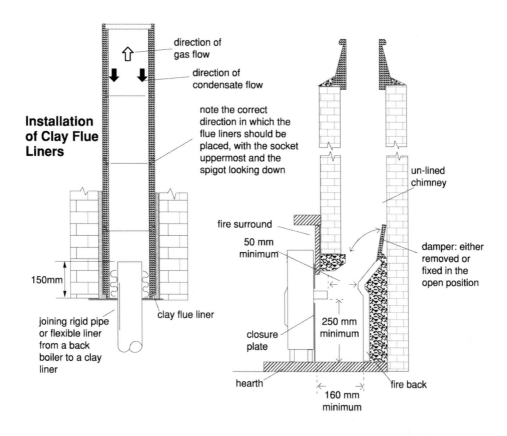

Installation of Clay Flue Liners

direction of gas flow

direction of condensate flow

note the correct direction in which the flue liners should be placed, with the socket uppermost and the spigot looking down

150mm

joining rigid pipe or flexible liner from a back boiler to a clay liner

clay flue liner

un-lined chimney

fire surround

50 mm minimum

closure plate

damper: either removed or fixed in the open position

250 mm minimum

hearth

160 mm minimum

fire back

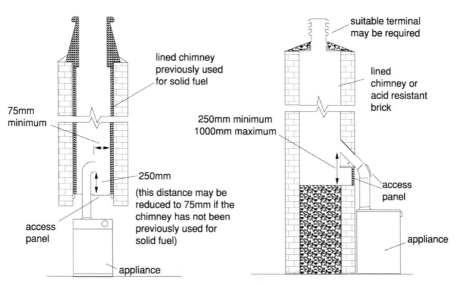

75mm minimum

lined chimney previously used for solid fuel

250mm minimum 1000mm maximum

suitable terminal may be required

lined chimney or acid resistant brick

250mm

(this distance may be reduced to 75mm if the chimney has not been previously used for solid fuel)

access panel

access panel

appliance

appliance

Connections to Brick Chimneys

6 Flues

Flexible Stainless Steel Flue Liners

These are specially designed, corrugated construction, flue liners which can be passed down through an existing flue to provide a sound flue system of the correct diameter. These liners must not protrude from the flueway and, where the appliance is external to the chimney, a flue pipe is required to extend to the flue. There should be no joints throughout the length of the flue liner, because it would not be possible to inspect them and, as a result, care needs to be taken in measuring for the length needed as the liner should be in one continuous piece. The installation is carried out as follows:

1. The existing chimney pot, where applicable, is removed. It may be left if an approved pot/liner plate is used.
2. The chimney needs to be thoroughly swept to remove any previously accumulated soot deposits and to ensure that it is clear.
3. A rope is dropped down the flue system. This is achieved by tying a weight to one end and passing it through the flue, from above.
4. With an end plug located on the liner end and tied to the rope, the liner is drawn up or down through the flue.
5. The top end is now secured with a clamping plate; a terminal is fitted and the whole area is suitably flaunched to drain the water away from the liner. *Note*: The terminal selected must discharge at a similar height to any adjoining terminals.
6. A second clamping or debris plate is affixed at the base of the chimney, securing the liner firmly. The annulus space at the base must be effectively sealed, using mineral wool if necessary, thus ensuring secondary flueing does not occur and allowing combustion products to pass up between the liner and flue. This plate also prevents debris falling on to the appliance.
7. The final connection is now made directly on to the appliance, where applicable, securing it with a series of self-tapping screws. Sometimes a flue pipe is run out from the chimney liner to the appliance, the liner being terminated in a socket joint. This joint must be made available for future inspection, therefore access will be needed.

Need for a flue liner

Appliance type	Flue length
Back boiler and gas fire	The flue needs to be lined for any length.
Gas fire (with or without a circulator)	>10 m external wall or >12 m internal wall
Other appliances	See manufacturers' instructions or BS 5440.

Existing Flexible Liners

Generally where an appliance is due for renewal and the installation has an existing stainless steel flue liner, it should preferably be replaced. The liner can rarely be guaranteed to last in good condition for longer than the life of an appliance and, in any case, it would be unlikely to last for sufficiently long to see out the life of the new unit. The liner could be left, providing the gas operative can be confident that it will last the expected 10–15 years that the new appliance may last, for example if it had only been in for a short time.

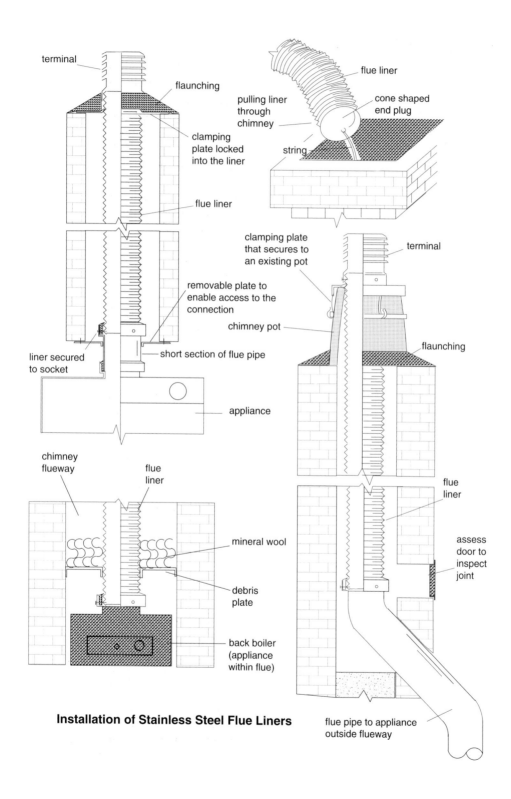

terminal

flaunching

clamping
plate locked
into the liner

flue liner

removable plate to
enable access to the
connection

liner secured
to socket

short section of flue pipe

appliance

chimney
flueway

flue
liner

mineral wool

debris
plate

back boiler
(appliance
within flue)

flue liner

pulling liner
through
chimney

cone shaped
end plug

string

clamping plate
that secures to
an existing pot

terminal

chimney pot

flaunching

flue
liner

assess
door to
inspect
joint

Installation of Stainless Steel Flue Liners

flue pipe to appliance
outside flueway

Pre-Cast Flue Blocks

Relevant Industry Documents
BS 5440 and BS 1289

The pre-cast flue block is designed to be incorporated within the structure of the building wall, forming an integral part and as such it is generally restricted to new buildings. The flue blocks are bonded into the block work of the wall during its construction and, as a result, allow for a greater volume within the living area. The flue is constructed from the base, where it is necessary to begin with the manufacturer's starter blocks, as shown. These are designed to provide the location for the appliance and support the chimney flue. Raking blocks are included where a vertical rise is not possible. However, they should be avoided where possible and in all cases the manufacturer's instructions should be followed. Invariably when the flue blocks reach the roof space the last section is run through this void using twin wall flue pipe, terminating at the ridge with a suitable terminal, the connection to the blocks being made using a specially manufactured transfer connector block. When laying the blocks the rebate needs to be at the top end, so that the spigot can sit inside, facing down. The joint needs to be made with the fireproof compound supplied or set within a fireproof mortar bed (see BS 5440). On completion joints require pointing and it is essential that any protruding compound or mortar inside the flueway is removed as construction continues.

The connection to the flue is made direct to the starter block, allowing for an appropriate catchment space (see page 292) or by means of a purpose designed ancillary component provided by the gas appliance manufacturer. Pre-cast block systems have limitations in that they should be run internally, any external runs being restricted to 3 m. It should be noted that a minimum vertical distance of 0.6 m is required above any appliance. The maximum input rating should not exceed 70 kW net. It is always necessary to follow manufacturers' instructions. They will state whether the installation is permitted, as not all appliances may be connected to a flue block system. For all new chimneys constructed in flue blocks the minimum cross-sectional area needs to be 165 cm² (16 500 mm²), with no dimension less than 90 mm. It should be borne in mind, however, that where the flue is to be used for a gas fire and the flue pipe is used within the roof void, this section needs to have a minimum flue diameter of 125 mm.

Temperature effects
Owing to the high temperatures that will be experienced on the wall surface directly above the appliance and the effect of any plaster surface cracking, the flue should be clad with either a row of bricks/blocks or a plasterboard facing, maintaining an air gap/insulating space. It is desirable to maintain a 50 mm space between the inner surface of the flue block and any structural timber. However, non-structural timbers, such as floorboards and picture rails, may be placed against the blocks.

The gas installer is unlikely to be involved with the construction of a flue block chimney but will ultimately be responsible for its safe use. Therefore, in addition to the usual flue flow testing, it is also advisable to undertake a soundness test of the system. This is fully explained on page 222.

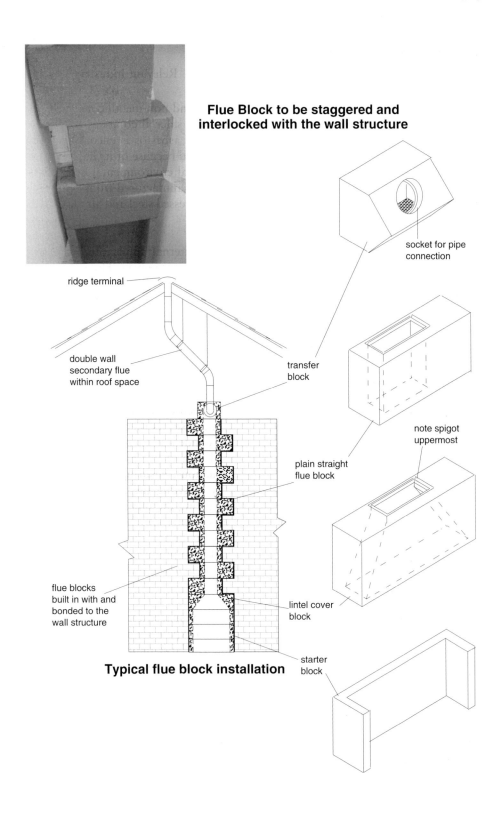

Flue Block to be staggered and interlocked with the wall structure

socket for pipe connection

ridge terminal

double wall secondary flue within roof space

transfer block

note spigot uppermost

plain straight flue block

flue blocks built in with and bonded to the wall structure

lintel cover block

Typical flue block installation

starter block

Flue Pipes

Relevant Industry Document
BS 715

Pipes will be found constructed of various metallic and non-metallic materials. Asbestos cement was once used extensively until several safer alternatives were found, but today these are no longer manufactured. Where asbestos is encountered with an existing installation great care needs to be observed because of its hazardous nature and information should be sought from the local environmental officer when removal is to be considered. Metallic flue pipes are manufactured from a variety of metals including stainless steel, cast iron, enamelled pressed steel and aluminium and are supplied as single or double (twin) walled.

Single walled pipes should no longer be used for external applications; the space within the roof void is also regarded as an external environment. However, an uppermost protrusion through the roof-line is permitted providing the distance is kept to a minimum. When making connections to single walled flue systems the method of jointing is often by means of a spigot and socketed joint. Should this be the case, it is essential that the socket is installed to the top end and that the spigot sits inside, facing down into the fitting. This is to ensure that any jointing medium used to assist in making the joint does not fall out and that any condensation flows back down the pipe internally.

Double walled flue pipes consist of two pipes, one secured inside the other, trapping an air space between the concentric void. This increases its thermal ability to keep the heat within and so overcomes many of the condensation problems associated with single walled flues. It is essential that this design of flue pipe be assembled in accordance with the manufacturer's fixing instructions, ensuring it is installed the correct way up. Connection is usually made by inserting the male coupling into the female coupling of the next component and giving the pipe a slight twist, thus interlocking the two sections. These flues may be used externally, however the distance is restricted to 3 m unless additional insulation is provided to overcome the problems associated with excessive flue cooling. It is possible to purchase double walled flue pipe for external applications that has an insulation material included.

Fire Precautions

As the flue passes up through the building it needs to be spaced a minimum distance of 25 mm from any surfaces to allow movement due to expansion and contraction and prevent any combustible material getting too hot. The basic requirement is to ensure that a combustible surface does not exceed 65°C. If using double wall flue pipe, the distance is measured from the inner wall. If it passes through combustible walls and floors, a non-combustible shield is required, again maintaining the 25 mm air gap. The space between the sleeve and flue pipe should be filled with a non-combustible material (e.g. mineral wool) and a ceiling/floor plate provided to ensure a suitable stop fire.

6 Flues

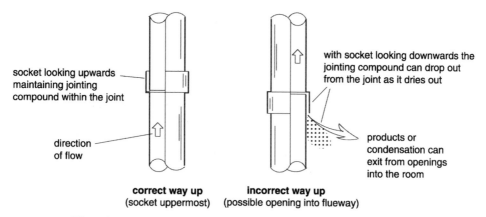

socket looking upwards maintaining jointing compound within the joint

with socket looking downwards the jointing compound can drop out from the joint as it dries out

direction of flow

products or condensation can exit from openings into the room

correct way up
(socket uppermost)

incorrect way up
(possible opening into flueway)

Direction of Flow for Single Walled Flue Pipe

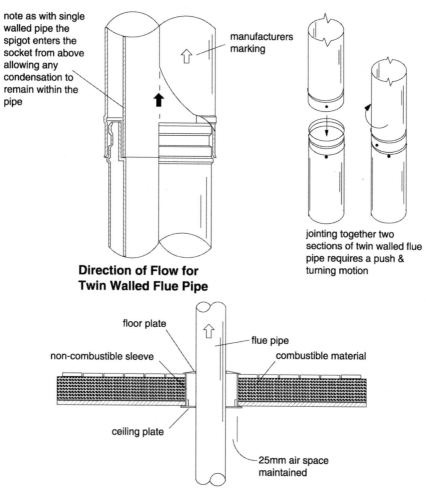

note as with single walled pipe the spigot enters the socket from above allowing any condensation to remain within the pipe

manufacturers marking

jointing together two sections of twin walled flue pipe requires a push & turning motion

Direction of Flow for Twin Walled Flue Pipe

floor plate

flue pipe

non-combustible sleeve

combustible material

ceiling plate

25mm air space maintained

Passing Flue Pipes Through Combustible Surfaces

Fan Draught Open Flue Systems

Unlike the natural draught open flue systems that work on convection currents, these systems have the additional motive force of a fan to assist in the extraction of flue gases from the building. The fan may be located either before the appliance (forced draught) or after it (induced draught). Where a fan has been located within the flue system away from the appliance it is invariably because there was some difficulty in designing the flue system to an acceptable standard and possibly the flue route fails to meet the required design criteria. The mechanical assistance of a fan can help overcome horizontal runs of flue and various restrictions such as flue diameter and an excessive number of bends.

Where any form of fanned draught is encountered it is essential that there is some form of automatic control that senses the pressure within the flue system and shuts down the appliance should insufficient draught be detected. Where shut down occurs there should be no automatic re-ignition and manual intervention will be required to investigate the cause of fan failure. This is generally achieved by the use of an air flow proving device or pressure switch.

Fan-flued systems have the following advantages over natural draught systems:

- There is greater freedom in the siting of the flue terminal.
- Wall termination is acceptable in most cases.
- Removal of combustion products is more positive, especially during cold spells.
- Flue outlet sizes may be smaller.
- There is no restriction on flue route except those applicable to condensation.
- The appliance design can allow for more efficiency, with the possibility of plastic flues being used.
- Greater flue dilution can be achieved.

The disadvantages include the following:

- There are more things to go wrong and additional safety features required.
- It can be more expensive.
- They tend to depressurise the room, causing products and smells to be drawn in from adjoining rooms.
- There is increased noise due to the operation of a fan.
- Additional maintenance to the fan unit is required.

Where a fan is to be selected for inclusion into a flue system, several factors need to be considered including the volume of flow to be allowed for, as well as the dilution air and the likely static pressure drop within the system, about which specialist advice should be sought. Some thought is also needed as to the location of the fan so that it is easily accessible for maintenance.

For fan draught open flue terminal locations see page 229.

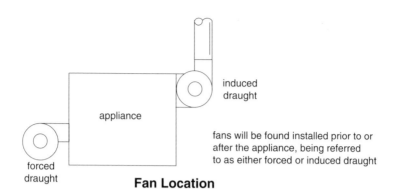

induced
draught

appliance

forced
draught

fans will be found installed prior to or
after the appliance, being referred
to as either forced or induced draught

Fan Location

**Fan Assisted Atmospheric
Open Flued Boiler**

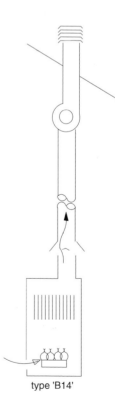

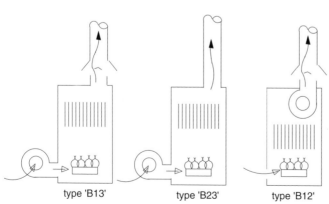

type 'B13'   type 'B23'   type 'B12'   type 'B14'

British Standard Flue Classification

Shared Open Flue Systems

Relevant Industry Documents
BS 5440, BS 6644 and IGE/UP/10

In order to accommodate several appliances, it is possible to join two or more appliances to the same flue system. However the following criteria apply:

- Each appliance, where applicable, requires its own draught diverter.
- Each appliance must have its own flame supervision device fitted.
- Each appliance must have a safety control incorporated to shut down the supply should there be a build up of combustion products within the room concerned.
- The flue needs to be suitably sized to take the combustion products from all appliances connected to the main flue and should have a minimum cross-sectional area of 400 cm^2(40 000 mm^2), which would be equivalent to a pipe diameter of about 230 mm.
- Access needs to be provided to the main flue system.
- Advice on installing several appliances to the same flue system should always be sought from the manufacturer.

Common Flue System: Appliances within the same room (modular flue)

With this system design all appliances are installed within the same room and have the same burner system, i.e. all are atmospheric or all forced draught. Gas burning appliances should not be discharged into flue systems used for solid fuel, however it is possible that a mix of gas and oil burning appliances are permitted together in the same system, providing this is in agreement with the manufacturer's instructions. Where one appliance is to operate for longer periods than the others it should be positioned nearest to the main flue. If the flue is to operate under natural draught the maximum number of appliances fed into the flue should be restricted to eight and, if the section includes a horizontal run, this should not exceed 2 m and is further limited to six appliances. Where the horizontal flue is included, the minimum height above the draught diverter needs to be 500 mm.

Branch Flue System: Appliances installed on different floors

This is a design of shared flue primarily restricted to natural draught appliances with a limited output such as a domestic boilers, water heaters or gas fires. The main flue must run up through the internal structure of the building, not form part of the external wall. Individual appliances must not discharge directly into the main flue but join via a subsidiary flue with a minimum height above the appliance of not less than 1.2 m or, for a gas fire, this distance is increased to 3 m. All appliances connected to this design of flue system must be of the same type (e.g. all gas fires) as indicated in the table opposite. Any appliances connected to such a system need to be suitably labelled to indicate that they are installed into a shared flue system.

Appliances discharging into the subsidiary flue of a branch flue system

Appliance type	Minimum cross-sectional area of main flue			
	>400 cm^2 and <620 cm^2		≥620 cm^2	
	Total input (kW)	Maximum appliances	Total input (kW)	Maximum appliances
Fire	30	5	45	7
Instantaneous water heater	300	10	450	10
Storage water heater, boiler or air heater	120	10	180	10

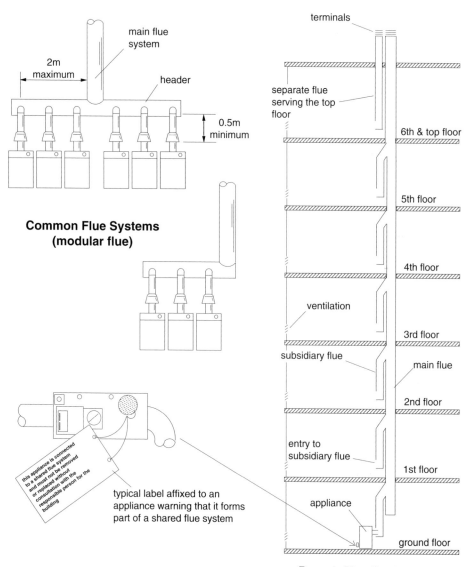

Common Flue Systems (modular flue)

Branch Flue System

Checking and Testing Open Flue Systems

Relevant Industry Document
BS 5440

The responsibility that the gas engineer takes for connections to any flue system feature highly in the Gas Regulations. As a consequence, any operative who fails to recognise fully the dangers of a poor flue system may find himself or herself in contravention of the law but, more importantly, they may install appliances that are potentially dangerous to the occupants of the building. The checks and test listed below need to be considered every time you work on an appliance and, where necessary, completed in full.

They include: visual inspection, soundness tests, flue flow tests and spillage tests (see also flue gas analysis on page 234).

Visual Inspection

A good visual inspection of a flue system invariably identifies potential problems before they are experienced. To assist in the completion of this task a schedule may be followed as shown on page 267, however the main points to look for include the following:

- That it is constructed of the correct materials and complete throughout its entire length, with the flue taking a suitable route.
- Internal inspection should reveal no dampers, unless fixed open, register plates or other potential blockages.
- That the flue serves only one location, unless it is a shared flue system and the rules on the previous page have been followed.
- That the flue outlet is correctly sited and a terminal fitted if applicable.
- That there is suitable ventilation.

Soundness test

This test will certainly not be required on every occasion and for many flue systems their integrity may be without question. However should you have any doubt as to the condition of a flue, e.g. a new pre-cast flue block system, installed by an unknown bricklayer, then this is a test that must be undertaken as follows:

1. With the flue open at the top and bottom and the flue warmed to assist in the creation of an up-draught, insert a smoke pellet/s at the base.
2. When smoke is seen to discharge from the top, while the pellet is still issuing smoke, temporarily close off both the base and top.
3. Any joints from which smoke can be seen to issue can now be identified as unsatisfactory.

Flue flow test (pull test)

This test determines whether the flue is capable of sustaining a suitable up-draught or pull. It is achieved using a smoke pellet that can produce a minimum volume of 5 m^3 of smoke in 30 seconds and, where necessary, more than one pellet can be used. The method to adopt for this test is as follows:

1. Close all windows/doors to the room in which the flue is to be checked.
2. Check with a smoke match to see if there is an evident updraft, if not introduce heat into the flue by means of a blowlamp for say 10 minutes.
3. Ignite a smoke pellet and place it in the base of the flue. In the case of a gas fire, if a closure plate is supplied this should be placed *in situ*.
4. The test can be deemed satisfactory if the smoke is seen to rise up into the flue, with no significant escape into the room, and can be seen clearly discharging only from the correct terminal outlet. In addition, during the discharge of smoke all accessible joints, including those in roof voids and cupboards, etc. should be examined for possible leakage.

Spillage Test

The spillage test is designed to see if the products are being safely carried to the outside under the worst possible conditions. It is carried out as follows:

1. Close all windows/doors to the room in which the appliance is installed.
2. Switch on any fans within the room, setting these to maximum. If the room contains any other open flue appliances, including solid fuel and oil, these should also be ignited as they also draw air from the room.
3. After the appliance has been alight for some 5–10 minutes, and with the appliance in operation, test for spillage in accordance with manufacturer's instructions. Where these are unavailable simply hold a lighted smoke match just inside the draught diverter and run it along its entire lower edge. All smoke, apart from an odd wisp, should be kept within the flue.
4. Should there be any fans in an adjoining room the test should now be repeated with the adjoining door open.

Where spillage occurs it may be possible to leave the appliance running for a further period to see if the problem rectifies as the flue heats up. It may be that the extractor fan is creating the lack of pull. This can be determined by turning off the fan, or opening a window slightly, to see if the flue now clears the smoke, in which case it suggests that more ventilation is needed.

Do not leave an appliance that fails the spillage test in operation. Treat the situation as immediately dangerous.

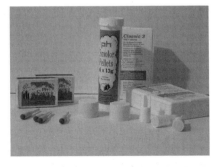

Smoke Matches and Smoke Pellets

Spillage Test to a Gas Fire
(spillage is occurring)

6 Flues

Room Sealed Flue System

The room sealed system, as previously stated, takes the air required for combustion from outside the room in which the appliance is situated. The room-sealed appliance may or may not be in balance. The term balanced flue refers to the position of the flue terminal in relation to the air supply inlet. For example, in the illustration opposite, the room sealed appliance is in balance, however the vertex flued appliance shown on page 227 is room sealed but it is not in balance because the air inlet is not directly adjacent to the flue outlet.

Natural Draught Appliances

The terminal outside a building is located in a positive or negative pressure zone, depending on the wind direction. The air inlet is also in this zone, with the pressure differential between each as zero. Thus the only motive force required to expel the flue gases from the appliance to the outside is that obtained via the heat rising due to convection currents. Natural draught systems have been used for many years because of the simplicity of their design. However, the need for improved efficiencies and the wish to site appliances away from the external wall systems have led to the incorporation of a fan to produce a more positive motive force and have made natural draught units less desirable. Also, where a natural draught appliance is used, the size of the flue terminal is much larger.

Balanced Compartments

This is a special arrangement in which an open flued appliance has been installed within a small room or compartment that is completely sealed from any adjoining room. All air openings to the compartment are taken from a point outside adjacent and within 150 mm of the flue terminal. The ventilation is taken to a high level only for appliances greater than 70 kW net input. For appliances below this input a low level opening is also required, achieved via a tee piece incorporated in the vent duct. For vent sizes see page 248. Access to the compartment is made via a secured panel or locked door, designed so that if it is opened the appliance will cease to operate, i.e. it would incorporate a switch and have a self-closing attachment. Draught excluders would also be fitted around the sides of the door, including the base, to seal further the space within. On the outside of the access door should be affixed a notice stating that the door is to be kept closed at all times. These compartments are generally large enough for the gas operative to work in for maintenance and commissioning. However if insufficient room is available then it is permissible to temporarily by-pass the door switch to complete the work. The access door to the compartment must not open into a bath/shower room and if the appliance is over 12.73 kW net input (14 kW gross), it also must not open on to sleeping accommodation.

This arrangement is particularly suitable for installations requiring greater heat inputs where availability or room sealed appliances are limited and long external runs of flue would be required.

225

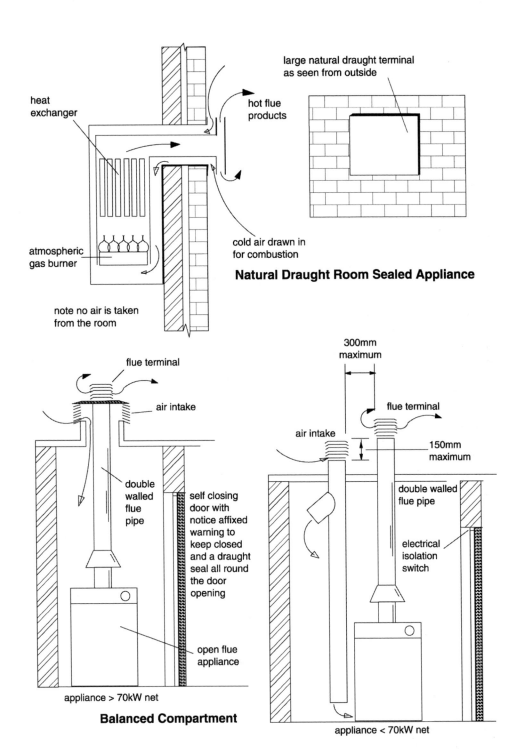

heat exchanger

hot flue products

large natural draught terminal as seen from outside

atmospheric gas burner

cold air drawn in for combustion

Natural Draught Room Sealed Appliance

note no air is taken from the room

flue terminal

air intake

300mm maximum

flue terminal

air intake

150mm maximum

double walled flue pipe

self closing door with notice affixed warning to keep closed and a draught seal all round the door opening

double walled flue pipe

electrical isolation switch

open flue appliance

appliance > 70kW net

Balanced Compartment

appliance < 70kW net

6 Flues

Fan Assisted Room Sealed Flue Appliance

Relevant Industry Document
BS 5440

Owing to the desire to have unrestricted siting of appliances, smaller termination outlets and the need for improved efficiencies, appliances need to rely on a fan to force the flue gases from the appliance and induce the air required for combustion to/from the external environment.

The fan may be located prior to the burner, giving a forced draught, creating a positive pressure on the appliance, or it may be placed after the heat exchanger causing an induced draught through the appliance, resulting in a negative pressure. Because a fan is incorporated with the system it is imperative that, should it malfunction or fail to create the necessary draught, the appliance must not be able to become established and burn the fuel. This is achieved by the use of a sensing tube, which detects a positive or negative pressure within the flue way, depending on the design of the appliance. This pressure is used to hold open a gas valve, where a zero governor is used, or allows electricity to flow to the gas valve via a pressure switch.

From the diagrams opposite it can be seen that there is a large range of designs in fan assisted appliances. In all cases the maximum flue distances and the total number of flue bends, etc. permitted will need to be found from the appliance manufacturer's instructions.

Vertex Flued System

This type of flue system, also manufactured under a trade name, 'solver' flue, falls within the category of C7 (see page 194 for flue classification). It is a strange mix in that the appliance is room-sealed. The flue system above the draught break in the roof void, however, is open-flued and as such a spillage test would be needed at this opening into the flueway.

Air is supplied to the roof void with a minimum free air size as indicated by the manufacturers of the appliance. This air is then drawn into the flue system via an opening in the flue-way, it can also enter a duct at this point and pass down via a concentric flue (air passing down the outside void) to the appliance in a room below. Thus no air is drawn from the room itself. The hot gases travel from the appliance, which is fan assisted, and pass up through the central flue to be expelled into the upper section of flue. Dilutent air is drawn in at this point and from here the flue gases pass up to the terminal.

The draught break located within the roof void must be a minimum distance above the insulation material of 300 mm and any bends above this point restricted to a minimum distance of at least 600 mm, allowing for a vertical rise.

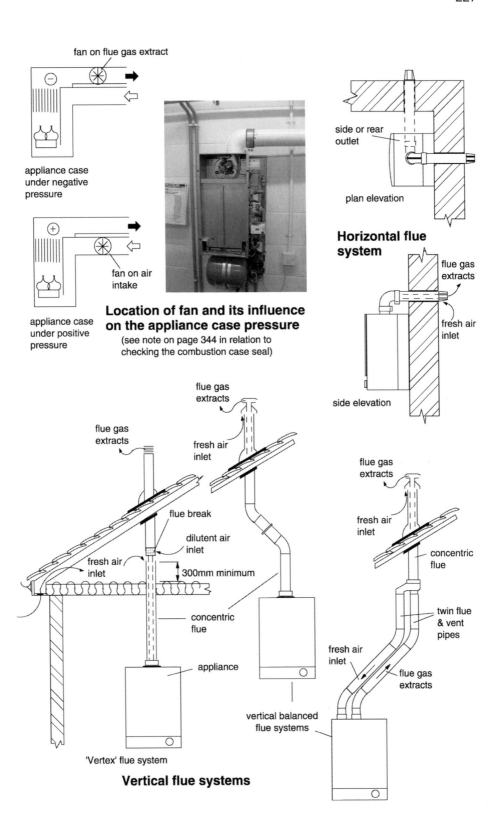

fan on flue gas extract

appliance case under negative pressure

fan on air intake

appliance case under positive pressure

Location of fan and its influence on the appliance case pressure

(see note on page 344 in relation to checking the combustion case seal)

side or rear outlet

plan elevation

Horizontal flue system

flue gas extracts

fresh air inlet

side elevation

flue gas extracts

fresh air inlet

flue gas extracts

flue break

dilutent air inlet

300mm minimum

fresh air inlet

concentric flue

appliance

vertical balanced flue systems

'Vertex' flue system

flue gas extracts

fresh air inlet

concentric flue

twin flue & vent pipes

fresh air inlet

flue gas extracts

Vertical flue systems

6 Flues

Room Sealed and Fan Flue Terminal Locations

Relevant Industry Document
BS 5440

The terminal location for a room sealed or open 'fan' flued appliance should be in accordance with the manufacturer's instructions. The table opposite gives sufficient information to enable you to comply with the appropriate standard.

Siting of Terminals in Relation to Boundaries

The combustion products should not cause a nuisance to adjoining or adjacent properties and it is recommended that any terminal from a fan flued appliance is not allowed to discharge combustion products across adjoining boundaries and, where it does occur, a distance of 2 m is maintained from any opening directly opposite. Permission from the local authority may be required if the terminal is to project into an area of public right of way.

Special caution needs to be considered in relation to possible future extensions to adjoining or adjacent properties, for example, if a neighbour extends their property to the boundary line it may unduly adversely affect the performance of your flue system. Thus the principle that needs to be adopted when selecting a suitable terminal position is to imagine that the neighbouring property has already been extended and therefore if it affects your minimum distance, as seen from the table opposite, the location should not be chosen.

Damage Due to Heat

Care needs to be taken to ensure that no heat damage will be caused. This would include locating a terminal guard around terminals that are within 2 m of any accessible surface. If the terminal is located close to any combustible surface, such as a gutter, soffit or fascia board, a heat shield minimum with a of 1 m length will be required to protect the material from damage. Sometimes the terminal is located where the external surface construction itself is of combustible material; if this is the case it will be necessary to protect the surface by using a non-combustible plate behind the terminal for a minimum distance of 25 mm beyond the external edges of the terminal.

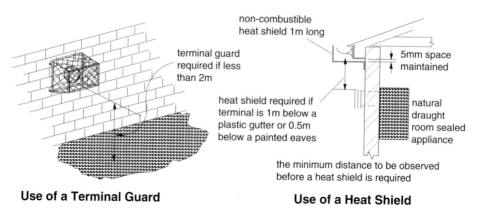

terminal guard required if less than 2m

heat shield required if terminal is 1m below a plastic gutter or 0.5m below a painted eaves

non-combustible heat shield 1m long

5mm space maintained

natural draught room sealed appliance

the minimum distance to be observed before a heat shield is required

Use of a Terminal Guard   **Use of a Heat Shield**

Terminal locations for room sealed and open fan flued appliances

Location		Heat input (net)	Room sealed		Open flued
			Nat. draught	Fan draught	Fan draught
A	Directly below an opening (e.g. window or air brick)	0–7 kW	300 mm	300 mm	300 mm
		7–14 kW	600 mm		
		14–32 kW	1500 mm		
		32–70 kW	2000 mm		
B	Directly above an opening (e.g. window or air brick)	0–7 kW	300 mm	300 mm	300 mm
		7–14 kW			
		14–32 kW			
		32–70 kW	600 mm		
C	Horizontally to an opening (e.g. window or air brick)	0–7 kW	300 mm	300 mm	300 mm
		7–14 kW	400 mm		
		14–32 kW	600 mm		
		32–70 kW			
D	Below gutters and pipes	0–70 kW	300 mm	75 mm	75 mm
E	Below eaves	0–70 kW	300 mm	200 mm	200 mm
F	Below balconies or car ports	0–70 kW	600 mm	200 mm	200 mm
G	Horizontally to vertical pipes	0–5 kW	300 mm	75 mm	150 mm
		5–70 kW		150 mm	
H	Horizontally to corners	0–70 kW	600 mm	300 mm	200 mm
I	Above ground or flat surface	0–70 kW	300 mm	300 mm	300 mm
J	Surface facing a terminal	0–70 kW	600 mm	600 mm	600 mm
K	Terminal facing terminal	0–70 kW	600 mm	1200 mm	1200 mm
L	Car port opening to inside	0–70 kW	1200 mm	1200 mm	1200 mm
M	Vertical to 2nd terminal	0–70 kW	1500 mm	1500 mm	1500 mm
N	Horizontal to 2nd terminal	0–70 kW	300 mm	300 mm	300 mm
O	Above the roof line	0–70 kW	See manufacturers' instructions		150 mm

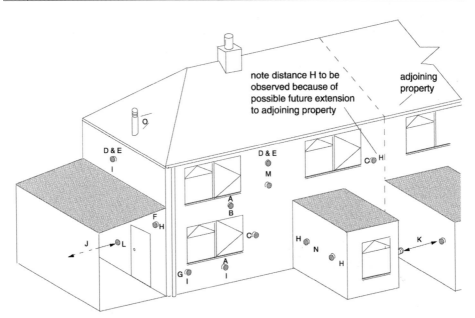

note distance H to be observed because of possible future extension to adjoining property

adjoining property

Minimum Fan Assisted Open Flue and Room Sealed Terminal Positions

Shared Room Sealed Flue Systems

Relevant Industry Document
BS 5440

There are two specific designs of flue that can be used so that several room sealed appliances can share a single flue system: the 'Se' duct and the 'U' duct. They are both used in multi-storey applications and boilers, water heaters and gas fires designed specially for this design of flue can be installed. Appliances operating on LPG should not be fitted.

Se-Duct System

The Se-duct consists of a vertical shaft run up through the building into which openings are provided at low level for the admittance of air and a suitable point of termination located at a high level. The openings at low level should be taken from two opposite sides of the building or from a single zone of neutral pressure, such as in a building supported on columns or that has a well ventilated sub-floor level.

U-Duct System

This design is a variation of the Se-duct and it is ideal where it is not possible to obtain suitable air from the base of the building. The U-duct has two vertical shafts that run through the building, both of which terminate above roof level. The air for combustion is drawn down through one leg into which it is essential that no appliances are connected.

Flue Design

Specialist advice will be needed at the design stage to include terminal selection. The terminal should be positioned away from all other structures on the roof and be at a suitable height as previously specified on page 206 for open flue terminals. The size of the duct is dependent on the appliances to be installed into the flue system. BS 5440 gives tables to assist in the design of new systems for this method of fluing.

Inspection and Maintenance

It is the responsibility of the building owner or landlord to ensure that the flue system remains in a safe condition and annual inspection and testing should be carried out in order to comply with the current Gas Regulations. Appliances connected to these flue systems will also require annual maintenance checks as inadequate installation may affect the safe operation of other appliances. *Note*: All appliances and ventilation systems must be suitably labelled, stating how they are part of a shared system.

Replacement Appliances

Replacement appliances must be suitable for connection to this design of flue system, with the manufacturer's instruction permission for installation and they must not be of a greater heat input than the appliance that has been removed. Any existing holes into the duct that are no longer required should be sealed with a plate made from a suitable non-combustible material. Any necessary flue inlet/outlet holes should be drilled through both the sealing plate and duct wall, taking care to prevent any rubble falling into the main flueway.

6 Flues

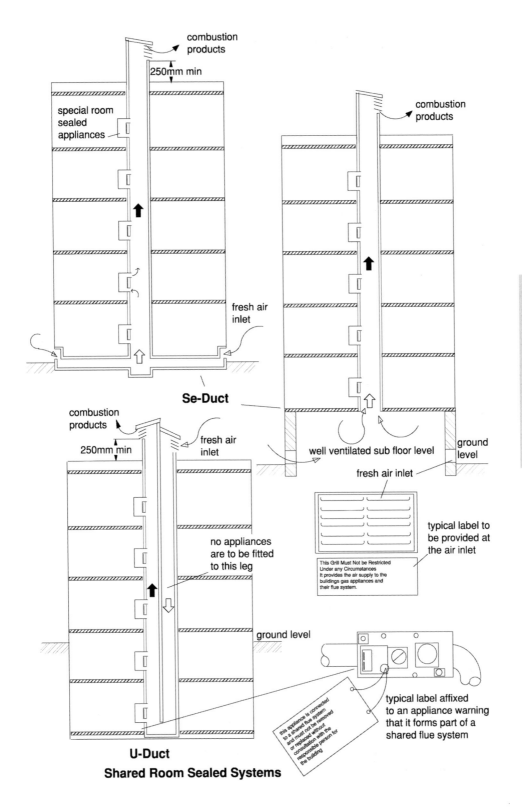

combustion products

250mm min

special room sealed appliances

combustion products

fresh air inlet

Se-Duct

well ventilated sub floor level

ground level

fresh air inlet

combustion products

fresh air inlet

250mm min

no appliances are to be fitted to this leg

typical label to be provided at the air inlet

This Grill Must Not be Restricted Under any Circumstances It provides the air supply to the buildings gas appliances and their flue system.

ground level

this appliance is connected to a shared flue system and must not be removed or replaced without consultation with the responsible person for the building

typical label affixed to an appliance warning that it forms part of a shared flue system

U-Duct

Shared Room Sealed Systems

Fan Dilution System

Relevant Industry Documents
BS 6644 and IGE/UP/10

The fan diluted flue system was developed to overcome the problem of discharging large volumes of flue products at a low level. The principle behind its design concept is that if sufficient fresh air could be mixed with the flue products it would dilute the final discharge down to an acceptable level. *Note*: The local environmental health officer should be consulted at the design stage. The combustion product level should not exceed the following values: $< 1\% \text{ CO}_2$; < 50 ppm CO and < 5 ppm NO_x.

Flue Discharge and Dilution Air Inlet

From the diagram one can see that the open flued appliances are connected to a horizontal flue, positioned above the appliances. This flue has an outlet discharge directly through the wall, with the outlet louvres diverted to direct the products away from the ground. The minimum height for this outlet should be 2 m to the lower edge of the grille for appliances up to 2 MW, with this distance increased to 3 m where the gross heat inputs exceeds 3 MW. Flue discharge grilles should not exhaust into areas, such as courtyards, where the products cannot readily disperse. The dilution air inlet should ideally be taken from the same wall as the flue extract, however it is possible to take this from another side wall but problems may be experienced from external wind conditions affecting the operation of the dilution fan. It is also possible, under exceptional circumstances, to take the dilutent air directly from the plant room; guidance from IGE/UP/10 should be sought here. The fan unit will be fitted within the flue system. This should be wired so that no appliance may be operated unless the fan has been proven to be operational. The fan and ductwork should be designed to provide a flow velocity of between 6 and 8 m/s and have a flow volume that meets the minimum combustion levels above. The volume per second can be determined from the following calculation:

$$Factor^* \times Net\ heat\ input \div 3600 = \text{m}^3/\text{s}$$

where $Factor^* = 10.8$ for natural gas and 12.8 for LPG.

Example Find the minimum volume flow rate where the net input of several natural gas appliances connected to a fan dilution system is 254 kW.

$$10.8 \times 254 \div 3600 = \underline{0.762\ \text{m}^3/\text{s}}$$

Within the inlet duct there should also be fitted a lockable damper that cannot be fully closed. It should be adjusted during the commissioning stage to a position that gives the desired combustion level. Any flue from an individual appliance should have fitted either a draught diverter or a stabiliser.

Plant-room Air Supply

These should be sized in accordance with the tables given on page 248 and ideally sited on a different wall from that of the flue discharge.

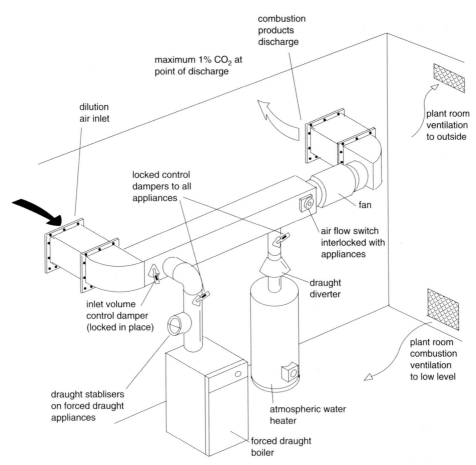

combustion
products
discharge

maximum 1% CO_2 at
point of discharge

dilution
air inlet

locked control
dampers to all
appliances

plant room
ventilation
to outside

fan

air flow switch
interlocked with
appliances

draught
diverter

inlet volume
control damper
(locked in place)

plant room
combustion
ventilation
to low level

draught stablisers
on forced draught
appliances

atmospheric water
heater

forced draught
boiler

6 Flues

Typical Arrangement of a Fan Dilution Flue System

Flue Gas Analysis

Flue gas analysis is a method employed to check the condition of combustion and is often used to make the necessary adjustment to the air intake, where applicable for a specific appliance, usually during commissioning. It should not be regarded as the ultimate in checking the condition of a flue system. When taking any flue gas reading, measurements can be made to ascertain the CO_2, CO levels, etc. and these may indicate that safe levels of gas are being produced. However, where an appliance is continuously spilling products into the room, the environment would soon become vitiated, resulting in a change in the combustion levels. Analysis does not determine the effectiveness of the flue, only that of the appliance, see also page 222.

Flue gas analysis generally aims to determine two specific qualities:

1. the efficiency of the appliance;
2. the CO/CO_2 ratio and therefore the level of safety.

Today most flue gas analysis is undertaken by the use of an electronic flue gas analyser, however older, more traditional, test apparatus may be selected as described over the following page.

Appliance Efficiency

Clearly an appliance that discharges heat out through the flue is discharging energy out from the appliance and as a result will not be realising its full potential. The efficiency of an appliance can be determined by comparing the flue gas temperature against the $CO_2\%$ level. As fuel is consumed, CO_2 is produced and potentially for an appliance that is 100% efficient this level would increase to a maximum of about 11.2% before dropping back, as it is diluted by the excess air drawn into the appliance.

Efficiency is calculated using the following calculation:

$$100 - [(0.343 \div CO_2\% + 0.009) \times (Flue\ temp. - Room\ temp.)°C + 9.78]\%$$

Example An appliance installed within a room has a flue temperature of 167°C and 6.5% CO_2. The room temperature is 21°C. Find the efficiency.

The efficiency of the appliance is:

$$100 - [(0.343 \div 6.5\% + 0.009) \times (167 - 21)°C + 9.78]\%$$
$$= 100 - [0.0618 \times 146 + 9.78]\%$$
$$= 100 - 18.8 = \underline{81.2\%\ efficient}$$

It is possible to complete the above calculation or the approximate efficiency can simply be obtained using a manufactured slide chart or by referring to the table opposite.

Typical combustion efficiency table

CO$_2$%	Net flue temperature °C (*Flue temperature – Room temperature*)											
	60	80	100	120	140	150	160	180	190	200	220	240
5	85.6	84.0	82.5	80.9	79.4	78.6	77.8	76.3	75.5	74.7	73.1	71.6
5.5	85.9	84.5	83.1	81.7	80.2	79.5	78.8	77.4	76.7	75.9	74.5	73.1
6	86.3	84.9	83.6	82.3	81.0	80.3	79.6	78.3	77.6	77.0	75.7	74.3
6.5	86.5	85.3	84.0	82.8	81.6	81.0	80.3	79.1	78.5	77.9	76.6	75.4
7	86.7	85.6	84.4	83.3	82.1	81.5	80.9	79.8	79.2	78.6	77.5	76.3
7.5	86.9	85.8	84.7	83.7	82.6	82.0	81.5	80.4	79.8	79.3	78.2	77.1
8	87.1	86.1	85.0	84.0	83.0	82.4	81.9	80.9	80.4	79.8	78.8	77.8
8.5	87.3	86.3	85.3	84.3	83.3	82.8	82.3	81.3	80.8	80.3	79.4	78.4
9	87.4	86.5	85.5	84.6	83.6	83.2	82.7	81.7	81.3	80.8	79.9	78.9
9.5	87.5	86.6	85.7	84.8	83.9	83.5	83.0	82.1	81.7	81.2	80.3	79.4
10	87.6	86.8	85.9	85.0	84.2	83.7	83.3	82.4	82.0	81.6	80.7	79.8
10.5	87.7	86.9	86.1	85.2	84.4	84.0	83.6	82.7	82.3	81.9	81.1	80.2
11	87.8	87.0	86.2	85.4	84.6	84.2	83.8	83.0	82.6	82.2	81.4	80.6

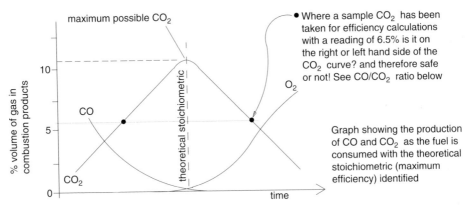

6 Flues

maximum possible CO$_2$

Where a sample CO$_2$ has been taken for efficiency calculations with a reading of 6.5% is it on the right or left hand side of the CO$_2$ curve? and therefore safe or not! See CO/CO$_2$ ratio below

Graph showing the production of CO and CO$_2$ as the fuel is consumed with the theoretical stoichiometric (maximum efficiency) identified

Stoichiometric Chart for Combustion of Gas

CO/CO$_2$ Ratio

The CO/CO$_2$ ratio gives a comparison between the CO and CO$_2$ levels recorded from an appliance. In general, recordings below 0.02 need to be achieved if the appliance is to be deemed to be working safely. When taking the CO$_2$ reading, to compare it against the appliance temperature and calculate the efficiency, one does not known which side of the CO$_2$ curve the reading is taken from. In the previous example, was the 6.5 CO$_2$% from the right or left-hand side of the curve? (See the stoichiometric chart above).Comparing it against the CO% would confirm whether it was on the side of safety (i.e. the right-hand side). The CO/CO$_2$ ratio is found by simply making the following calculation: CO% $\div$ CO$_2$%. *Note*: If the CO is recorded in ppm divide it by 10 000 to convert it to a percentage.

In the previous example if a CO% of 3.5% is recorded at the same time as the 6.5% CO$_2$, it would be found that the CO/CO$_2$ ratio is 3.5 $\div$ 6.5 $=$ 0.54. Therefore, although the appliance indicates an efficiency of 81.2% it is unsafe and more air would be required to achieve complete combustion.

Test Equipment for Flue Gas Analysis

Most flue gas analysis work today is undertaken using an electronic flue gas analyser. However, it is possible to use the more traditional manual test pump apparatus such as the 'Fyrite' and 'Draegar' analyser in conjunction with a thermometer or pyrometer, as shown on page 332. A pyrometer is an instrument used for measuring high temperatures.

Electronic Flue Gas Analyser

There are several manufacturers who specialise in this kind of equipment and the range of functions and capabilities vary from analyser to analyser and with the manufacturer's design. Most give a digital readout, with a facility to record and store the information or give a printout. Electronic analysers are sophisticated microprocessors and as such compute all the data sensed by the unit to produce a collection of readings including:

- O_2 content;
- CO_2 content;
- CO content;
- CO/CO_2 ratio;
- ambient temperature;
- flue temperature;
- efficiency;
- net temperature (*Flue temperature – Ambient temperature*).

The analyser can be set and used with the various fuels and is not restricted for use with gas. The biggest drawback with the electronic analyser is the short working life of the cells used in the unit to measure the flue gases; in most cases the cells last not much more than 12 months, necessitating continual expensive replacements. In addition to this is the annual cost of re-calibration, which is required in order to ensure the accuracy of the readings.

Fyrite Flue Gas Analyser

This device is used to sample the CO_2 content in the flue products. It consists of a clear plastic tube that contains a liquid capable of absorbing CO_2. When a sample of flue gas is drawn into the device by sucking it in through a non-return valve, and the unit inverted several times, the CO_2 is absorbed into the liquid. This creates a partial vacuum and a diaphragm in the base of the unit flexes up, pushed by atmospheric pressure. This causes the liquid to rise up in the tube, whereupon the $CO_2\%$ can be read off. The device can also be used to sample O_2, however a different liquid, capable of absorbing O_2, is required.

Draegar Analyser

This device consists of a small bellows hand pump into which a glass sampling tube is inserted. It operates in a similar way to the breathalyser. Within the glass tube are crystals capable of absorbing all kinds of different gases, depending on the tube selected. Generally tubes capable of sampling CO are selected, this allows comparison with the CO_2 found using the Fyrite analyser above to give the CO/CO_2 ratio (see previous page). In order that the hot flue gases do not give a false reading, it is essential that the gases are cooled prior to entry into the glass tube. This is usually achieved by passing the products through a copper cooling tube. *Note*: A new glass tube is used for each occasion.

Flue sampling is undertaken by inserting the sampling probe into a temporary hole drilled into the primary flue pipe, which is plugged on completion. This location is chosen because no false reading of temperature, due to the infrared heat from the combustion chamber, is then recorded and no dilutent air has been drawn in via a draught diverter. Prior to taking any readings, the probe and connecting tube need to be purged so that a good sample of flue products can be obtained.

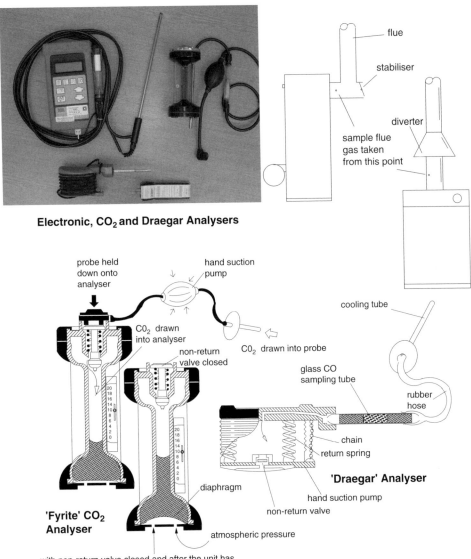

Electronic, CO$_2$ and Draegar Analysers

6 Flues

'Fyrite' CO$_2$ Analyser

with non-return valve closed and after the unit has been inverted a few times atmospheric pressure passes through holes in base of unit to flex diaphragm up, causing liquid to lift

'Draegar' Analyser

Part 7

Ventilation

Need for Ventilation

Relevant Industry Documents
BS 5440, 6644 and 6230 and IGE/UP/10

Ventilation is required for several reasons, including:

- to provide fresh air to breathe;
- to refresh the condition of the air, e.g. removing smells;
- to provide a means of removing high levels of water vapour;
- for combustion of fuel;
- for cooling down an environment, e.g. compartment or plant room.

A great deal of the air supply enters the building through adventitious means, which is basically through cracks in the window and door openings. The gas operative is particularly interested in the last two points listed above – air for combustion and cooling purposes.

Combustion Air

This is needed for combustion to take place without using up all the oxygen within a room and causing the air to become what is referred to as vitiated (i.e. lacking oxygen) resulting in dangerous gases being produced. The amount of air needed depends on the appliance type, location and size.

Air for Cooling Purposes

In order to prevent the risk of fire and avoid unacceptable temperature conditions it is essential to observe the temperature levels within a room or compartment and ensure that they do not exceed the following:

- 40°C air temperature at the ceiling or up to 100 mm below ceiling level;
- 32°C air temperature at 1.5 m above floor level;
- 25°C air temperature at floor level or up to 100 mm above the floor level;
- less than 65°C at surfaces in close proximity to the appliance.

The effectiveness of an air supply is something that must be considered on every occasion, not just when installing an appliance. If there is insufficient air available the consequences may be fatal.

Natural and Mechanical Ventilation

The method of ventilation selected would depend on the ability to provide air by natural means. It may be possible to use some form of mechanical ventilation, however this would normally be interlocked with the gas supply so that if the fan failed no gas would be allowed to flow to the appliance. Where mechanical extraction is chosen for a room, gas interlock is generally not required. However it is essential to ensure that the air intake exceeds the volume of the extracted air, otherwise a negative zone may be created, causing open flued appliances to spill their products back into the room. Catering establishments use the concept of a negative pressure environment to prevent smell accumulation, but this is a special case and no open flued appliances would be located within this negative zone. See Ventilation in Commercial Kitchens on page 394.

**Large Commercial
Ventilation Grill**

Domestic Sized Grills

Plant Room Ventilation at High and Low Level

Ventilation Location

The location of an air vent should be carefully selected to ensure that a suitable air supply is maintained. Positions where leaves or similar debris may block the vent should be avoided, as should areas liable to flooding. Where an air vent is to be located in close proximity to a flue terminal, the minimum distances identified in the following table should be observed.

Minimum distance between open flued/room sealed terminal and air vent

Appliance net heat input	<7 kW	>7–14 kW	>14–32 kW	>32 kW
Fan draught appliance In any direction	300 mm	300 mm	300 mm	300 mm
Natural draught appliance Above terminal Horizontally to terminal Below a terminal	300 mm 300 mm 300 mm	600 mm 400 mm 300 mm	1500 mm 600 mm 300 mm	2000 mm 600 mm 600 mm

Where an air vent into a particular room is located directly outside the building, air may enter the room at any position, either at a high level or at a low level, and ideally close to the appliance, to reduce the amount of draught. Ducted ventilation from above should be treated with caution as pressures within the room can restrict a good flow of air.

An appliance may need the ventilation to be supplied either directly from outside or, in some cases, it may be permitted to be taken from another room. However care should be observed as in many situations, for example flueless appliances or a commercial plant room, the supply must be taken directly from outside and not from an adjoining location. Where it is possible to take a supply of air from another room, the other room must have been supplied with the fresh air. The communicating grille in such circumstances between each room must be at low level, at a distance no greater than 450 mm above floor level, in order to prevent the spread of smoke in the event of a fire. Should the air supply need to pass through several rooms, the vent needs to be 50% larger than the outside grille in order to take account of the flow resistance. Taking a supply of air from a roof void or an under-floor space may be permitted in certain circumstances, providing that the roof or floor void itself is adequately ventilated and that this space does not communicate with an adjoining property.

Where air for cooling is required two grilles are needed: one at low level and one at high level, allowing for good convection currents to be set up, circulating the air through the compartment. It will be seen (over the page) that for open flued appliances the lower vent is twice as big as the high level vent. This is because some of the air is used in the combustion process and as a result passes up into the flue system.

Special Precautions where Radon Gas is a Problem
Radon gas is a colourless, odourless gas that is radioactive and is found where uranium or radium is present, such as in certain parts of South-West England. Where identified as a problem, below floor ventilation should not be used.

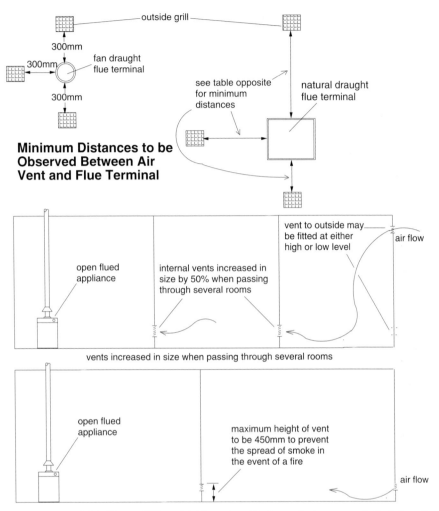

outside grill

300mm

300mm fan draught
flue terminal

300mm

see table opposite
for minimum
distances

natural draught
flue terminal

**Minimum Distances to be
Observed Between Air
Vent and Flue Terminal**

vent to outside may
be fitted at either
high or low level

air flow

open flued
appliance

internal vents increased in
size by 50% when passing
through several rooms

vents increased in size when passing through several rooms

open flued
appliance

maximum height of vent
to be 450mm to prevent
the spread of smoke in
the event of a fire

air flow

Location of Air Vents When Taking the Air Supply from an Adjoining Room

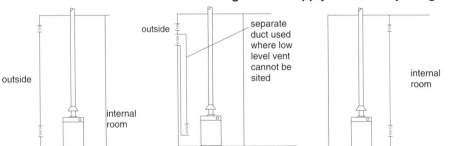

outside

outside

separate
duct used
where low
level vent
cannot be
sited

internal
room

internal
room

High and Low Level Compartment Ventilation

7 Ventilation

Effective Ventilation

The size of a ventilation grille for a specific purpose is calculated over the page. However, when calculating the vent size it is essential to know what is needed in terms of 'effective free area'. This is the actual size of the opening through which air can pass. The size of the actual grille has no bearing on the amount of air that might pass through. For example, the terracotta grille shown on page 241 would not allow the same throughput of air as a plastic or pressed metal grille of the same overall dimensions.

The size of air grille specified is usually in cm², therefore a randomly chosen grille or one found installed within a wall may physically measure 12 cm × 28 cm and therefore takes up 336 cm² of wall space, however this would not be the effective free area of air flow that could pass into the building. This area is determined by calculating the size of one hole and then multiplying this figure by the number of holes in the grille.

Example In the diagram of the grille opposite there are 15 holes, each individual hole measuring 7 mm × 80 mm. Therefore the effective size would be:

$$7 \times 80 \times 15 = 8400 \, mm^2$$

To convert mm² to cm² simply divide by 100:

$$\therefore 8400 \div 100 = \underline{84 \, cm^2}$$

For a vent to be effective, each individual hole needs to be small enough to stop vermin getting through, but not so small that it becomes blocked by dust, flies and general lint. British Standards gives the aperture as between 5 mm and 10 mm. Therefore, where a fly screen is incorporated, the grille should not be used except where permitted, such as in a few special cases – where food preparation and food hygiene are of high importance, etc.

Other specific points in air vent design include the following.
- Vents with manual closing devices should not be used.
- The air vent when used in a cavity wall would need to be ducted fully across the cavity void. In order to cut down on draughts and noise transmission from the outside it is possible to use special ducts that divert the air flow around a series of baffles.
- Any duct used must not have a cross-sectional area that is less than the effective free area of the ventilation grille.
- Ducts over 3 m in length should be avoided, without increasing in size.
- The number of 90° bends should be restricted to a maximum of two.
- Ducts intended to convey the air flow downwards should be avoided.

Intumescent Vents
These are special ventilator grilles that are designed to close and prevent the spread of smoke in the event of a fire. They generally employ a lattice arrangement that expands and closes off the holes in the event of extreme heat.

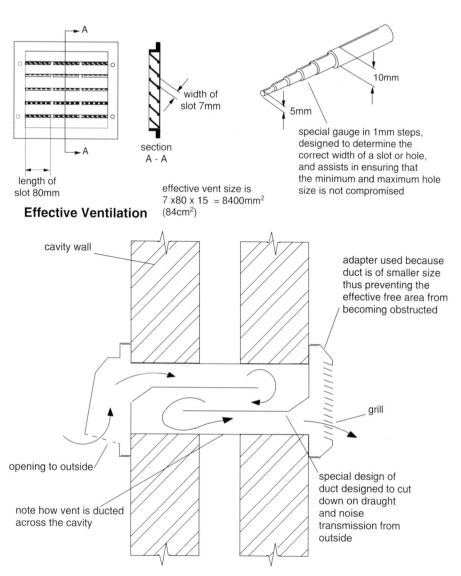

A

width of
slot 7mm

section
A - A

10mm

5mm

special gauge in 1mm steps,
designed to determine the
correct width of a slot or hole,
and assists in ensuring that
the minimum and maximum hole
size is not compromised

length of
slot 80mm

effective vent size is
7 x80 x 15 = 8400mm²
(84cm²)

Effective Ventilation

cavity wall

adapter used because
duct is of smaller size
thus preventing the
effective free area from
becoming obstructed

grill

opening to outside

note how vent is ducted
across the cavity

special design of
duct designed to cut
down on draught
and noise
transmission from
outside

7 Ventilation

Design of Grill to Cut Down on Draughts

Inappropriate Closeable Grill for Ventilation

Ventilation Sizing and Tables

Relevant Industry Documents
BS 5440, 6644 and 6230 and IGE/UP/10

This chapter deals with ventilation sizing for ventilation within permanent buildings for both natural gas and LPG. However, reference also needs to be made to Part 11, which deals with non-permanent dwellings and commercial catering (page 394), where specific requirements are described. See page 258 for a domestic ready-reckoner that can be used for open flued appliances. See also page 31, which explains why natural gas and LPG systems can have the same size air grille.

Size of Air Grille Needed

The size of an air vent into a building for combustion or cooling purposes can be calculated; it depends on several factors including appliance type, size, location and fuel. There may be more than one appliance in a room and these may be of different designs and burn different fuels. The air supply requirement for a particular appliance is calculated from specific data that can be obtained from various industry documents and standards. This data has been collected in the following tables (shown opposite and over the page).

Caution: Changes to the Standards

Unfortunately, at the time of publication, changes are being made to BS documents to update them and bring them in line with current EC directives. This has resulted in some confusion in that net input is referred to in BS 5440 (Domestic Appliances) and gross inputs are referred to in BS 6644 (Commercial Boilers) and 6230 (Commercial Air Heaters).

It is therefore very important to note, when calculating the ventilation requirements for a particular appliance, whether the *net or gross* heat input is quoted.

- To convert gross into a net heat input divide by 1.1.
- To convert net into gross multiply by 1.1.

Example 60 kW gross input $= 60 \div 1.1 = \underline{54.5 \text{ kW net input}}$.

In order to standardise the approach to completing the calculations, this book uses only net inputs, and has made a simple conversion from gross input to net input of BS 6644 and 6230, rounding the 54.5 kW conversion to 54 kW, thus mirroring IGE/UP/10. The revised British Standards, however, may do far more than this and therefore may need to be referred for further study.

It needs a little practice to use and understand the tables given in this book. Several worked examples are given for both domestic and commercial premises on the following pages.

To find the specified size of an air grille the following procedure needs to be adopted.

1. Select the correct *table*, based on the room location and flue type.
2. Choose the correct *row* based on the net input or appliance type.
3. Undertake the calculation or select a size as defined by the table.

Multiple Appliance Installations

Sometimes you will encounter a multiple appliance installation in which several appliances are installed within the same location. In this case you need to make a judgement to ensure that sufficient ventilation is provided. The following guide from BS 5440, covering domestic ventilation, suggests selecting a vent size based on the largest of the following:

1. the total maximum flueless space heating input*, or
2. the total maximum flued space heating input*, or
3. any individual rated input from any other appliance.

*Where 'space heating' refers to a gas fire; central heating boiler or air heater/convector.

Notes for ventilation tables given below

Note: All tables are based on *net heat input*.

[$\leq$ = less than or equal to; $<=$ less than; $>=$ greater than]

Natural Ventilation for Flued Appliances Within Rooms
(Rooms other than plant rooms or compartments)

Table 1 Natural ventilation for open flued appliances (IGE/UP/10)

Appliances type (net input)	Ventilation requirements	
Open flued appliances <7 kW	No additional ventilation is required.	
Open flued appliances >7 kW ≤70 kW	5 cm^2 per kW in excess of 7 kW	
Indirect fired air heaters or boilers with natural draught >70 kW	No additional ventilation is required if the air change rate is >0.5 per hour or the volume of room containing appliance is >5.2 m^3 per kW.	Where additional ventilation is required it would need to be: 2.5 cm^2 per kW in excess of 54 kW gross plus 270 cm^2.
Indirect fired air heaters or boilers with forced or induced draught >70 kW	No additional ventilation is required if the air change rate is >0.5 per hour or the volume of room containing appliance is >2.44 m^3 per kW.	
Room sealed appliances	No additional ventilation is required.	
Decorative fuel effect gas fire	100 cm^2 for appliances ≤20 kW	

7 Ventilation

Ventilation Tables for Flued Appliances Within Compartments and Plant Rooms

Table 2a Natural ventilation in enclosures or compartments*

Appliance Type (net input)	Low level opening		High level opening	
	To Outside	Into room	To outside	Into room
Open flued $\leq$70 kW	10 cm^2 per kW	20 cm^2 per kW	Half that of low level	Half that of low level
Open flued > 70 kW	10 cm^2 per kW	Not permitted to be installed	Half that of low level	Not permitted to be installed
Room sealed appliance	5 cm^2 per kW	10 cm^2 per kW	Same size as low level	Same size as low level

*A room not large enough to enter and perform work other than maintenance

Table 2b Natural ventilation in plant rooms

Appliance type/input (net input)	Low level opening		High level opening	
	To Outside	Into room	To outside	Into Room
Open flued >70 kW $\leq$1.8 MW	5 cm^2 per kW in excess of 54 kW plus 540 cm^2	Not permitted to be installed	Half that of low level	Not permitted to be installed
Room sealed appliance	Suitable ventilation to maintain the minimum temperature of: 25°C @ 100 mm off the floor; 32°C @ mid position and 40°C @ 100 mm off the ceiling			

Table 2c Natural ventilation in balanced compartments

Appliance input (net input)	Method of ventilation	Ventilation size
$\leq$70 kW	Ducted to low level	7.5 cm^2 per kW
	Permanent opening at high level	12.5 cm^2 per kW
>70 kW $\leq$1.8 MW	Permanent opening at high level Low level openings not permitted	* 6.25 cm^2 per kW in excess of 54 kW plus 675 cm^2

*Ventilation size suggested using IGE/UP/10.

Table 2d Mechanical ventilation for plant rooms or compartments

Appliance type	Flow rate per kW of net heat input	
	Low level inlet air m^3/s	High level extract air m^3/s
Natural draught appliance	0.0012	0.0005
Forced/induced draught appliance	0.001	0.0007

Note 1: Automatic control required to shut down appliance in the event of fan failure.
Note 2: Air intake must always exceed air extract.

7 Ventilation

Ventilation Tables for Flueless Appliances

Table 3a Natural ventilation direct to outside (domestic flueless appliances)

Appliance type	Net max heat input	Room volume	Minimum vent size (cm^3)	Openable window*
Domestic oven, hob or grill	No maximum	< 5m^3	100	Yes
		5–10	50 (If door direct to outside 0)	Yes
		>10	0	Yes
Instantaneous water heater	11 kW	<5	Not permitted to be installed	Yes
		5–10	100	Yes
		>10–20	50	Yes
		>20	0	Yes
Tumble dryer	N/A	<10	100	Yes
	N/A	>10	0	Yes
Refrigerator	N/A		0	No
Natural gas space heater	45 W/m^3		55 cm^2 for every kW in excess of 2.7 kW plus 100 cm^2	Yes
	90 W/m^3		27.5 cm^2 for every kW in excess of 5.4 kW plus 100 cm^2	Yes
LPG space heater to EN449	45 W/m^3		27.5 cm^2 for every kW in excess of 1.8 kW plus 50 cm^2	Yes
	90 W/m^3		13.7 cm^2 for every kW in excess of 3.6 kW plus 50 cm^2	Yes

*Alternative acceptable opening includes adjustable louvres of hinged panels.

Table 3b Un-flued radiant heaters

Natural air supply change rates	Natural ventilation size	
	Low level opening	High level opening
Unknown air change rate or air change ≤ 0.5/h	46 cm^2 per kW	Same size as low level
Air change > 0.5	Make the following calculation and locate the grilles at high and low level: [33 − (Air change rate × Room volume ÷ Max. kW)] × 1.4 × max. kW = cm^2	

Table 3c Direct fired air heaters installed within a compartment/plant room

Net input	Low level opening	High level opening
≤ 55 kW	5 cm^2/kW	Same size as low level
>55 kW	2.5 cm^2 per kW in excess of 55 kW plus 270 cm^2	Same size as low level

Note: The carbon dioxide (CO_2) levels within the heated space must not exceed 0.28%, equivalent to 2800 p.p.m. at any position where the products are likely to be inhaled.

7 Ventilation

Ventilation Calculations 1

Worked Examples for Domestic Premises

The tables on the previous two pages cover a whole range of gas installations from domestic to commercial applications and include LPG in permanent buildings and at first they may seem a little daunting. In order to find the required size of air grille, the following procedure needs to be carried out.

1. Select the correct table, based on the room location and flue type.
2. Choose the correct row based on the net input or appliance type.
3. Undertake the calculation or select a size as defined by the table.

The calculations provide the suggested minimum size of ventilation needed for a particular appliance but it must be remembered that where spillage occurs the only cure may be to use a larger vent size!

Example 1: A 22 kW net input open flued warm air unit has been installed within a compartment and its ventilation has been taken directly from outside. What is the required minimum size of grille?

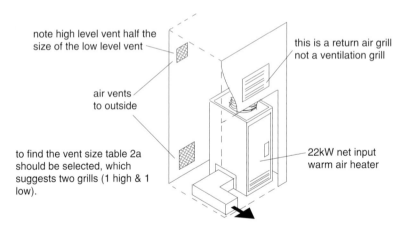

note high level vent half the size of the low level vent

this is a return air grill not a ventilation grill

air vents to outside

to find the vent size table 2a should be selected, which suggests two grills (1 high & 1 low).

22kW net input warm air heater

Step 1: Review the situation and note that the appliance is within a compartment, therefore Table 2a should be selected.
Step 2: The appliance is opened flued and is less than 70 kW, therefore the first row is selected.
Step 3: The table suggests two grilles: one at high level and one at low level, the high level grille being half the size of the low level grille.

Thus the calculation is as follows:

For the low level grille $= 10 \times 22 = \underline{220 \text{ cm}^2}$
and for the high level grille $= 220 \div 2 = \underline{110 \text{ cm}^2}$

Example 2: A 24.2 kW gross input open flued boiler in a garage has been found, previously installed, with no ventilation grille fitted, what size should the grille be?

First the gross input needs to be converted to a net input:

$$Net\ input = 24.2 \div 1.1 = 22\ kW$$

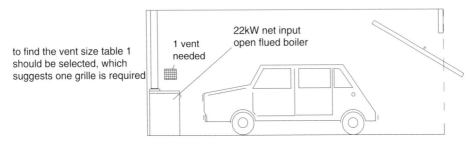

to find the vent size table 1 should be selected, which suggests one grille is required

1 vent needed

22kW net input open flued boiler

Step 1: With this example the appliance is installed in a garage. Table 1 should be selected because the appliance is not within a compartment.

Step 2: From the table it can be seen that the second row needs to be selected because the appliance is open flued, operating at 22 kW and clearly falls within the range $>7\,kW \leq 70\,kW$.

Step 3: The table suggests one grille of the following size: 5 cm² per kW in excess of 7 kW. Therefore first the 7 kW is subtracted from the 22 kW appliance, then the following calculation is made: $(22 - 7) \times 5 = \underline{75\ cm^2}$. The grille may be located at any suitable position into the room.

Example 3: A domestic cooker of 17.2 kW is to be installed within a kitchen that measures 3 m × 3 m × 2.8 m. There is an outside window. What ventilation is needed?

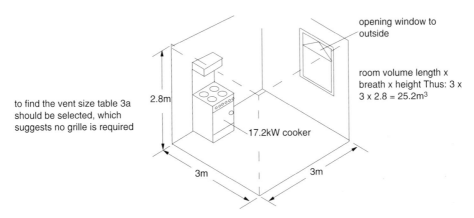

to find the vent size table 3a should be selected, which suggests no grille is required

opening window to outside

room volume length x breath x height Thus: 3 x 3 x 2.8 = 25.2m³

2.8m

17.2kW cooker

3m   3m

Step 1: Table 3a is selected as the oven is flueless and of a domestic type.

Step 2: The first row is selected; the heat input is irrelevant.

Step 3: The room volume denotes the size of the room, which is calculated as 25.2 m³. Therefore it is >10 m³ so <u>no grille is required</u>. An opening window is, however, required for the installation that is in place.

Example 4: A 14 kW net, open flued water heater has been installed in a kitchen in which there is also a cooker. The room volume is 8 m³. There is no door to outside. What ventilation is needed?

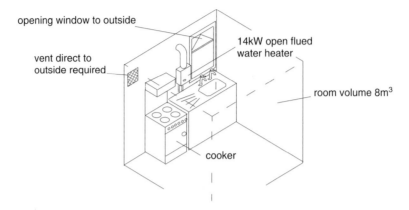

Step 1: Tables 1 and 3a are both applicable. Because it is a multi-appliance installation, the size of the vent selected should be based on the largest individual requirement (see Multiple Appliance Installations, page 247). So initially the requirements for both appliances need to be worked out.

Step 2: This is completed for each appliance as with previous examples.

Step 3: With the two calculations completed it is found that the water heater needs $(14 - 7) \times 5 = 35$ cm² and, from the table, the cooker requires 50 cm².

As the cooker has the larger requirement, a <u>50 cm²</u> vent is fitted.

Example 5: A conventional open flued gas fire of 5.4 kW is to be installed in a sitting room. What ventilation is needed?

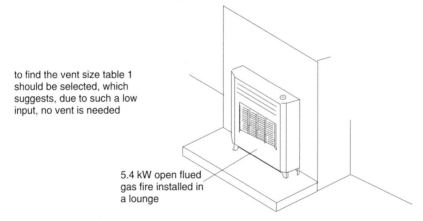

Step 1: Table 1 is selected.

Step 2: The first row is chosen as the appliance is less than 7 kW.

Step 3: Because of the low input, sufficient adventitious air can enter the building to support combustion, therefore <u>no vent is needed.</u>

Example 6: A gas fire of 5 kW net has be installed in conjunction with a back boiler that has a maximum net heat input of 17 kW, yet is only set to run at 14 kW. No ventilation has been provided.

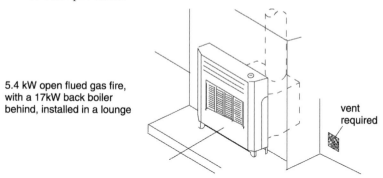

5.4 kW open flued gas fire, with a 17kW back boiler behind, installed in a lounge

vent required

As with Example 4, this is a case where two appliances are installed, namely a boiler and a fire, within the same room. Because these are both flued space heating appliances, the heat inputs must be added together. The fact that the boiler is set to run at a lower heat input is irrelevant and its ventilation requirements need to be based on the maximum heat input. Thus the fire is 5 kW and the boiler 17 kW, therefore the required ventilation is based on a total input of 22 kW.

Following the steps previously described, Table 1 is selected. A vent is required of size:

$$(22 - 7) \times 5 = \underline{75 \text{ cm}^2}$$

Example 7: A 22 kW net input open flued boiler has been installed within a compartment and its ventilation has been taken directly from the room.

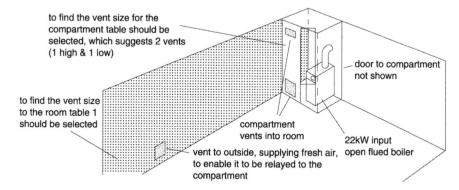

to find the vent size for the compartment table should be selected, which suggests 2 vents (1 high & 1 low)

to find the vent size to the room table 1 should be selected

door to compartment not shown

compartment vents into room

22kW input open flued boiler

vent to outside, supplying fresh air, to enable it to be relayed to the compartment

In this last domestic situation, it will be seen that two separate tables need to be selected, first Table 2a, which is used to determine the vent sizes for the compartment itself, and then Table 1 needs to be consulted in order to find the size of the vent needed to serve the room in which the appliance is situated. The fact that it is in a cupboard is irrelevant. Thus the vent sizes would be:

Compartment low level: $20 \times 22 = \underline{440 \text{ cm}^2}$ and high level $440 \div 2 = \underline{220 \text{ cm}^2}$

Room ventilation: $(22-7) \times 5 = \underline{75 \text{ cm}^2}$

7 Ventilation

Ventilation Calculations 2

Worked Examples for Commercial Premises

Example 8: A 92 kW net input forced draught open flued warm air heater is to be installed within a workshop with a room volume of 196 m³. What ventilation would be required?

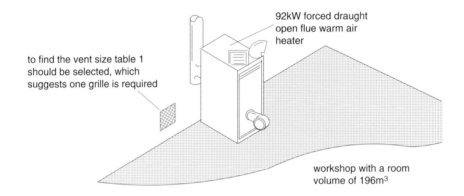

92kW forced draught
open flue warm air
heater

to find the vent size table 1
should be selected, which
suggests one grille is required

workshop with a room
volume of 196m3

Step 1: With this example the appliance is installed in a workshop. Table 1 should be selected because the appliance is not within a compartment.

Step 2: From the tables for open flued appliances and the example it can be seen that the appliance is operating at 92 kW and clearly falls within the band >70 kW. The fourth row is finally selected because the appliance is of a forced draught design.

Step 3: The table gives options and suggests that a grille is required only if:

1. the air change rate is <0.5 room volumes/h (in this example the air change rate is unknown), <u>or</u>
2. if the room volume is smaller than a certain minimum requirement (which must be calculated based upon information from Table 1).

The appliance requires a minimum room size of: 2.44 × 92 = <u>224.48 m³</u>. The room volume is only <u>196 m³</u>, therefore additional ventilation is required.

The vent size can be calculated as: 2.5 cm² per kW in excess of 54 kW + 270 cm². Therefore (92 – 54) × 2.5 + 270 = <u>365 cm²</u> of effective free air will be needed.

Example 9: A 104 kW net input forced draught open flued warm air heater is to be installed within a workshop with a room volume of 400 m³. What ventilation will be required?

This example follows the same format as the previous example, therefore Steps 1 and 2 are as before. When tackling Step 3 and calculating the minimum room size to be considered before a ventilator grille is required, it will be seen that the appliance requires a minimum room size of: $2.44 \times 104 = \underline{253.76 \text{ m}^3}$. Therefore no additional ventilation is required because the room volume is 400 m³ and is therefore large enough to accommodate the appliance.

Example 10: A 92 kW net input forced draught open flued boiler is to be installed within a compartment. What ventilation is required?

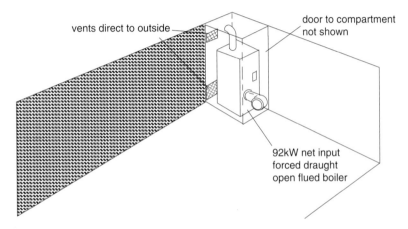

Step 1: With the appliance located in a compartment Table 2a should be selected.
Step 2: The appliance is opened flued and has input greater than 70 kW net, therefore the second row is selected.
Step 3: The table suggests two grilles: one at high level and one at low level, both taking their air from outside. The high level grille is half the size of the low level grille.

Thus the calculation is as follows:

low level grille = $10 \times 92 = \underline{920 \text{ cm}^2}$ and high level grille = $920 \div 2 = \underline{460 \text{ cm}^2}$.

If the same appliance had been installed within a plant room, Table 2b would need to be selected, giving the ventilation grille sizes as:

low level grille = $(92 - 54) \times 5 + 540 = \underline{730 \text{ cm}^2}$ and high level grille = $730 \div 2 = \underline{365 \text{ cm}^2}$.

Thus, as can be seen, smaller grilles are required. This is because a plant room is larger than a small compartment and not subject to such a large heat gain.

7 Ventilation

Example 11: Two 230 kW net input forced draught open flued boilers are installed within a plant room. What ventilation would be required?

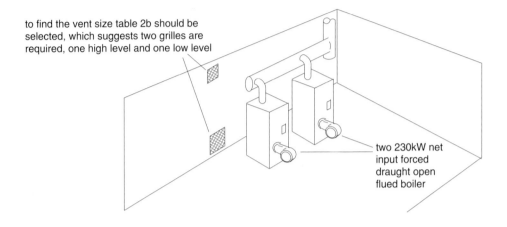

to find the vent size table 2b should be selected, which suggests two grilles are required, one high level and one low level

two 230kW net input forced draught open flued boiler

Step 1: The appliances are located in a plant room, therefore Table 2b is selected. They are of the same type, therefore it is best to take the sum total of the two appliances in determining the ventilation needs. Thus $2 \times 230 = 460$ kW is allowed for.

Step 2: The appliances are opened flued and greater than 70 kW net, but less than 1.8 MW, therefore the first row is selected.

Step 3: The table suggests two grilles: one at high level and one at low level, both taking their air from outside. The high level grille is half the size of the low level grille.

Thus the calculation would be as follows:

low level grille $= (460 - 54) \times 5 + 540 = \underline{2570 \text{ cm}^2}$ and high level grille $= 2570 \div 2 = \underline{1285 \text{ cm}^2}$.

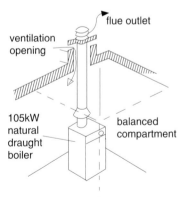

flue outlet

ventilation opening

105kW natural draught boiler

balanced compartment

Example 12: A 105 kW net input natural draught open flued boiler is installed within a balanced compartment.

The size of the ventilation for this will be found in Table 2c. The ventilation should only be at high level and the size would need to be: $(105 - 54) \times 6.25 + 675 = \underline{993.75 \text{ cm}^2}$.

Example 13: A series of un-flued radiant heaters are found installed in a warehouse. The total net heat input makes a sum total of 74 kW net. The size of the heated space is 800 m² and the rate of air change is unknown, What ventilation would be required?

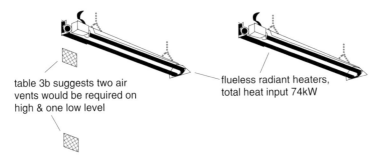

table 3b suggests two air vents would be required on high & one low level

flueless radiant heaters, total heat input 74kW

Step 1: As these are flueless appliances, Table 3b needs to be consulted.

Step 2: Because the rate of air change is unknown, the first row is selected. The volume of the warehouse has no bearing on the ventilation requirements.

Step 3: The table suggests two grilles: one at high level and one at low level, both of the same size. Thus the calculation would be as follows:

low level grille = 46 × 74 = 3404 cm² and the high level grille is also 3404 cm².

Referring back to Example 13, if the rate of air change had been known to be, say, 2 volumes per hour, a calculation could be made according to the second row of Table 3b. Thus the ventilation would now be:

$$[33 - (Air\ change\ rate \times Room\ volume \div max.\ kW)] \times 1.4 \times max.\ kW = cm^2$$

giving $[33 - (2 \times 800 \div 74\ kW)] \times 1.4 \times 74\ kW = 1179\ cm^2$ low level and $1179\ cm^2$ high level.

Example 14: There is a natural draught open flued appliance installed within a plant room and mechanical ventilation is required. The maximum net input is 154 kW.

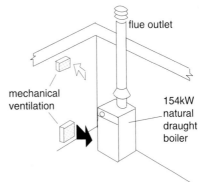

flue outlet

mechanical ventilation

154kW natural draught boiler

Table 2d suggests a low level air intake of 0.0012 × 154 = 0.1848 m³/s and a high level extract of 0.0005 × 154 = 0.077 m³/s.

Domestic Ventilation Ready Reckoner

Domestic ventilation ready reckoner

			Compartment ventilation					
							Room sealed	
							Vent to outside	Vent to room
Net input kW	Gross input kW	Air supply to room	Open flued				High and low	High and low
			Vented to outside		Vented into room			
			Low	High	Low	High		
1	1.1	0	10	5	20	10	5	10
2	2.2	0	20	10	40	20	10	20
3	3.3	0	30	15	60	30	15	30
4	4.4	0	40	20	80	40	20	40
5	5.5	0	50	25	100	50	25	50
6	6.6	0	60	30	120	60	30	60
7	7.7	0	70	35	140	70	35	70
8	8.8	5	80	40	160	80	40	80
9	9.9	10	90	45	180	90	45	90
10	11	15	100	50	200	100	50	100
11	12.1	20	110	55	220	110	55	110
12	13.2	25	120	60	240	120	60	120
13	14.3	30	130	65	260	130	65	130
14	15.4	35	140	70	280	140	70	140
15	16.5	40	150	75	300	150	75	150
16	17.6	45	160	80	320	160	80	160
17	18.7	50	170	85	340	170	85	170
18	19.8	55	180	90	360	180	90	180
19	20.9	60	190	95	380	190	95	190
20	22	65	200	100	400	200	100	200
21	23.1	70	210	105	420	210	105	210
22	24.2	75	220	110	440	220	110	220
23	25.3	80	230	115	460	230	115	230
24	26.4	85	240	120	480	240	120	240
25	27.5	90	250	125	500	250	125	250
26	28.6	95	260	130	520	260	130	260
27	29.7	100	270	135	540	270	135	270
28	30.8	105	280	140	560	280	140	280
29	31.9	110	290	145	580	290	145	290
30	33	115	300	150	600	300	150	300
31	34.1	120	310	155	620	310	155	310
32	35.2	125	320	160	640	320	160	320
33	36.3	130	330	165	660	330	165	330
34	37.4	135	340	170	680	340	170	340

Net input kW	Gross input kW	Air supply to room	Compartment ventilation				Room sealed	
			Open flued				Vent to outside High and low	Vent to room High and low
			Vented to outside		Vented into room			
			Low	High	Low	High		
35	38.5	140	350	175	700	350	175	350
36	39.6	145	360	180	720	360	180	360
37	40.7	150	370	185	740	370	185	370
38	41.8	155	380	190	760	380	190	380
39	42.9	160	390	195	780	390	195	390
40	44	165	400	200	800	400	200	400
41	45.1	170	410	205	820	410	205	410
42	46.2	175	420	210	840	420	210	420
43	47.3	180	430	215	860	430	215	430
44	48.4	185	440	220	880	440	220	440
45	49.5	190	450	225	900	450	225	450
46	50.6	195	460	230	920	460	230	460
47	51.7	200	470	235	940	470	235	470
48	52.8	205	480	240	960	480	240	480
49	53.9	210	490	245	980	490	245	490
50	55	215	500	250	1000	500	250	500
51	56.1	220	510	255	1020	510	255	510
52	57.2	225	520	260	1040	520	260	520
53	58.3	230	530	265	1060	530	265	530
54	59.4	235	540	270	1080	540	270	540
55	60.5	240	550	275	1100	550	275	550
56	61.6	245	560	280	1120	560	280	560
57	62.7	250	570	285	1140	570	285	570
58	63.8	255	580	290	1160	580	290	580
59	64.9	260	590	295	1180	590	295	590
60	66	265	600	300	1200	600	300	600
61	67.1	270	610	305	1220	610	305	610
62	68.2	275	620	310	1240	620	310	620
63	69.3	280	630	315	1260	630	315	630
64	70.4	285	640	320	1280	640	320	640
65	71.5	290	650	325	1300	650	325	650
66	72.6	295	660	330	1320	660	330	660
67	73.7	300	670	335	1340	670	335	670
68	74.8	305	680	340	1360	680	340	680
69	75.9	310	690	345	1380	690	345	690
70	77	315	700	350	1400	700	350	700

7 Ventilation

Part 8

Gas Installer Responsibility

Commissioning Gas Installations/Appliances

Commissioning

Once a gas appliance e.g. cooker, boiler, fire, etc. has been installed, or for a whole system of pipework, once the installation checks have been completed to ensure that all manufacturers' instructions have been complied with, the installation/appliances need to be commissioned. Commissioning is the checking of the installation to make sure that it is working safely; where this is not completed, the appliance **must** be disconnected from the gas supply. Commissioning a system will flag up any operational defects and invariably if this turns out to be, for example, a poor flame picture, ineffective flue or under-pressure, then it would be an installation defect, rather than a fault in the appliance itself. However, if an appliance is faulty, stripping it down to find the cause of the problem may well invalidate any warranty or guarantee, therefore the manufacturer should be contacted.

Commissioning is not restricted to a new appliance and, following any work on an old appliance, 're-commissioning' would be needed, because it is the last person to have worked on an appliance who deems whether the appliance is safe to use.

What is deemed as working on the appliance or system?

Many of the Gas Regulations imply that various checks and tests need to be undertaken. However, it would not be possible to trade within the industry if you took ownership and responsibility for every part of a gas installation that you worked on. So what are you responsible for? Well, first you have a duty of care. This means that if you know or suspect that a system or appliance is operating incorrectly you must take reasonable steps to make the installation safe. This includes all your observations on the system, not only the part on which you have been working. This is discussed later in the book (Gas Industry Unsafe Situations, page 274). How is the system or appliance on which you are actually working defined? Regulation 2 of the Gas Safety (Installation and Use) Regulations clearly defines 'work' and includes any of the following activities:

- installing, or re-connecting;
- repairing, maintaining or servicing;
- disconnecting;
- purging (when simply reinstating the gas supply, purging through an appliance is not deemed as work, unless it affects safety).

Further study of the Regulations identifies the section of the system that you will be deemed to have ensured is gas tight as that nearest to the valve upstream and downstream of the section on which you are working. For example, if you visit a property to service a boiler in a garage, all the work could be undertaken without entering the property as the local isolation valve will be adjacent to the appliance. In addition, most of the work could be undertaken at the boiler; the only other gas pipework you may expect to see would be at or around the gas meter. Thus, should there subsequently be an incident in relation to a gas appliance in the house, you could not be expected to have known of it and therefore are not responsible for it.

Record of Work Undertaken

To ensure that nothing is left to chance, the gas engineer may well follow a checklist, such as the one shown on page 266, filling in the details that are applicable to the work undertaken. These records of work are not just a tool to assist the engineer to complete the work; they also provide evidence, should it ever be required, of the actual work completed by the operative at any particular time. Completing a Record of Work also enhances the perception of the client for work undertaken and may lead to more work at a later date. The record of work illustrated is generic, covering a general wide spectrum, and is one of many found throughout this book. Other Records relate to specific appliances (for example see 'Commissioning and Servicing Cookers', pp. 316 and 317). These records should be copied or tailored, if required, to suit your needs, inserting your company details and logo, and then used every time that a system or appliance is worked on. Alternatively, organisations such as CORGI produce pre-designed duplicate forms for purchase.

Gas Work Notification

From 1 April 2005 when a gas appliance is installed or exchanged in a residential dwelling it is a requirement that CORGI is notified. In retrun they will send a declaration of safety to the homeowner.

Explanation of Commissioning Tasks Listed over

The certificate or Record of Work shown overleaf is reasonably self-explanatory. However, the rationale as to why a test is undertaken and how each task is executed, if required, is also given here, along with a page reference for further study. *Note:* The checklist shown allows for only one appliance, therefore where several appliances are to be commissioned several forms will be needed.

The commissioning checklist can be indefinite in its design and, clearly, the version shown on page 265 only illustrates the advantage of using some form of guide to ensure that you have undertaken all that is possible to make sure that the gas appliance/s are working safely. Without a checklist it is easy to miss something. The manufacturer's instruction booklet will list other items to be checked and will not include all of the above.

One final point to consider where commissioning or undertaking any service work is 'Put down a dust sheet'– it does no harm and gains the operative a lot of respect!

8 Gas Installer Responsibility

Rationale to checklist opposite and when to undertake checks/test	Page Ref
Data badge details *These should always be entered as they indicate the appliance on which you have been working on and therefore allows your findings to be cross-checked with the appliance data.*	–
System tightness test *This would be undertaken in any situation where the gas pipework has been broken into, allowing the pressure to be lost.*	162–171
Standing pressure *When undertaking the testing of a new installation or at times when the manometer has been connected to the gas meter to confirm the regulator locks up at less than 30 mbar for natural gas, 47 mbar for propane and 38 mbar for butane gas.*	42, 46 and 84
Purging *When supplying gas for the first time or reinstating the supply that has been depressurised.*	167—168
Visual inspection of pipework *This must be undertaken on all occasions, inspecting work within the section on which you are working.*	–
Gas suitable for use An important check to be made when a new appliance has been installed. It has not been unknown for the wrong appliance to be sent to the job!	–
Appliance complying with manufacturers' instructions This is clearly one of the most important checks that need to be undertaken and should be seen as a major requirement to ensure safe operation. The manufacturer also identifies specific checks that may need to be undertaken.	–
Appliance level, secure and seals maintained This is self-explanatory as no appliance can be expected to be safe where inappropriate movement is allowed.	–
Preliminary electrical checks Five statutory electrical checks have been included here and must only be undertaken by competent operatives. Where gas installers do not possess the knowledge and skills of this essential electrical work they should have this work inspected by another individual.	434
Appliance operating pressure This is a mandatory check that should be undertaken every time you work on an appliance.	84 and 46
Total heat input Not a mandatory check if you have undertaken the previous pressure check, however it does allow you to confirm that the correct injector has been installed in relation to the pressure.	56
Main burner flame picture good A check that allows you to flag up any problems, generally a stable blue flame is sought.	36
Pilot flame correct For appliances with a permanent pilot flame this check looks to see that the flame is playing on to the correct location allowing for smooth ignition. It also confirms that it is not unnecessarily large, so wasting fuel and prematurely shortening the life of the thermocouple.	–
Flame supervision device operational Another mandatory check confirming that the appliance would shut down in the event of the flame being lost within the combustion chamber.	98
Thermostat operational Again another mandatory check to confirm that the appliance goes off as the temperature reaches its designed safe limit.	100
Ignition devices Spark ignition devices etc. should be checked to ensure trouble free operation, this includes electrodes and leads.	106
Appliance tightness test A test, often overlooked, on the gas pipework and controls down stream of the on/off valve. The test is achieved by the use of leak detection fluid, or a gas detector, when the gas is flowing to the burners. A typical example would be the compression joint located at the back of a free standing cooker, supplying the grill.	–

Rationale to checklist opposite and when to undertake checks/test	Page Ref
Ventilation Where appropriate the size and location needs to be investigated. Included are checks for combustion and cooling air.	246–249 and 258
Visual inspection of flue system Many items need to be reviewed here including the materials used, flue route, termination location, fire hazards, openings into the flue system etc. The list is so great that in itself warrants its own checklist and a checklist such as that illustrated opposite may prove useful.	–
Flue flow performance A must on every occasion where an open flued appliance is in use. The flue flow test gives a good indication that the flue draught is sufficient to meet the needs of the appliance. It also helps identify any potential leaks. However it should not be confused with a flue soundness test, where the smoke is trapped into the flue system.	222
Spillage checks Again a must where working on an open flued appliance. It differs from the flue flow in that the heat generated to send the smoke out through the terminal is generated only by the appliance combustion process; the small quantity of smoke would not be seen outside therefore the flue flow confirms its path is good.	223
CO/CO$_2$ ratio Not an essential test, however it is a test that could be undertaken to confirm the safe operation of a flue system. It must be understood, however, that the reading is only a window event, and in a poor installation the reading may alter for the worst in the fullness of time as the air within the room becomes vitiated.	235
Appliance efficiency Again for the domestic appliance this is not an essential test. For the commercial appliance it may form part of the commissioning procedure to ensure correct appliance set up.	234
Meter working pressure This test should be undertaken when installing a new appliance to the system to ensure that sufficient gas is available. It is often undertaken to investigate a problem where insufficient gas is available at the appliance inlet.	46 and 84–86
Working pressure drop across system As above, undertaken when installing a new appliance to confirm that the pipework is of adequate size.	42–46

Gas Installation Commissioning Checklist		
Gas Installer Details	**Client Details**	**Appliance Date Badge Details**
Name :	Name :	Model/Serial No:
CORGI Reg. Nº :	Address :	Gas Type: Natural ☐ LPG ☐
Address :		Heat Input: max. . . kW min. . . kW
		Burner Pressure Range: . . .–. . . mbar
		Gas Council Nº/ CE Nº:................
Preliminary System Checks: Tightness test, to include let by		PASS ☐ FAIL ☐ N/A ☐
Standing pressure of system (Lock-up < ....mbars)		PASS ☐ FAIL ☐ N/A ☐
Appliance/System is purged of air		PASS ☐ FAIL ☐ N/A ☐
General visual inspection of pipework		PASS ☐ FAIL ☐
Gas suitable for use		PASS ☐ FAIL ☐
Appliance complying to manufacturer's instructions		PASS ☐ FAIL ☐
Clearance from combustible materials		PASS ☐ FAIL ☐
Appliance level, secure and seals maintained		PASS ☐ FAIL ☐
Preliminary Electrical Checks: Conductors secure		PASS ☐ FAIL ☐ N/A ☐
Earth continuity and bonding maintained		PASS ☐ FAIL ☐ N/A ☐
Polarity correct		PASS ☐ FAIL ☐ N/A ☐
Insulation resistance (>0.5M Ohm)		PASS ☐ FAIL ☐ N/A ☐
Fuse rating:.........amps		PASS ☐ FAIL ☐ N/A ☐
Gas Utilisation Checks: Appliance operating pressure		mbars....... PASS ☐ FAIL ☐
Total heat input		kW.......PASS ☐ FAIL ☐ N/A
Main burner flame picture good		PASS ☐ FAIL ☐ N/A ☐
Pilot flame correct		PASS ☐ FAIL ☐ N/A ☐
Flame supervision device operational		PASS ☐ FAIL ☐ N/A
Thermostat operational		PASS ☐ FAIL ☐ N/A ☐
Ignition devices		PASS ☐ FAIL ☐ N/A ☐
Appliance tightness test		PASS ☐ FAIL ☐
Flue and Ventilation Checks: Combustion ventilation grille size		cm^2.........PASS ☐ FAIL ☐ N/A ☐
Cooling ventilation grill sizes High....... cm^2. Low........ cm^2		PASS ☐ FAIL ☐ N/A ☐
Visual inspection of flue system		PASS ☐ FAIL ☐ N/A ☐
Flue flow performance		PASS ☐ FAIL ☐ N/A ☐
Spillage checks		PASS ☐ FAIL ☐ N/A ☐
CO/CO_2 ratio		PASS ☐ FAIL ☐ N/A ☐
Appliance efficiency		PASS ☐ FAIL ☐ N/A ☐
Safe operation of appliance explained to customer		YES ☐ NO ☐
Post System Checks: Meter regulator adequately sealed		PASS ☐ FAIL ☐ N/A ☐
Meter working pressure (.........mbars)		PASS ☐ FAIL ☐ N/A ☐
Working pressure drop across system (max. ......mbar)		PASS ☐ FAIL ☐ N/A ☐

Recommendations and/or Urgent Notification   Benchmark Logbook completed YES ☐ NO ☐

Date:........	**Appliance Safe to Use** YES ☐ NO ☐	**Service Due Date:........**
Installer's Signature:		**Customer's Signature:**

Flue Inspection and Testing Checklist

Gas Installer Details	Date:
Name:	
CORGI Reg N°:	**Description of Flue Type and Location**
Address:	....
	....
	....
	Notes

Chimney notice plate correctly located	Yes ☐ No ☐ N/A ☐	
Catchment space correct size	Yes ☐ No ☐ N/A ☐	
Hearth construction	Suitable ☐ Unsuitable ☐ N/A ☐	
Materials used and jointing method	Suitable ☐ Unsuitable ☐	
Liner correctly sealed	Yes ☐ No ☐ N/A ☐	
Flue route, including bends	Suitable ☐ Unsuitable ☐	
Openings into the flue system	Suitable ☐ Unsuitable ☐	
Is the flue continuous throughout?	Yes ☐ No ☐	
Does the flue only serve one appliance?	Yes ☐ No ☐	
Flue supports	Suitable ☐ Unsuitable ☐	
Terminal position	Suitable ☐ Unsuitable ☐	
Terminal, where fitted	Suitable ☐ Unsuitable ☐	
Terminal guard required	Yes ☐ No ☐ N/A ☐	
Signs of spillage	Yes ☐ No ☐ N/A ☐	
Protection from combustible materials	Yes ☐ No ☐	
Maintained fire stop	Yes ☐ No ☐	
Fan flow proving device	Suitable ☐ Unsuitable ☐ N/A ☐	
Flue soundness test	Pass ☐ Fail ☐ N/A ☐	
Flue flow test	Suitable ☐ Unsuitable ☐ N/A ☐	
Spillage test	Suitable ☐ Unsuitable ☐ N/A ☐	
CO_2%	.........%	
CO%	.........%	
CO_2/CO ratio	Suitable ☐ Unsuitable ☐ N/A ☐	
Net temperature (*Flue temp. – Room temp.*)	.........°C	
Appliance efficiency	.........%	

Installer's Signature: **Customer's Signature:**

Servicing Gas Installations/Appliances

Servicing

In general the maintenance or servicing of any gas appliance should be undertaken at 12-monthly intervals. However, it is essential to observe the manufacturers' instructions, which may give further guidance. It would also be well to consider the environment into which the appliance has been installed: for example, how often is the appliance used; what is the state of the surrounding environment, are there old or young people within close proximity; is there a large volume of dust, etc. produced within the environment or is there a presence of 'black dust' (see below). Factors such as these may well lead you to recommend more frequent inspection and maintenance.

Servicing of an appliance can be described as examining the working of the components of an appliance and ensuring that it is operating as the manufacturer intended, by appropriate cleaning, re-greasing, replacing and exchanging of components not fit for use. Invariably, where they are available, the manufacturer's instructions will give specific details of the servicing for an appliance. The checklist opposite may act as a guide to the many tasks to review. The service concludes with the re-commissioning of the appliance as previously described.

Unfortunately the servicing of many appliances is often neglected and often the customer only calls out the gas engineer to sort out a fault/problem. With this in mind, the operative should try to ascertain the general performance of the appliance prior to any work and, above all, check its full operation before starting. Failure to do so may leave the engineer arguing with the customer over a fault that the customer claims was not there before.

Why service an appliance?

There are many reasons for undertaking a service including: maintaining maximum efficiency; ensuring continued trouble free operation and ensuring continued safe operation. Air currents, whether caused by natural convection or by forced draught systems, may cause dust, animal fur, lint, etc. to partially block the passages through which the air, gas and flue products pass. If these spaces are restricted, the way in which the gas and products flow is affected. For example, if the primary airway to a burner becomes blocked, insufficient air will be drawn in to support complete combustion. If the gas injector is blocked, the gas rate will be reduced and primary air intake increased, again affecting combustion.

Black dust

Where high levels of hydrogen oxide, which is found naturally in gas, is present in the gas supply it can attack the metal pipework. Where copper is attacked, a black film of copper sulphide is produced on its surface. Where this black dust, as it is commonly called, flakes off from the surface it can clog injectors and valves and, in extremely bad cases, it can cause a blockage, reducing the appliance heat input.

Gas Service Checklist

Gas Installer Details	Client Details	Appliance Date Badge Details
Name :	Name :	Model/Serial No:
CORGI Reg. No° :	Address :	Gas Type: Natural ☐ LPG ☐
Address :		Heat Input: max... kW min... kW
		Burner Pressure Range: ... – ... mbar
		Gas Council N°/ CE N°:...............

Component checked, dismantled and/or cleaned   **Remedial Work Required/Notes**

	Yes	No	N/A
General condition of pipework/appliance	☐	☐	☐
Air intake/lint arrestor free from dust etc	☐	☐	☐
Primary air port clear	☐	☐	☐
Main injectors correct size and undamaged	☐	☐	☐
Pilot injectors correct size and undamaged	☐	☐	☐
Burner surfaces free from blockage	☐	☐	☐
Burner surfaces free from damage	☐	☐	☐
Control taps and valves working freely	☐	☐	☐
Ignition devices, electrodes and leads	☐	☐	☐
Filters	☐	☐	☐
Fan louvers free from lint	☐	☐	☐
Fan motor, air tubes	☐	☐	☐
Pressure switches and connections	☐	☐	☐
Combustion chamber and seals effective	☐	☐	☐
Heat exchanger	☐	☐	☐
Refractory plaques	☐	☐	☐
Appliance and flue seals	☐	☐	☐
Flue inspection/testing check completed	☐	☐	☐
Debris collection space	☐	☐	☐
Fan flow and proving switches	☐	☐	☐
Draught stabiliser	☐	☐	☐
Thermostats	☐	☐	☐
Flame supervision device	☐	☐	☐
Oxygen depletion system or ASD	☐	☐	☐
Remaining appliance controls	☐	☐	☐
Other/external controls	☐	☐	☐
Electrical wiring	☐	☐	☐
Supply and cooling ventilation paths	☐	☐	☐

Recommendations and/or Urgent Notification

Date:........   **Commissioning Completed** YES ☐ NO ☐   **Next Service Due Date:**........

Installer's Signature: **Customer's Signature:**

Installation of Second-Hand Appliances

The Gas Appliance (Safety) Regulations, as described on page 14, are not applicable to second-hand gas appliances. There is applicable legislation, the Gas Cooking Appliance (Safety) Regulations and the Heating Appliance (Fireguards) Regulations, but no such Regulations are as important as the Gas Safely (Installation and Use) Regulations. These place an obligation on the installer to ensure that the appliance is **safe to use.**

The supply and installation of second-hand appliances often pose a problem because it is often difficult to assess the condition of the appliance until the commissioning stage. The manufacturer's instructions supplied with any appliance are possibly the best guide to the installation of any appliance. However, these may be out of date in terms of compliance with current Gas Regulations. For example, when a gas fire was installed in a bedroom 15 years ago, the manufacturer's installation instructions were valid. However current Regulations require some form of atmospheric sensing device to be incorporated, and an older appliance may not have one. With the introduction of Part L of the Building Regulations most new boilers installed will be required to meet SEDBUK rating of category A or B, see page 334.

The Regulations require the operative to check the physical condition of a previously used gas appliance before it is installed. This is because it may not be possible to check it completely once it is installed.

The term second-hand refers to all appliances even those moved from one location to another within the same room. An appliance that has been disconnected from the gas supply should be treated as second-hand, and requires the operative on removing it to check it over before reinstating it. An appliance removed for servicing or repair, however, is not considered to be second-hand.

Points to consider where second-hand appliances are concerned include the following:

- Are manufacturers instructions available?
- Is the data badge intact and legible?
- Does it meet current standards (e.g. ODS or ASD incorporated)?
- Have any modifications been made and is it complete?
- Are spare parts available?
- Is the appliance suitable for the gas being used?
- Is its general condition good?
- Has it been tested before and after installation?
- Will it operate safely?

If in doubt do not fit the appliance!

Gas Fire Service

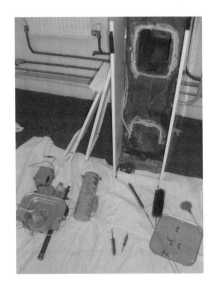

**Domestic Boiler With
Atmospheric Burner**

**Commercial
Appliances
With Forced
Draught
Burners**

**Various Appliances Opened To
Enable Internal Inspection and Servicing**

Gas Safety and Landlord Inspections

Under part E of the Gas Safety (Installation and Use) Regulations it is a requirement that the landlord of any domestic property ensures that all the gas appliances and flues under their control are maintained in a safe condition and that suitable records are provided, and that these are kept for a minimum period of two years. It should be noted that it is not a requirement to check the installation pipework as it is only a recommendation that tightness testing and visual inspection be made. The landlord has no duty to inspect appliances 'owned' by the tenant, such as a cooker or gas fire and, where such a fire is solely connected to a flue system, this equally needs not to be inspected. When a tenant vacates a property the landlord needs to ensure that all gas pipework, appliances and flues are safe before re-letting. Occupants may have removed appliances unsafely or left their own appliance in place. Property types requiring an annual inspection include:

- leased accommodation where the lease does not exceed 7 years;
- rented accommodation, both private sector and local authorities;
- part of a building such as where a room is let out;
- hostels and bed-sits;
- holiday cottages, chalets, flats and caravans;
- hired out narrow boats used on inland waterways.

In addition, safety checks apply to additional appliances such as water heaters and central heating boilers serving a relevant property above, although the appliance may not itself be located in the tenant's accommodation.

The Gas Safety Record form may be purchased from an organisation such as CORGI or one may produce it oneself, as in the example shown opposite, provided that it contains the following minimum information and the information is contained on a form that is numbered and can be tracked:

1. Date of appliance inspection.
2. Address of the property.
3. Name and address of landlord.
4. Description and location of each appliance checked.
5. Any identified defects and remedial action taken.
6. Confirmation that the flue and air supply is effective, the operating pressure and/or heat input is correct and that the appliance is operating safely.
7. The name, signature and CORGI company registration number of the individual undertaking the checks.

If a room that contains a gas appliance is converted into sleeping accommodation, any existing gas appliance needs to be inspected to ensure that it complies with current standards and regulations in force and, where this is not the case, the appliance needs to be replaced. However, this does not apply to accommodation existing prior to the current Gas Safety Regulations.

Home Owner/Landlord Gas Safety Record

Gas Installer Details	Tenant/Home Owner Details	Client/Landlord Details
Company Name:	Name :	Name :
CORGI Reg. Noº :	Address :	Address :
Address :		
Tel:	Tel:	Tel:

Details	Appliance 1	Appliance 2	Appliance 3	Appliance 4
Type				
Location				
Make				
Model				
Flue design	OF ☐ RS ☐ FL ☐	OF ☐ RS ☐ FL ☐	OF ☐ RS ☐ FL ☐	OF ☐ RS ☐ FL ☐
Owner	Tenant ☐ Landlord ☐	Tenant ☐ Landlord ☐	Tenant ☐ Landlord ☐	Tenant ☐ Landlord ☐
Inspected	Yes ☐ No ☐	Yes ☐ No ☐	Yes ☐ No ☐	Yes ☐ No ☐
Operating pressure	..............mbar	..............mbar	..............mbar	..............mbar
Heat Input if applicable	..............kW	..............kW	..............kW	..............kW
FSD operational	Yes ☐ No ☐ Na ☐	Yes ☐ No ☐ Na ☐	Yes ☐ No ☐ Na ☐	Yes ☐ No ☐ Na ☐
Thermostat operational	Yes ☐ No ☐ Na ☐	Yes ☐ No ☐ Na ☐	Yes ☐ No ☐ Na ☐	Yes ☐ No ☐ Na ☐
Ventilation satisfactory	Yes ☐ No ☐ Na ☐	Yes ☐ No ☐ Na ☐	Yes ☐ No ☐ Na ☐	Yes ☐ No ☐ Na ☐
General condition	Satisfactory Yes ☐ No ☐	Satisfactory Yes ☐ No ☐	Satisfactory Yes ☐ No ☐	Satisfactory Yes ☐ No ☐
Visual flue condition	Pass ☐ Fail ☐ Na ☐	Pass ☐ Fail ☐ Na ☐	Pass ☐ Fail ☐ Na ☐	Pass ☐ Fail ☐ Na ☐
Flue flow	Pass ☐ Fail ☐ Na ☐	Pass ☐ Fail ☐ Na ☐	Pass ☐ Fail ☐ Na ☐	Pass ☐ Fail ☐ Na ☐
Spillage	Pass ☐ Fail ☐ Na ☐	Pass ☐ Fail ☐ Na ☐	Pass ☐ Fail ☐ Na ☐	Pass ☐ Fail ☐ Na ☐
Termination	Pass ☐ Fail ☐ Na ☐	Pass ☐ Fail ☐ Na ☐	Pass ☐ Fail ☐ Na ☐	Pass ☐ Fail ☐ Na ☐
Service undertaken	Yes ☐ No ☐	Yes ☐ No ☐	Yes ☐ No ☐	Yes ☐ No ☐
Appliance safe to use	**Yes ☐ No ☐**	**Yes ☐ No ☐**	**Yes ☐ No ☐**	**Yes ☐ No ☐**
Defects and remedial work undertaken				

Installation pipework satisfactory Yes ☐ No ☐   Tightness Test Satisfactory Yes ☐ No ☐ Na ☐

Recommendations and/or Urgent Notification

Issue Number:......
Issue Date:............   **NEXT SAFETY INSPECTION DUE WITHIN 12 MONTHS OF DATE OF ISSUE**

Installer's Name:	Customer's Name:
Installer's Signature:	Customer's Signature:

Gas Industry Unsafe Situations

Under part E of the Gas Safety Regulations the gas operative working in a property has an obligation to take steps and make safe any appliance or installation that they know or have a reason to suspect is unsafe.

Depending on the severity of the apparent danger, the operative needs to categorise the situation as either At Risk (AR) or Immediately Dangerous (ID). Both these actions will have the same effect: as a minimum, to ensure that the appliance or system is turned off but, in the case of an ID situation, corrective action needs to be undertaken immediately or the appliance/system must be disconnected from the supply. Where an appliance/system is found to be 'Not to Current Standards (NCS)' and, providing there is no immediate danger to life or property, the situation may remain in operation.

Examples of 'Immediately Dangerous' (ID) Installations

- Gas leaks, e.g. failed tightness test.
- Failed spillage test or signs of spillage.
- Excessive gas pressure, e.g. due to the absence of a primary gas regulator.
- Open ended pipework.
- Leaking product of combustion from a flue into a building.
- Flues termination within the building, e.g. conservatory. Should it terminate into a roof space it is deemed AR.
- Absence of a flue where it is required.
- Inoperative safety devices, e.g. FSD inoperative.
- Appliances being supplied with the wrong gas.

Examples of 'At Risk' (AR) Installations

- Poor meter installation, e.g. incorrect location or with signs of damage.
- Incorrect pipework materials, e.g. plastic fittings or hosepipe, etc.
- Pipework with signs of corrosion or damage, raising concerns for safety.
- Pipework grossly undersized, affecting the operation of the appliance.
- Flexible connections to flued domestic appliances.
- Unstable or insecure appliances.
- Lack of air supply where required for appliance.
- Evidence of scorching to adjacent combustible materials, e.g. gas fire burner less than 225 mm above carpet and damage evident.
- Warm air unit in compartment without positive return air connection or warm air unit in compartment with unsealed plenum.
- Builder's opening inadequately sealed, or inadequate catchment space.
- Manual damper not fixed within a flue system.
- Flueless or open flued appliance in a bath or shower room.
- Flueless or open flued water heater or space heater greater than 14 kW gross input in a bed-sitting room (installed after 31.10.98).

- Flueless or open flued space heater less than 14 kW gross input in a bed-sitting room without a suitable ASD (installed after 31.10.98).
- Flueless or O/F water heater without ASD (installed after 31.10.98).
- Damaged flue or incomplete, e.g. lack of fixings.
- Natural draught open flued appliance with the terminal simply pointing horizontally out through the wall. All natural draught O/F terminations in pressure zones need to be treated as suspect.
- Fan assisted fluing/ventilation system not interlocked with gas supply.
- No low operating pressure protection where boosted gas supply is used.
- High energy/limit stat inoperative or not fitted for specific appliances.
- Absence of powered extract system for a commercial kitchen.
- LPG appliance with automatic ignition installed below ground level.
- Propane cylinders used internally or four or more cylinders without OPSO protection.

Examples of 'Not to Current Standards' (NCS) Installations

- Pipework not sleeved or unsealed when passing through a wall. (Providing no signs of corrosion)
- Pipework not adequately supported.
- Pipework inadequately protected from corrosion.
- Pipework installed within an unventilated duct or cavity wall.
- Freestanding cooker without stability device or restraint chain.
- Incorrect location or use of a flueless appliance, however see AR locations above.
- Lack of terminal guard to appliance less than 2 m above access level.
- Insufficient or incorrect air vent fitted, e.g. undersized, or closeable.
- Open flue with less than 600 mm vertical rise before first bend.
- Open flue with horizontal runs or 90° bends.
- Incorrect use of flue liner.
- Undersized flue.
- Unsuitable chimney pot or terminal.
- Commercial appliance connected to flue system with both forced and atmospheric burners fitted.
- Overhead radiant heater fitted at an incorrect height.

Clearly the situations that may be regarded as an NCS sometimes give cause for concern and in many cases the situation will need to be investigated to ensure that the situation is not AR or ID. It would also be appropriate to consider the occupants, for example are there young children about where the flue terminal is without a guard? The gas operative should advise the customer of any such NCS situation, giving them clear guidance of any corrective action that may be required and they should record any defects on their appropriate work records, keeping a copy on file.

Where two or more NCS ventilation or flue defects are encountered on the same natural draught flueless or open flued appliance, it may be advisable to treat the installation as At Risk (AR) as the combined effects of the faults may lead to an unsafe situation.

8 Gas Installer Responsibility

Dealing with Unsafe Situations

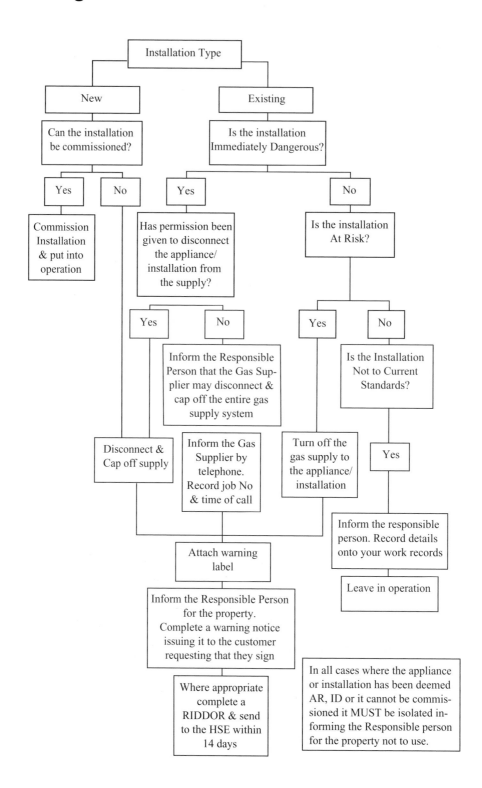

Installer Ref No [ ]

GAS SAFETY INSPECTION
This form is not to be used as a Landlord's Gas Safety Record

CORGI

Registered Business: _____
Address: _____

CORGI Registration No [ ][ ][ ][ ][ ][ ]

Postcode: _____
Tel No: _____

Gas User (Mr/Mrs/Miss/Ms): _____
Address: _____

GAS PIPEWORK/TIGHTNESS
Is the i... tight?
Have the ...materials ...in th...
Is the installation ...ed?
Where ...ha... n elect... nding be...

EMERGENCY CONTROLS
Is the emergency control correctly positioned/accessible...
Is the emergency control labelled?

APPLIANCES
Is flue system satisfactory? Yes/No
_____ [ ]
_____ [ ]
_____ [ ]
_____ [ ]
_____ [ ]
_____ [ ]

GAS PIPEWORK/TIGHTNESS
Is the gas installation safe for use?
If No, has a Warning/Advice Notice (CP14) been raised...
Has warning labels been affixed?
Is any remedial work required?

Details of remedial work required:
_____
_____
_____

Key: Top Copy – Gas User Second Copy – Gas Operative
Copyright © CORGI The format and layout of this docume...

SAMPLE

WARNING/ADVICE NOTICE
This form should be completed in accordance with the requirements of the current CORGI Gas Industry Unsafe Situations Procedure*

CORGI

Installer Ref No [ ]

Registered Business: _____
Address: _____

CORGI Registration No [ ][ ][ ][ ][ ][ ]
CORGI ID card serial No _____
Date of issue: _____
Time of issue: _____

Postcode: _____
Tel No: _____

Issued by: _____
Print Name: _____

Gas User (Mr/Mrs/Miss/Ms): _____
Address: _____

Postcode: _____
Tel No: _____
Rented Accommodation: Yes/No

THE INFORMATION WITHIN THIS BOX CONCERNS YOUR SAFETY

GAS ESCAPE
An escape of gas has been detected on the installation and the supply has been turned off and disconnected. [ ]

THE GAS APPLIANCE/GAS INSTALLATION PIPEWORK (delete as appropriate)
(Make) _____ (Mo... ...be) _____ (Serial...)
Location (Position... _____

IS IMMEDIATELY ...NGE...S* beca... **SAMPLE** and,

A. with your permission it has been disconnected from the GAS SUPPLY and a WARNING LABEL attached, or [ ] A

B. as you have refused to allow it to be made safe, a WARNING LABEL has been attached. [ ] B

PLEASE NOTE: CORGI registered gas installers are required by law to report cases where they are refused permission to disconnect an IMMEDIATELY DANGEROUS gas installation to the Gas Emergency Service Call Centre for Natural Gas, or for Liquefied Petroleum Gas (LPG), the Gas Supplier. All Gas Transporters (GT's) operate a gas emergency service and have powers under the Gas Safety (Rights Of Entry) Regulations to visit properties and disconnect unsafe gas appliances/installations.*

IS AT RISK* (AR) because:
_____
and has been turned off and a WARNING LABEL attached. [ ]

FOR YOUR OWN SAFETY, 'IMMEDIATELY DANGEROUS' AND 'AT RISK' GAS APPLIANCES/INSTALLATIONS SHOULD NOT BE USED

THE GAS APPLIANCE/GAS INSTALLATION PIPEWORK (delete as appropriate)
(Make) _____ (Model) _____ (Type) _____ (Serial No.) _____
Location (Position/Room) _____
IS NOT TO CURRENT STANDARDS (NCS)* because: _____

The appliance/installation is currently operating safely and does NOT constitute either an Immediately Dangerous or At Risk situation. The defect(s) do not present a gas safety hazard at this time. However, in the interests of safety, it is recommended that work is carried out to upgrade the installation to current requirements.

I confirm that I have received this Warning/Advice Notice concerning the safety of the gas installation. I understand that the use of the appliance/installation in the case of an IMMEDIATELY DANGEROUS or AT RISK installation, could present a hazard and could place me in breach of the Gas Safety (Installation and Use) Regulations.

Signed: _____ Print Name: _____ Date: _____

The gas user was not present at the time of the visit, and where appropriate, (an IMMEDIATELY DANGEROUS (ID) or AT RISK (AR) situation) the installation has been made safe and this notice left on the premises.

* See overleaf for definitions   Key: Top Copy – Gas User Second Copy – Gas Operative   To re-order quote Ref. CP14
Copyright © CORGI The format and layout of this document may not be reproduced in any manner without the prior written consent of CORGI

Example Inspection and Warning Forms

Example Labels
Forms such as those above can be obtained from CORGI, (address on page 2).

Gas Escape Procedures

Responsibilities of the Gas User (Responsible Person for the Premises)

Where the responsible person for the premises knows or suspects that there is a gas leak or fumes are escaping into the building, they must act immediately by turning off the appliance and contacting a gas engineer. Where the suspected gas leak is on the system, they should turn off the gas supply at the emergency control valve. If the smell of gas persists after shutting off at the emergency control valve, then they must report the gas escape immediately to the supplier, or to one of the National Gas Emergency call centres listed opposite.

Responsibilities of the Gas Operative (not on site)

When a gas operative is advised of a potential gas leak and they are not on site the following advice needs to be given to the responsible person:

- Turn off the gas immediately at the inlet emergency control valve.
- Turn off all sources of ignition, do not operate any light switches and do not smoke.
- Ventilate the building by opening doors and windows.
- Ensure access is made to the premises for the gas operative or emergency help provider.
- Instruct the responsible person to report the escape to the supplier, where applicable.

Responsibilities of the Gas Operative (on site)

Where the gas operative is on site it is clearly possible to instigate the above actions either directly or indirectly with the responsible person's permission. It must be fully understood, however, that the gas operative, although fully responsible for their own actions, may not have permission to shut off the gas supply. This is usual in a commercial situation where interrupting the gas supply to an industrial process may itself lead to a dangerous situation, possibly resulting in a third party claim against the gas operative for losses and damage. Where permission is not given to make the situation safe, the gas operative must treat the situation as immediately dangerous and follow the procedure previously described under 'Gas Industry Unsafe Situations' on page 276.

When reporting a gas escape to the emergency service/supplier be prepared to provide the following information:

- Address of the property and where the smell is most noticeable.
- Address and telephone number of the gas user.
- Address and telephone number of the person reporting the gas escape.
- Time the gas escape was first discovered.
- Whether the emergency control is turned off or not. If not, the reason why. Also whether the gas can still be smelt and if it is inside or outside.
- Any special circumstances such as access problems or infirm residents.

Upon reporting the gas escape, the service provider will give you the following information, which you should keep for reference:

- The job number.
- Date and time of the report.
- The name of the person to whom you reported the incident.

National Emergency Service Call Centre Telephone Number

- England, Scotland and Wales: 0800 111 999.
- Isle of Man: 01624 644 444.
- Northern Ireland: 0800 002 001.

Note: In the case of LPG the supplier details should be on the bulk tank or can be found under GAS in the local telephone directory.

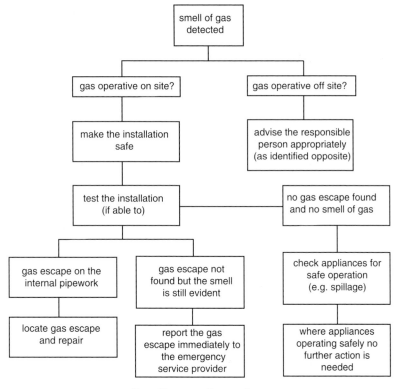

Gas Escape Procedure

Reporting of Injuries, Diseases and Dangerous Occurrences Regulations (RIDDOR)

RIDDOR is applicable to all aspects of health and safety. The notes here focus on those areas that affect gas safety. Where a situation is encountered that involves the discovery of a dangerous gas installation the installer has to decide if the situation needs to be notified to the Health and Safety Executive (HSE). The work may be put right on discovery, however the incident still needs to be reported and failure to do so may result in prosecution. The HSE is not interested in every incident, for example where signs of spillage are found, unless due to poor installation practices. If the spillage were the result of, say, a nesting bird, the incident does not need reporting.

Examples of Reportable Situations

- Poor installation work, resulting in a major gas escape.
- Signs of spillage or combustion problems resulting from bad practices.
- Inappropriate materials being used.
- Flued appliances not being flued correctly.
- Appliances being supplied with the incorrect gas, e.g. LPG or natural gas.
- Instances where a safety device has been made inoperative.
- If an appliance has become dangerous as a result of faulty servicing.

Some gas fittings/installations may not be installed in accordance with current standards or legislation. However, unless there is a good reason to believe that they are dangerous, there is no need to report them to the HSE. It is about questioning the competence of the gas operative who completed the installation in the first place and whether they should be allowed to operate un-checked.

When and What to Report

By law, a RIDDOR report must be sent in within 14 days of the discovery of a dangerous gas installation or fitting. The report should be completed using form F2508G2. An example of this form is shown opposite. However, forms are available from HSE books or the report can be completed on-screen over the Internet from the Web address below. When completing the report it is essential to stick to the questions asked on the form and not send photographs and other pieces of evidence as they tend to get lost. Keep these until they are requested. The completed form should be sent to:

Health and Safety Executive
The Incident Contact Centre
Caerphilly Business Park
Caerphilly CF83 3GG
Wales
Telephone: 0845 3009923   Website: http://www.riddor.gov.uk

Part D

About the dangerous gas fitting

1. What was the main fault?
   - [ ] gas leak
   - [ ] inadequate flue
   - [ ] inadequate ventilation
   - [ ] other

2. What type of appliance was involved?
   - [ ] boiler
   - [ ] instant water heater
   - [ ] combined fire and boiler
   - [ ] warm air unit
   - [ ] decorative fire
   - [ ] non-decorative fire
   - [ ] convector
   - [ ] cooking appliance
   - [ ] other appliance

3. What type of gas was involved?
   - [ ] natural gas
   - [ ] liquid petroleum gas (LPG)
   - [ ] LPG/air

4. Was the appliance:
   - [ ] flueless
   - [ ] open-flued
   - [ ] room-sealed
   - [ ] other (eg closed flue)

5. Who last serviced the appliance (if k

6. What date was the appliance install
   / /

7. Was the appliance bought second h
   - [ ] no
   - [ ] yes
   - [ ] don't know

8. What is the name of the installer (if i

9. What is their address and postcode

10. What is their telephone number?

For official use
Client number   Loca

Part E

Summary of the dangerous gas fitting

Please say how dangerous you consider it to be, and why, and what action you have taken to make things safe by repairing faults at the time, disconnecting the gas supply, or advising occupiers (or the landlord or managing agent for the property) of the faults you are reporting.

Health and Safety at Work etc Act 1974
The Reporting of Injuries, Diseases and Dangerous Occurrences Regulations 1995

HSE
Health & Safety
Executive

Report of a dangerous gas fitting

Explanatory notes

1. This form should be used to report to HSE gas appliances and installations using either natural gas or liquefied petroleum gas (LPG) that have been examined or tested and regarded to be dangerous*, but have not actually caused any injuries.

   * To be regarded as "dangerous" there must be a serious fault in either the design or construction of the gas fitting (including any flueing or ventilation provided for appliances), or in the way the initial installation was carried out or later serviced or modified. The fault must be so serious that people are likely to suffer death, or major injury from the acute effects of carbon monoxide poisoning or the effects of fires or explosions following gas escapes.
2. Form F2508G1 should be used to report actual incidents that have led to death or major injuries to gas consumers from the use of faulty installations.
3. Form F2508 should be used to report any deaths or major injuries arising from the use of gas that involve persons whilst at work.

Part A

About you and your organisation

1. What is your full name?

2. What is your job title?

3. What is the name of your organisation or company?

4. How can we contact you if we need more information?
   Your address and post code

   Your telephone/fax number

Part B

Some general details

1. When was the dangerous fitting found?
   / /

2. What was the address and postcode at which it was found?

3. Was it in a building? [ ] no yes
   What type of building?   What type of room?
   - [ ] house
   - [ ] flats (four storeys or less)
   - [ ] flats (more than four storeys)
   - [ ] bungalow
   - [ ] maisonette
   - [ ] other
   - [ ] kitchen
   - [ ] bathroom
   - [ ] bedroom
   - [ ] lounge
   - [ ] dining room
   - [ ] other room

4. Was the fault repaired at the time?
   - [ ] yes
   - [ ] no

5. If not, was the situation made safe by disconnection, or contact with BG Transco's emergency centre for them to disconnect?
   - [ ] yes
   - [ ] no

Part C

About the person

1. What was the name of the person living in the premises? (If they cannot be contacted, give the name address and telephone number of a relative or friend)

2. Are the premises rented?
   - [ ] no
   - [ ] yes – what is the name, address and telephone number of the landlord/managing agents

3. Was the landlord (or the managing agent for the premises) notified about the faults.
   - [ ] yes
   - [ ] no

F2508G2 (rev 04.99)

continued overleaf

8 Gas Installer Responsibility

RIDDOR Form

Part 9

Domestic Appliances

Gas Fires and Space Heaters

Relevant ACS Qualification
HTR1

Relevant Industry Document
BS 5871

There is a whole range of gas fires and wall, or free-standing heaters. These appliances may be either flueless, open flued or room sealed, each designed to provide a level of comfort and appearance to suit the environment in which they are installed. A range of appliances that fall within this spectrum of the gas industry are shown opposite. The term 'space heater' refers to an appliance designed to heat a 'space' or individual room. A space heater can provide heat to a room where it is required, unlike a large central heating system that takes time to warm up many rooms. The heat is primarily distributed from these appliances by convection currents. Radiant heat is also given off, especially from appliances that use radiants, to transmit the infra-red heat rays freely from the appliance. Convection is described on page 198. Radiation is further identified on page 378.

Radiant Heaters

These appliances are usually mobile; no heat exchanger is incorporated in this design and all heat is obtained directly from the effect of burning the fuel on a fireclay plaque, which glows red-hot.

Convector Heaters

These appliances warm the environment by convection currents. Any radiants incorporated are purely there for decorative purposes and are not incorporated to improve the efficiency. Apart from the decorative convector heater, they are generally quite plain in appearance, comprising a metal case with louvres at the top through which the hot air passes out into the room. The cold air is drawn into the heater at a low level below the heat exchanger.

Radiant Convector Heaters

These appliances operate as above, but they include radiants as part of their design and, as such, generally prove to be the most efficient form of space heating within this category of appliances.

Decorative Heaters

There are several designs of decorative heaters, including the inset live fuel effect (ILFE), decorative fuel effect (DFE) and the heating stove, to name a few. These heaters have been developed to create the illusion of a wood or coal-burning appliance, without the mess and inconvenience of using solid fuel.

Condensing Heaters

These gas fires operated with efficiencies of up to 90%, unlike the previous heaters. However, due to the high cost of manufacture, they have fallen out of favour and are now no longer produced.

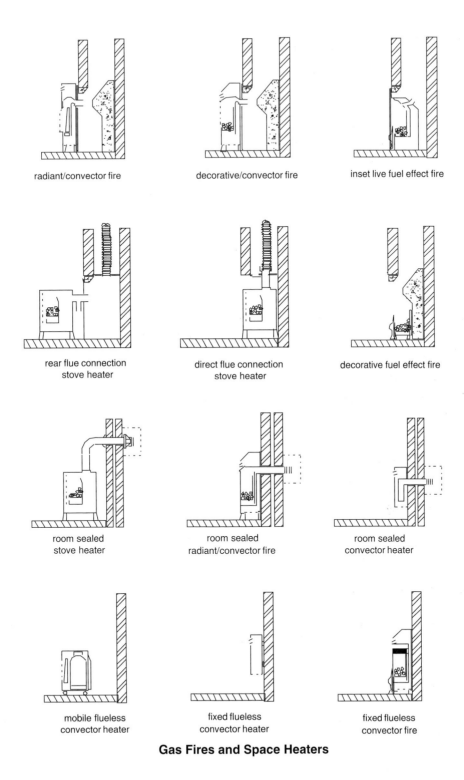

radiant/convector fire

decorative/convector fire

inset live fuel effect fire

rear flue connection
stove heater

direct flue connection
stove heater

decorative fuel effect fire

room sealed
stove heater

room sealed
radiant/convector fire

room sealed
convector heater

mobile flueless
convector heater

fixed flueless
convector heater

fixed flueless
convector fire

Gas Fires and Space Heaters

9 Domestic Appliances

Component Parts of a Gas Fire

A typical gas fire consists of the following components:

Outer case
This is an enamelled pressed steel surround, possibly with some other material attached to give the desired decorative finish; it also serves to protect the user from the hot surfaces. A fireguard or glass panel, designed to prevent anyone from touching the red-hot radiants, is attached to the front. A chromium plated, or stainless steel reflector is often incorporated. This serves two purposes: to enhance the appearance of the fire and to assist in reflecting the heat energy away from the case and floor.

Firebox
This is the metal box in which the combustion process takes place. There are radiants and firebricks in the firebox. The top or sides of the firebox usually forms the draught diverter. It also directs the combustion products to the heat exchanger and flue.

Radiants
A radiant is a fireclay, ceramic or volcanic lava rock that is incorporated in order to increase the efficiency and/or improve the appearance of the appliance. When surrounded by the gas flame the radiant glows to a white or red-hot temperature of around 900°C. The radiants need to be located in accordance with the manufacturer's instructions to ensure that incomplete combustion and severe sooting does not occur. The combustion products pass up through the radiants and pass out through the heat exchanger and flue. During maintenance work it is often good practice to rotate the bricks to enhance the life of the appliance.

Firebrick
This is a refractory lining, sometimes located behind the radiants or coals. It is designed to prevent the appliance from becoming overheated and cracking. It also prevents the heat from being lost into the chimney.

Burners
General burner design has been described on page 38. However, in gas fires, in order to allow a pair of radiants to glow independently from the full set of four, the injectors are sometimes positioned in an arrangement referred to as a duplex design. Therefore, for clarity, if only one injector is used, it is referred to as a simplex burner; where more than one injector is used, the arrangement is called a duplex burner (see diagram opposite).

Heat exchanger
This consists of a flat metal enclosure located above the firebox. Cool air enters the base of the fire and passes over the flat surface, quickly heating up before passing out through louvres in the top front edge of the casing. It is essential to check the condition of the heat exchanger thoroughly, as they are prone to fatigue cracking, as a result of the huge heat transference.

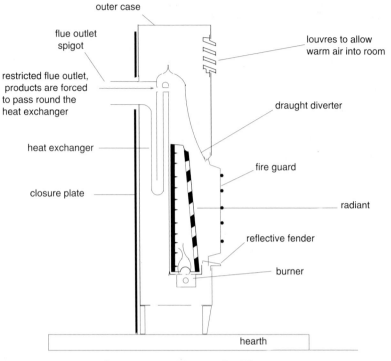

outer case

flue outlet spigot

restricted flue outlet, products are forced to pass round the heat exchanger

louvres to allow warm air into room

draught diverter

heat exchanger

fire guard

closure plate

radiant

reflective fender

burner

hearth

Component Parts of a Fire

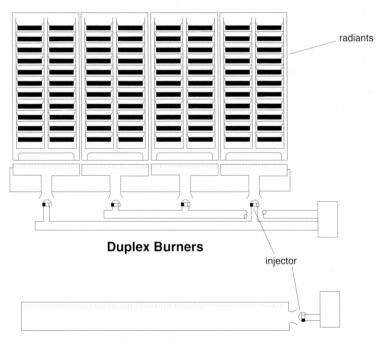

radiants

Duplex Burners

injector

Simplex Burner

Open Flued Radiant Convector Gas Fires

This design of fire takes the air for combustion from inside the room. The products of combustion are expelled from the room into a chimney or system of flue pipework of no less than 125 mm diameter, usually by means of convection currents. Owing to the comparatively small amount of fuel consumption, i.e. less than 7 kW, no additional ventilation is generally required to support the combustion process. The air in the room is replaced by adventitious means, through cracks in window frames, etc. These fires have typically been found to be around 65% efficient.

The design of the fire has changed over the years to encompass several variations since the traditional design of the 1950s, which typically had four fireclay radiants positioned at the front of those that give the effect of live solid fuel, i.e. coal or wood, being burnt. Some fires incorporate a glass front, allowing the solid fuel burning effect to be seen behind the glass. This prevents material falling on to the exposed flame and can also have the advantage of increasing the overall efficiency.

Cool air enters the base of the fire, passes round through the heat exchanger and out through a series of louvres located at the top of the fire. At the same time radiant heat is directed into the room to be heated.

To enable connection to the builder's opening at the base of a chimney and to allow access to the flueway for servicing purposes a closure plate is used. The closure plate is just a thin sheet of metal about 0.5 mm thick, supplied with the appliance, in which there is a slot through which the flue spigot of the fire can be located. The size of the slot controls the amount of air that can be drawn in from the room without creating too great a draught and also maintains a ventilation rate of about 60–70 m³/h, equivalent to approximately two air changes per hour. The plate must be sufficiently sealed to the opening, along all edges, with a heat-proof tape or other suitable sealing material capable of withstanding temperatures up to 100°C. This must be flexible enough to seal along uneven surfaces, such as stonework. Failure to provide a suitable seal may lead to too much air passing into the flue due to a strong pull, starving the burner of sufficient combustion air, and causing the flames to be drawn back downwards away from their intended direction.

A typical domestic dwelling has a flue draught of about 0.1 mbar pressure, which is more than sufficient to pull the products of combustion through the heat exchanger without pulling too much through the appliance. If the flue draught is too great the additional air pulled through the heat exchanger has the effect of cooling it down, thereby reducing the overall efficiency. To overcome this problem some manufacturers recommend the use of a spigot restrictor, which creates a resistance to flow through the appliance.

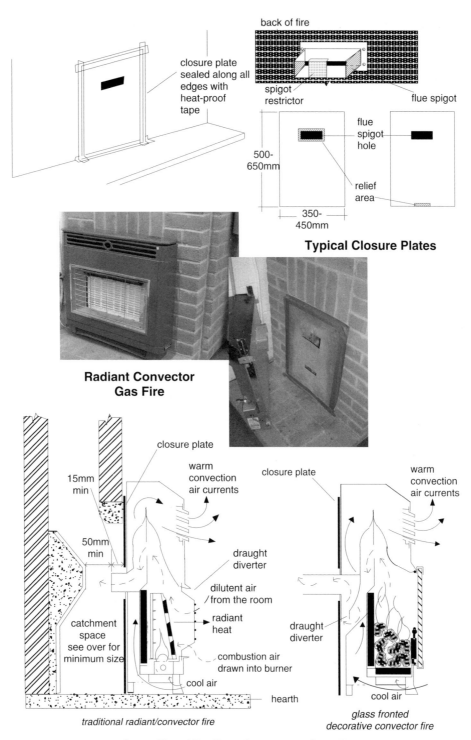

back of fire

closure plate
sealed along all
edges with
heat-proof
tape

spigot
restrictor

flue spigot

flue
spigot
hole

500-
650mm

relief
area

350-
450mm

Typical Closure Plates

**Radiant Convector
Gas Fire**

closure plate

warm
convection
air currents

closure plate

warm
convection
air currents

15mm
min

50mm
min

draught
diverter

dilutent air
from the room

radiant
heat

draught
diverter

catchment
space
see over for
minimum size

combustion air
drawn into burner

cool air

hearth

cool air

traditional radiant/convector fire

*glass fronted
decorative convector fire*

Open Flued Radiant Convector Gas Fires

Open Flued Solid Fuel Effect Fires and Heaters

These heaters, because they give the illusion of an exposed flame, similar to that given off by solid fuel, often have lower efficiencies than the more conventional radiant convector heater; efficiencies of not much greater than 45% can be expected. The appliances have a firebox on which imitation logs or coals are placed and through which the gas can pass to burn to give the desired flame pattern. *Note*: The logs/coals must be positioned strictly in accordance with the manufacturer's instructions. Failure to do so may lead to high levels of CO and soot being produced; for the same reason it is essential that the customer is advised how the appliance operates and is instructed not to throw paper and cigarette ends, etc. on to the burning flame as the accumulating ash will have a detrimental effect. Designs include the ILFE and heating stove, which connect to the flue system. However, other gas fires including open flued radiant convector heaters, DFE fires and flueless gas fires also give a solid fuel effect (see neighbouring pages).

Inset Live Fuel Effect Gas Fire (ILFE)
This fire comes supplied with its own fire back and heat exchanger, which is located either fully or partially within the builder's opening. It must be understood that the whole fire, including this fire back, will need to be removed for servicing purposes and therefore it is essential that any frame front can be detached from the fire surround. Heat passing from the burner rises up through the heat exchanger and discharges into the flue. Cold air also enters the base of the fire, passes up through an air passage behind the burner round the heat exchanger to pass out through the top front edge, thus allowing the room to be warmed by convection currents. The amount of radiant heat is limited to a small quantity from the incandescent coals and black-coloured fire back.

The minimum flue size where an ILFE fire is installed is 125 mm. Additional ventilation may or may not be required, depending on the input rating of the appliance; the manufacturer's instructions should be consulted.

Heating Stove
Two basic designs of heating stove are available: those with a vertical flue pipe and those with a rear flue spigot that passes through a closure plate, as with the radiant convector gas fire. For the stove with a flue pipe connection, the flue will need to have a minimum diameter equal to the size of the appliance outlet. As with the gas fire, unless the manufacturer specifies differently, a heating stove will need to be positioned on a suitable hearth as described in the next section. It should be noted that when installing a heating stove it is not acceptable to make the connection to an unlined chimney and either a clay or metal liner is needed. This would need to be suitably sealed at the base and, where necessary, an inspection hatch provided.

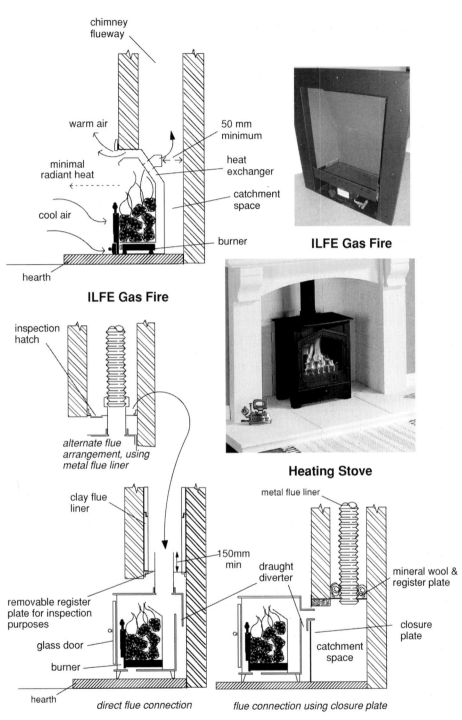

chimney
flueway

warm air

50 mm
minimum

minimal
radiant heat

heat
exchanger

catchment
space

cool air

burner

hearth

ILFE Gas Fire

ILFE Gas Fire

inspection
hatch

*alternate flue
arrangement, using
metal flue liner*

Heating Stove

clay flue
liner

metal flue liner

150mm
min

draught
diverter

mineral wool &
register plate

removable register
plate for inspection
purposes

closure
plate

glass door

catchment
space

burner

hearth

direct flue connection

flue connection using closure plate

Heating Stove

9 Domestic Appliances

Installation of Open Flued Gas Fires 1

Relevant ACS Qualification
HTR1

Relevant Industry Document
BS 5871

Flue Termination

It is not normally necessary to fit a terminal to any flue system where the flue is greater than 170 mm in diameter. Where a terminal does become necessary, the design will have to meet the requirements of the manufacturer and follow the recommendations in Part 6, Flues.

Connecting Fires and Heaters to an Existing Flue

Generally in a traditionally designed dwelling, due to the low heat input of a gas fire, there is no requirement to line the chimney unless the flue length exceeds the distance identified in the table on page 212. However, the manufacturer's instructions will need to be consulted for clarification. An existing flue will need to be inspected fully to ensure that it is suitable for the installation and of the correct material. It will be necessary to sweep the flue, especially if it was previously used for appliances burning fuels other than gas. Older chimneys will need to be inspected to ensure that there is no evidence of restriction within the flue, such as a damper plate or an old restrictor. Where these are evident, they will need to be removed. In the case of a damper, securing it in the open position will suffice. It is possible to leave old block, back boilers within the opening, providing they do not restrict the void (see below). However, where the boiler has been drained of water, drilling a small 6 mm hole in its base will prevent any pressure from building up inside and causing it to become unsafe. Any existing ventilation from openings into the builder's opening will need to be sealed off as this may affect the operation of the appliance. The flue will need to be inspected throughout its entire length to assess its condition and to ensure that it only serves the one location. A flue flow test will need to be completed successfully to ascertain its condition. Flue flow testing was described on page 222. Following a check on the condition of the flue, the catchment space and fireplace opening will need to be inspected to ensure that they are adequate for its purpose.

Catchment Space

This is the opening located behind an appliance, usually as part of the builder's opening or flue box, in which any debris that falls down the chimney can collect. DFE appliances do not have a catchment space, however conventional radiant convector fires, ILFE and heating stoves do. The size of this void is determined by the flue system. For unlined flues or flues that have been previously used with other fuels, the space should have a minimum depth of 0.25 m and volume of 0.012 m^3.

For example, if the void is 0.25 m high, 0.16 m deep and 0.3 m wide, the volume will be $0.25 \times 0.16 \times 0.3$ m $= 0.012$ m^3 (12 litres). The catchment space can be reduced if the flue system has never been used by other fuels, to 75 mm high with a volume of 0.02 m^3 (2 litres), however, the brick chimneys would still need to be lined.

Hearth Construction

Where a hearth is required, it must be of fireproof construction and have a minimum thickness of 12 mm. For DFE and ILFE gas fires an additional up-stand is required along the front and side edges of 50 mm, to discourage the placing of rugs on the hearth. This may be achieved by the use of a fender plate. Where a hearth is located at the base of a builder's opening it will be necessary to extend it forward a minimum distance of 300 mm in front of the fire surround and a distance of 150 mm to each side, beyond the edge of a naked flame or incandescent material. It will also be necessary to extend the hearth beneath any naked flame. See page 197 for details of the hearth notice plate.

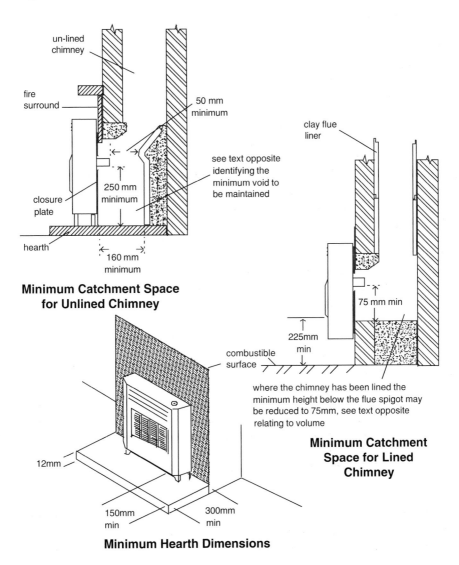

Minimum Catchment Space for Unlined Chimney

Minimum Catchment Space for Lined Chimney

Minimum Hearth Dimensions

Installation of Open Flued Gas Fires 2

Fire Protection

Rear-wall Where the fire is to be located against a surround, additional protection should be provided to minimise the effects of heat transmission unless the manufacturer states otherwise. No combustible material should be placed within the fireplace or builder's opening.

Side-wall If the fire is to be located in a corner, it is essential where the construction includes combustible material to ensure that a minimum distance of 500 mm is observed between the wall and the flame or incandescent material.

Shelf An appliance may only be located below a combustible shelf where the manufacturer's instructions clearly specify that it is permitted.

Wall-mounted fires Where a fire is mounted above floor level into a wall, no hearth is needed, providing a distance of 225 mm is maintained between any combustible floor covering and the flame or incandescent material; see diagram on previous page.

Connecting a Radiant/Convector or Heater to the Flue Opening

Using a secured closure plate, as described on page 288, the flue passes through the slot or hole provided. Where an existing fire back or chair-brick is in place, it is essential to ensure that a minimum distance of 50 mm is maintained and that the void provides the required catchment space as mentioned. Sometimes, due to the design of the fire surround, it will be necessary to fit a flue spigot extension piece as shown or, where the flue spigot passes into a pre-cast flue block system, a cooler device may be required.

Oversized Openings and Voids

If a fire or heater is to be put up to an opening that is larger than the closure plate is to be put in, it will be necessary to reduce the opening size accordingly. This may be achieved by simply bricking up the opening using non-porous bricks, alternatively it is acceptable to use a panel of non-combustible material. Materials such as wood cannot be covered in cladding and used as the heat within the opening may result in a fire. As stated above, there should be no combustible material in the opening. Whatever method is used, the final opening should be small enough that the closure plate fits over it and *not* the size of the fire-flue spigot. Should the builder's opening or the void itself be too large, as identified by the manufacturer (normally more than 800 mm high, 650 mm wide and 475 mm deep), it may be possible to reduce the size using non-porous bricks or blocks. Insulation blocks, such as 'Celcons', should not be used as they will absorb the water vapour from the combustion products and slowly deteriorate. Alternatively, it may be possible to fit a flue-box.

Flue-Box or Collector

These metal boxes have been especially designed for the installation of gas fires and heaters where there is no chimney or the existing chimney is unsuitable. The box must be secured in accordance with the manufacturer's instructions and be positioned on a non-combustible base. When in position, the box can be treated in much the same way as the builder's opening, with the closure plate secured as necessary. The flue box is connected to a lined chimney or, where a false chimney and breast is proposed, a rigid flue pipe could be used.

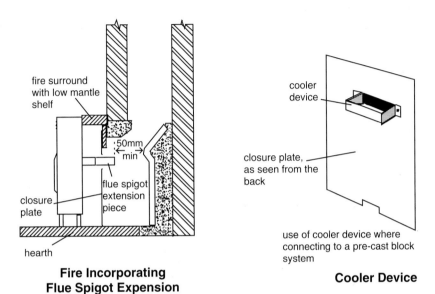

**Fire Incorporating
Flue Spigot Expension**

Cooler Device

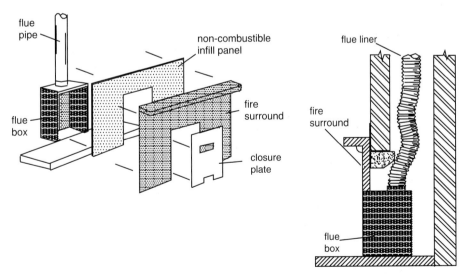

Flue Box Installation

Installation of Open Flued Gas Fires 3

Openings into the Fireplace

It is absolutely essential that there are no openings into the builder's opening except that allowed for in the closure plate, as described on page 288. In all cases any additional air that may be drawn into the fire via this unintended route may lead to poor combustion and adversely affect flue performance. Specific points to look for include:

- poorly fitted proprietary fire surrounds or infill panels;
- gaps in pre-cast flue blocks or where they may adjoin a dry-lined wall;
- unsealed redundant pipes that enter the void;
- poorly sealed and secured closure plates;
- wrongly sized closure plates.

Direct Flue Connection

In the last few years, fires and heaters with direct flue connections have come on to the market. Where these are to be fitted, it is essential that they are installed to a lined or twin wall flue system and treated similarly to a boiler connection and not connected simply through a closure plate. They may look like traditional gas fires or heating stoves, but they have their own installation requirements as well as alternative fixing methods.

Securing the Fire

The appliance should be positioned on a sound level stable base and, in general, the fire or heater is freestanding, requiring no additional securing. However, if stability is in question, additional fixing should be undertaken and, where necessary, the appliance should be secured to the wall to ensure no undue movement. Where the appliance is to be wall mounted, all of the manufacturer's fixing points should be used and give sufficient hold. Silicon sealant should on no account be used for securing purposes unless the manufacturer approves this method. Some ILFE fires require the use of a restraining cable to prevent the fire tipping; where this is the case care needs to be given to the fixing method used in securing the cable in the catchment space as plastic type wall plugs may melt due to the heat transfer along the cable. Fibre plugs may prove a better alternative.

Gas Supply

The gas pipework to the heater must be a permanently fixed rigid supply and incorporate a readily accessible local isolating control tap to facilitate removal of the fire for servicing and maintenance purposes. It is permissible to conceal the pipe that runs to the fire by running it below the floor or through the wall into the chimney recess, providing the pipe takes the shortest possible route and is adequately protected against corrosion. All connections must be such that the fire can be removed for periodic inspection.

Ventilation

All gas appliances need adequate supplies of air for combustion and the gas fire is no exception. However, it has been seen that the adventitious ventilation (air which enters through cracks in doors and window openings, etc.) is sufficient to supply an open flued radiant/convector or ILFE appliance of up to 7 kW input, providing the manufacturer's instructions do not state otherwise. For fires in excess of 7 kW see the notes in Part 7, Ventilation. A DFE fire generally requires 100 mm^2 of additional ventilation, see over.

Restricted Locations under current Gas Safety Regulations

The installation of open flued appliances is prohibited in *showers and bathrooms*.

Appliances installed in *bedrooms and bed-sits* or, in an adjoining cupboard, over 12.7 kW net (14 kW gross) input must be room sealed. Alternatively, an open flued appliance of less than 12.7 kW net input may be installed, providing a safety control, such as a vitiation sensing device (see page 108) is incorporated.

Basements and cellar locations have restrictions on the use of LPG appliances.

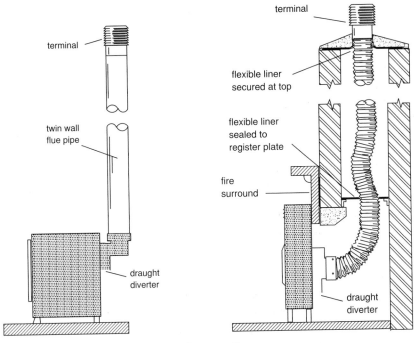

Direct Flue Connections

Decorative Fuel Effect Gas Fires (DFE)

This heater consists of no more than a firebox and burner. The unit is simply located within the builder's/fireplace opening, flue box or under an independent canopy and no direct connection is made to the flue. Because there is no heat exchanger, warmed convection currents passing from the appliances are minimal and the only heat supplied to the room is the radiant heat from the coals. Generally these appliances are used as a focal point and the room will have to been heated by some other means.

Flue Size

The minimum flue size to be looked for when installing these appliances is 175 mm across the axis (the old class 1 flue). Where installing a DFE, the notes from the previous section, dealing with the installation of open flued fires should be referred to, as these details also apply. The chair-brick, where fitted, assists in directing the products into the flueway. It is not necessary to alter the size if connecting into an existing chimney, providing the flue system has been proved to work safely. However, it must not be assumed that a DFE appliance positioned below a flue system will work simply because the hot products of combustion would be expected to rise upwards.

The fireplace opening in relation to the flue height needs to be considered to work out a minimum cross-sectional flue dimension, as a diameter greater than 175 mm may be required. For example, where the height of the fireplace opening is 0.75 m and the width 0.6 m, the opening area will be $0.75 \times 0.6 = 0.45$ m^2. If this is to be installed to a flue of height 6 m, from the chart opposite it will be seen that a flue diameter of either 225 mm or 250 mm will be required. Large diameter flues, particularly when they are short, can be prone to downdraught, therefore the smaller diameter of 225 mm should be selected.

Installations Below Independent Canopies

Where an independent canopy is to be used to collect the flue products it should be sited no more than 400 mm above the fire bed. The outer edges of the canopy should extend by the amounts shown in the diagram. The flue connection should be positioned at the top of the canopy with no further opening into the flue system. The angle of the canopy opening should be no less than 45°.

If a DFE fire is freestanding beneath a canopy or similar flue, the hearth will need to be extended 300 mm beyond the fire in all directions and, where such a fire is adjacent to a wall within this 300 mm space, the wall will need to be adequately protected.

Ventilation

The ventilation should be no less than 100 mm^2 for appliances up to 20 kW input unless the manufacturer states otherwise.

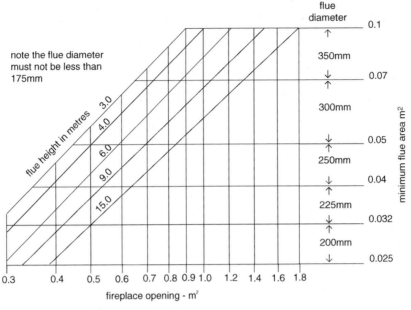

Flue Sizing Chart

note the flue diameter must not be less than 175mm

flue diameter

flue height in metres — 3.0, 4.0, 6.0, 9.0, 15.0

fireplace opening - m^2

minimum flue area m^2

opening 0.6 x 0.75 = 0.45m^2

Example:
should the fire place opening area be 0.45m^2 as with this example, & the flue height was 6m, by referral to the chart a flue size of approximately 0.04m^2 would be needed and where a pipe is to be used 225mm dia should be the minimum size selected

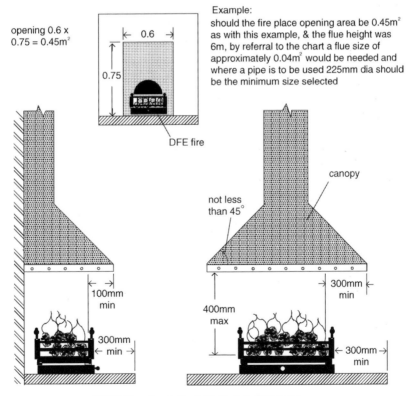

DFE fire

canopy

not less than 45°

100mm min

300mm min

400mm max

300mm min

300mm min

DFE Fire Installed Below a Canopy

9 Domestic Appliances

Fan Flued and Room Sealed Heaters

Fan Flued Systems

Fans are sometimes used to assist in the removal of combustion products with a space heater, particularly in the case of installations that incorporate a false chimney breast. Usually the appliance has a fan included as part of the manufacturer's design, however, occasionally such units have a fan fitted at the point of the chimney termination in place of a chimney pot. One example would be where a DFE appliance has been installed in a pub or restaurant and, due to the nature of the extracting system and ventilation changes, additional draught is needed. Should this be the case, it is essential that the unit is prevented from working unless the fan has been proved to be functional by a flue flow sensing device. For proprietary fan draught systems, discharging at low level, the terminal usually needs to be positioned so that there is a free passage of air across its surface. However, the appliance manufacturer's instructions need to be checked for compliance in terms of maximum flue lengths and terminal positions.

Many heaters have a fan fitted as part of the manufacturer's design; these are often wrongly seen as being room sealed but, in fact, they take the air from the room in which the fire is positioned. The Gas Safety Regulations permit the installation pipework to a fanned draught living flame effect fire to be installed in a cavity wall. However, the pipe must be enclosed within a gas-tight sleeve and take the shortest possible route. The sleeve will need to be sealed at the point at which it enters the fire.

Room Sealed Heaters

There are two distinct designs of room sealed unit: those that have a glass fronted panel, designed to provide a radiant element and/or characteristic solid fuel effect and those that are just a unit from which convected heat is distributed. The room sealed heater needs to be installed with reference to the manufacturer's instructions, in particular the termination requirements and location. See page 229 for suitable terminal positions. It should be noted that with the room sealed heater no ventilation is taken from the room in which it is situated and therefore no additional ventilation to the room will be required. The heat resistant glass panel fitted to these units is normally removable for servicing, etc., in which case it is essential that the condition of the seal used is maintained, thus preventing any combustion products entering the room.

With both the fan assisted open flued appliance and the room sealed appliance, when the terminal is less than 2 m above ground a terminal guard is required and other building openings and obstructions must be considered. Heaters that use a fan to assist the removal of the combustion products, unlike many of the other designs of gas fire/heaters, require an electrical connection, which is usually made via a fixed fused spur.

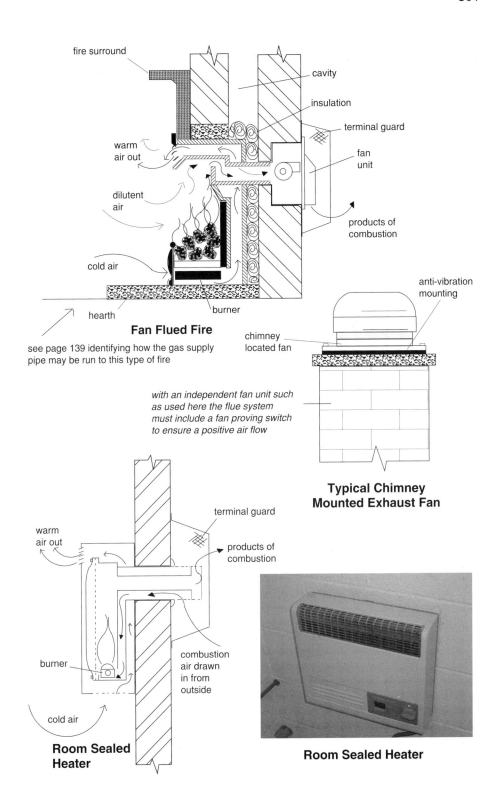

fire surround

cavity

insulation

terminal guard

warm
air out

fan
unit

dilutent
air

products of
combustion

cold air

hearth

burner

Fan Flued Fire

see page 139 identifying how the gas supply
pipe may be run to this type of fire

*with an independent fan unit such
as used here the flue system
must include a fan proving switch
to ensure a positive air flow*

anti-vibration
mounting

chimney
located fan

**Typical Chimney
Mounted Exhaust Fan**

warm
air out

terminal guard

products of
combustion

burner

combustion
air drawn
in from
outside

cold air

**Room Sealed
Heater**

Room Sealed Heater

9 Domestic Appliances

Flueless Space Heaters

These may be either fixed or mobile. Flueless heaters discharge their products directly into the room in which they are installed, therefore they are 100% efficient as none of the heat is lost into a flue system. However, because they discharge their products into the room it is essential that additional ventilation is provided, as well as some form of openable window to ensure complete combustion of the fuel and to overcome the increased problems of condensation, resulting from the combustion process. Refer to the table on page 249 for the size of ventilation grill for a specific appliance. The size of the room would also need to be considered as a room smaller that 24 m^3 would be too small to accommodate such an appliance. These heaters are generally quite low in output and therefore should be used in conjunction with some other form of heating.

Fixed Flueless Heaters

Traditionally, fixed heaters were installed in hallways. However, with the development of the room-sealed heater, they have now fallen out of general use. Currently, however, there is a revival with a design that incorporates a sealed combustion chamber and catalytic converter. These heaters are proving popular where a solid fuel effect fire or stove is wanted and no chimney or flue system is available.

The catalytic converter is a device that converts the poisonous gases, such as carbon monoxide and aldehydes into less harmful emissions, such as carbon dioxide. It works by passing the products of combustion through an inner honeycomb section, usually made of a ceramic structure coated in a metal such as platinum, palladium and rhodium. This causes a chemical reaction in the flue gases and speeds up the process of changing the products to less harmful ones.

Mobile Flueless Heaters

Flueless heaters are not restricted to fixed appliances and there are mobile heaters that use a butane gas cylinder housed within the case. Propane must not be used with these mobile heaters because of the higher cylinder pressures. When using a mobile cabinet heater that uses a series of three radiant plaques, allowing one, two or all three burners to be lit, it should be noted that the ventilation requirements alters for each input setting and having the appliance on full may be permitted in one room but not in another, where only one radiant may be lit, due to room size and available ventilation. The size of vent required is given on page 249 (Table 3a). These heaters generally have a label inside the case, usually where the cylinder connects, warning of the requirements. As these heaters run on LPG they are restricted from use in cellars and basements where any escape of gas could prove disastrous.

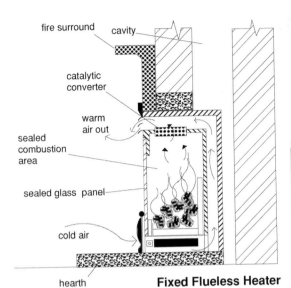

fire surround

cavity

catalytic
converter

warm
air out

sealed
combustion
area

sealed glass panel

cold air

hearth

Fixed Flueless Heater

Fixed Flueless Heater

**Looking Up from the Base
of the Fire to the
Catalytic Convertor**

Mobile Heater

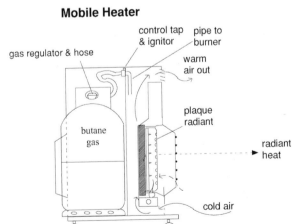

control tap
& ignitor

pipe to
burner

gas regulator & hose

warm
air out

plaque
radiant

butane
gas

radiant
heat

cold air

Mobile Heater

Mobile Heater

Commissioning and Servicing Space Heaters

Commissioning and Servicing Work Record

The Work Record opposite is given purely as a guide to the many tasks to be undertaken when servicing/maintaining and commissioning a space heater. In order to complete the form you may need to refer back to Part 8, page 264, where many of the tasks are explained in greater detail. However, the additional 'appliance specific' checks on space heaters include some of the following.

Visual inspection of flue, catchment space, damper plate and hearth

You cannot assume that a flue system is working effectively when you start work on a gas fire. The flue system or chimney into which the fire is connected is regarded as part of the installation and therefore its safe operation rests in the hands of the gas engineer working on the fire, So, for example, where an ILFE gas fire is being worked on, it is essential that the whole unit is removed from the builder's opening, allowing internal inspection of the flue and catchment space. The pipe to the appliance may enter the catchment area and special care needs to be given at inspection for corrosion problems and its effective seal into the chamber. The Flue Inspection and Testing Checklist on page 267 may prove useful in completing this task.

Flue/Hearth notice plate located and correctly filled in

A plate should be in place that gives details of the flue, location and hearth (see page 197). This is a comparatively new requirement in this country.

Metal fatigue of the heat exchanger

Owing to the high intensive heat generated within the fire box, sometimes cracks will appear behind the radiant and fire brick; these can only be seen by completely removing them from the appliance. These cracks are the result of the continued expansion and contraction of the metal. Where the heat exchanger is damaged in this way the fire needs to be condemned as the combustion products may be drawn in through these cracks to be discharged around the room, giving rise to the production of carbon monoxide.

Fireguard in place

A fireguard may be required in order to comply with the Heating Appliances (Fireguards) (Safety) Regulations; this is particularly important where the elderly, infirm or children may be put at risk from the open flames or hot surfaces. The fireguard needs to be secured to prevent its removal.

Specific notes applicable to mobile cabinet heaters

Because the supply of gas is via a butane gas bottle, a check needs to be made on the condition of the hose connections and regulator to make sure they are in a sound working order.

Certificate/Record of Space Heater Service/Commission

Gas Installer Details	Client Details	Appliance Date Badge Details	
Name :	Name :	Model/Serial No:	
CORGI Reg. Nº :	Address :	Gas Type: Natural ☐ LPG ☐	
Address :		Heat Input: max…kW min…kW	
		Burner Pressure Range: …–…mbar	
		Gas Council Nº/ CE Nº:………………	

Date: | **Appliance Location:**………………………………
Install/Commission ☐   **Service/Commission ☐**

Preliminary System Checks Compliance with manufacturer's instructions	PASS ☐ FAIL ☐
General visual inspection of pipework	PASS ☐ FAIL ☐
Clearance from combustible materials	PASS ☐ FAIL ☐
Visual inspection of flue, catchment space, damper plate and hearth	PASS ☐ FAIL ☐ N/A ☐
Flue/hearth notice plate located and correctly filled in	PASS ☐ FAIL ☐
Closure plate sealed along all edges	PASS ☐ FAIL ☐ N/A ☐
Flue flow performance test	PASS ☐ FAIL ☐ N/A ☐
Appliance level and secure	PASS ☐ FAIL ☐
Electrical connections	PASS ☐ FAIL ☐ N/A ☐
Bonding maintained	PASS ☐ FAIL ☐ N/A ☐
Fuse rating:……….amps	PASS ☐ FAIL ☐ N/A ☐
System tightness test; to include let by	PASS ☐ FAIL ☐ N/A ☐
Standing pressure of system	PASS ☐ FAIL ☐ N/A ☐
Appliance/System is purged of air	PASS ☐
Service/Commission Checks Clean primary air ports and lint arrestor	PASS ☐ FAIL ☐ N/A ☐
Clean/Check condition of injectors; burners and radiants/coals	PASS ☐ FAIL ☐ N/A ☐
Check condition of heat exchanger, examining for metal fatigue	PASS ☐ FAIL ☐
Radiants/Coals correctly located and aligned	PASS ☐ FAIL ☐ N/A ☐
Check for easy operation and grease, if necessary, control taps	PASS ☐ FAIL ☐ N/A ☐
Ignition devices effective including condition of electrodes, leads and battery	PASS ☐ FAIL ☐ N/A ☐
Clean and check operation of fans	PASS ☐ FAIL ☐ N/A ☐
Pilot flame correct	PASS ☐ FAIL ☐ N/A ☐
Burner pressure………….mbar	PASS ☐ FAIL ☐
Total heat input………….kW	PASS ☐ FAIL ☐ N/A ☐
Flame picture good	PASS ☐ FAIL ☐ N/A ☐
Flame supervision device operational	PASS ☐ FAIL ☐ N/A ☐
Condition of frame and combustion seals effective	PASS ☐ FAIL ☐ N/A ☐
Appliance tightness check	PASS ☐ FAIL ☐ N/A ☐
Spillage tests	PASS ☐ FAIL ☐ N/A ☐
Operating thermostat correct	PASS ☐ FAIL ☐ N/A ☐
Flue guard fitted to low level terminals	PASS ☐ FAIL ☐ N/A ☐
Additional ventilation grill, where necessary………….cm^2	PASS ☐ FAIL ☐ N/A ☐
Fireguard in place in presence of elderly, infirm or children	YES ☐ NO ☐
Post System Checks Meter working pressure (……….mbar)	PASS ☐ FAIL ☐ N/A ☐
Working pressure drop across system (max………mbar)	PASS ☐ FAIL ☐ N/A ☐
Safe operation of appliance explained to customer	YES ☐ NO ☐

Recommendations and/or Urgent Notification

Appliance Safe to Use   YES ☐ NO ☐   **Next Service Due:**…………

Installer's Signature:	**Customer's Signature:**

It is generally recommended that all flexible hoses are replaced every five years, however any hose that shows signs of fatigue needs replacing. A good test is to bend the hose back on itself, forming a very tight bend; this will expose any hair-line cracks that may be developing. As it is difficult to test the atmosphere sensing device (ASD) located at the pilot flame on mobile heaters it is advisable to replace it every five years. The radiant plaques should be inspected and it is recommended that damaged or cracked plaques be replaced. When re-bedding the plaque, the fire cement should be allowed to dry for at least 24 hours before lighting the fire. Where a catalytic panel has been incorporated with the heater this should be inspected for wear, which is often seen as bald patches or holes. This may need replacing and care needs to be observed as catalytic heaters made before 1983 may include asbestos. Where the heater is mounted on rollers, the casters should be generally examined to ensure smooth running and the control tap must operate freely.

Fault Finding
The typical faults with gas fires are also common to any burner or control device in a gas appliance. However, the list opposite may help to identify specific problems.

Photograph showing the radiants removed from a gas fire and the crack formed in the heat exchanger behind

Fault Diagnosis Chart

Fault	Possible cause
Pilot will not light	• Gas supply turned off • Air in pipe • Pilot injector blocked or incorrect size • Incorrect spark gap • No spark or lead not connected properly
Poor pilot flame and will not stay alight	• Thermocouple connection loose • Pilot flame too small or pilot tube blocked • Faulty thermocouple of thermo-electric valve • Inadequate ventilation • ODS defective or blocked
Poor heat output and lack of warm air to the room	• Poorly sealed closure plate, allowing heat to be lost up the flue • Inadequate gas pressure • Blocked gas injectors • Flame reversal, resulting from incorrect or poorly sealed closure plate • Is a spigot restrictor needed? • Faulty thermostat, where fitted • High curb or up-stand at the lower edge of the fire preventing cool air from circulating through the heat exchanger
Staining to outer case or wall areas surrounding the fire	• Inadequate catchment space, causing spillage • Inadequate ventilation, causing spillage • Blocked or restricted flue, causing spillage
Poor combustion, e.g. yellow flames and soot	• Burner and primary air intake blocked • Incorrectly positioned radiants or coals
Uneven flame pattern or ghosting where the flames begin to emerge from the top of the radiant	• Faulty or linted burner • Insufficient air • Damaged radiants, try swapping about • Blocked injector, where duplex injectors used • A thick rug restricting air flow

9 Domestic Appliances

Domestic Gas Cookers

Relevant ACS Qualification	Relevant Industry Document
CKR1	BS 6172

The domestic gas cooker consists of three separate components, namely the grill, oven and hob/hotplate. These may be purchased separately or as one complete unit and are designed to be installed with the associated kitchen cupboards or as a free-standing appliance.

The Grill

The grill is used for toasting, grilling or browning previously cooked food. It works by directing radiant heat that has been produced on the surface of a red-hot fret, mesh or gauze surface, on to the food. There are two designs of grill.

First, those referred to as *conventional grills*, as they have been around the longest. The grill consists of a pressed steel burner, fed via an injector at one end and located beneath an expanded metal fret. The flames, on leaving the burner, heat the metal fret and cause it to glow red-hot. The combustion products rise by convection through holes in the canopy top. The biggest problem with this design is the problem of uneven cooking, a result of the burner failing to heat the whole surface of the fret uniformly.

In the second design of grill, known as a *surface combustion grill*, the injector feeds gas into the primary air intake, this entails some 80–90% of that required for combustion. The mixed gas and air now travels to the centre of a sealed chamber, which has a fine metal mesh burner surface where the ignition probe is located and where combustion takes place, the air/gas mixture clings to its face, producing an evenly heated radiant surface across the entire burner.

The Oven

The oven is used for warming, roasting and baking. It works by surrounding the food with hot convection currents. The oven temperature is adjusted by a thermostat that varies the amount of heat surrounding the food to be cooked. The grading ranges from a simmer or economy setting of around 100°C through to gas mark 9, which generates temperatures of around 245°C. Most domestic ovens found in the UK work by natural convection currents. The burner located at the rear of the base creates a circulation of hot gases that eventually discharge from the rear of the oven by means of a flue. Owing to the relatively slow circulatory motion, temperature zones invariable develop in the oven, with the hottest region at the top.

In another design of oven, the gas burner is located outside the food compartment and hot air is allowed to enter via various ports, so producing a more even spread of heat temperature throughout the oven. This design is often used with an additional fan to give improved efficiency.

Built-In Oven and Hob

Freestanding Cooker

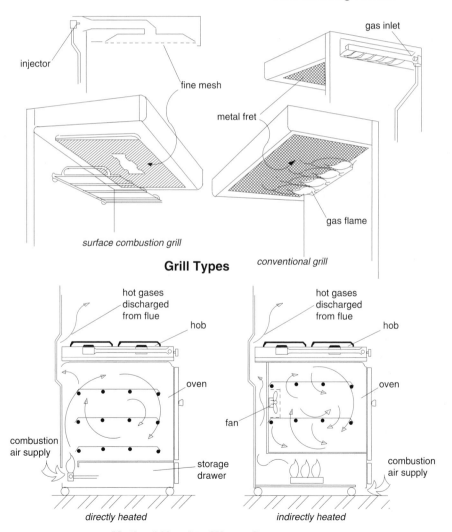

injector

fine mesh

gas inlet

metal fret

gas flame

surface combustion grill

Grill Types

conventional grill

hot gases
discharged
from flue

hob

oven

combustion
air supply

storage
drawer

directly heated

hot gases
discharged
from flue

hob

oven

fan

combustion
air supply

indirectly heated

Method Used to Warm Ovens

The Hob or Hotplate

The hob is used for boiling, frying, steaming, simmering and braising. It primarily works by the conduction of heat from the flame through the pan surface. However, a certain amount of radiant heat is also generated. A large volume of primary air is provided so that the pan can be located as close as possible to the burner head and so gain the heat from the flame. The flame, however, should not be allowed to lick up the side of the pan, as heat will be wasted. The hotplate is usually a solid steel flat plate or ceramic surface, on which the pan sits, or it can take the form of a griddle, which is used for dry-frying foods such as eggs, bacon, hamburgers, etc. The flame is located beneath the metal and heats it to the desired temperature.

The heat setting of a hob or hotplate often has a grading, such as simmer, medium and high. The flame does not adjust to the heat requirements of the food being cooked and, if set too high, the contents of the pan may boil over or burn. However, having said that, some appliances have thermostatically-controlled burners. They work using a contact sensor that is filled with a volatile fluid that, when heated, expands forcing the gas valve to close down to a by-pass rate.

Hobs are often found with a glass drop-down lid that is designed to provide a smoother pleasing appearance. Where these are incorporated it is essential to understand the additional safety feature, which is incorporated to prevent gas passing to the burners when the lid is closed. This works in a number of ways, one of which, as shown, allows a pin to be pushed in to open the valve when the lid is raised. *Note*: When tightness testing this valve must be open otherwise several joints will remain unchecked, therefore the lid must be in the upright position.

With free standing cookers all control taps and the oven thermostat are located on a pipe, named the float rail, which is located behind the control knob fascia. During any maintenance work all these gas connections should be sprayed with leak detection fluid to ensure they are gas tight.

The gas burners used with the associated parts of the cooker are generally unprotected, in terms of flame failure. This means that should the flame go out, gas is allowed to discharge freely into the room. The oven will be found to have a flame failure device fitted, usually of the liquid vapour type, which will shut down the main flow of gas but will still allow a discharge through the by-pass. Cookers supplied by LPG, however, are controlled and all control taps generally require the operation of a thermo-electric FSD before gas is allowed to flow freely. These control devices, the gas taps, thermostats and associated ignition devices themselves are described in Part 3, Gas Controls, which should be referred to for further information.

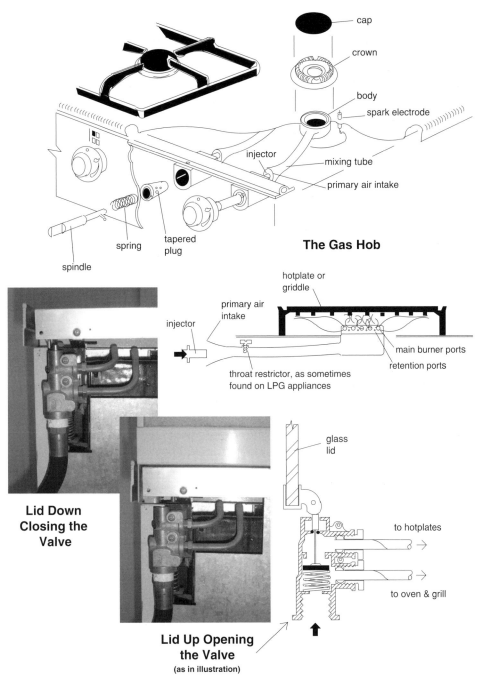

cap

crown

body

spark electrode

injector

mixing tube

primary air intake

spring

tapered plug

spindle

The Gas Hob

hotplate or griddle

primary air intake

injector

main burner ports

retention ports

throat restrictor, as sometimes found on LPG appliances

glass lid

Lid Down Closing the Valve

to hotplates

to oven & grill

Lid Up Opening the Valve
(as in illustration)

Cooker Drop Down Lid Safety Cut-Off Device

Installation of Cooking Appliances

Location
A cooker is not allowed to be installed in the following areas:

- bathroom or shower room;
- bed-sitting room less than 20 m^3 volume (unless a single hotplate burner); and
- for LPG – below ground or basement type areas.

The room into which the appliance is to be installed will require an openable window or similar adjustable opening. In addition, if the room is small, additional ventilation may be required as listed below.

Room volume m^3	Ventilation free area cm^2
Less than 5	100
Between 5 and 10	50*
Greater than 10	Nil

*If there is a door to outside no additional vent is required.

Siting Requirements
The position of a cooker in a room will depend on several factors. It should not be affected by draughts from windows nor must it affect adjacent appliances such as a refrigerator. It may be necessary to have an adjacent electrical point and a cooker will generally need to be near other appliances, work tops, etc. for convenience. However, wherever it is sited, care will need to be taken to ensure that materials in close proximity will not be in danger of catching fire. The appliance manufacturer generally gives clear guidance as to the correct siting, however the dimensions given opposite can be taken as suitable minimum recommendations.

Gas Connections
This may be by either a rigid connection, with an accessible isolation valve, as found in ranges or where a gas hob is installed directly into a kitchen unit or, as is often the case, by a flexible connector and self-sealing bayonet valve. The valve must be accessible for disconnection purposes and the flexible hose should hang freely down, thus avoiding any undue stress to the rubber (see diagram). *Note*: LPG connectors are marked with a red band. If excessive heat, greater than 70°C, is anticipated a flexible connector should not be used.

Cooker Stability
With a free-standing cooker it is possible that when the oven door is opened and a minimal weight applied, that the cooker might tip forward. This could have disastrous effects, especially if a pan of boiling water is on the hob. For this reason a stability bracket or chain must be installed as shown.

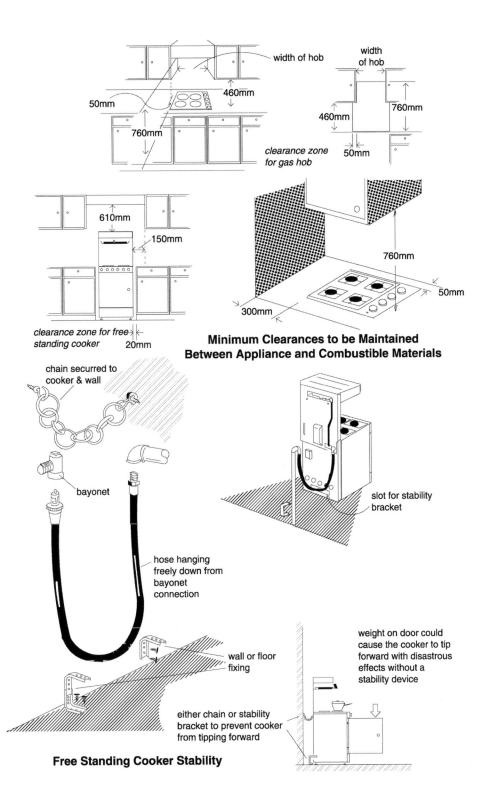

width of hob

width of hob

460mm

50mm

760mm

clearance zone
for gas hob

460mm

760mm

50mm

610mm

150mm

clearance zone for free
standing cooker   20mm

760mm

50mm

300mm

**Minimum Clearances to be Maintained
Between Appliance and Combustible Materials**

chain securred to
cooker & wall

bayonet

slot for stability
bracket

hose hanging
freely down from
bayonet
connection

wall or floor
fixing

either chain or stability
bracket to prevent cooker
from tipping forward

weight on door could
cause the cooker to tip
forward with disastrous
effects without a
stability device

Free Standing Cooker Stability

Domestic Flued Cooking Range

Relevant ACS Qualification
CKHB1

This is a cooker that is based on the traditional cast iron design of a solid fuel or oil burning range. There are two designs: those with atmospheric burners and those that use a forced draught burner. These appliances invariably have more than one function, such as heating water for central heating and domestic purposes, and so require a flue system, which may be either of open flue or balanced flue design.

Those ranges designed with an atmospheric burner often include a maintained flame to keep the appliance at a constant temperature, thus mirroring the traditional design. Invariably two separate ovens and two large hotplates are incorporated, operating at differing temperatures with graduated zones.

Some designs of this type cooker are too heavy to be supplied as a complete unit and it is the specialist installer's responsibility to ensure that all the internal parts are correctly located to ensure safe and adequate heat transfer. Owing to the weight, which may be several hundred kilogram, a suitable non-combustible hearth with a minimum thickness of 12 mm needs to be constructed.

The gas connection to a range needs to be of the rigid type because it is a flued appliance and so a flexible connection as used for the normal freestanding cooker must not be used. Following the isolating valve, a disconnecting joint, such as a union connector, is required for servicing purposes.

Fluing

The flue design depends on the appliance. However, where an open flued appliance has been chosen, care should be take to observe the manufacturer's instructions and a flue should never be less than the appliance flue spigot size. Fan assisted open flued models are available where obtaining a natural draught is a problem. Existing chimneys need to be lined with a double skin liner, certified as being suitable for use with solid fuel and the void between the liner and chimney filled with a suitable insulating material such as vermiculite. The terminal would require an effective free area opening of at least twice that of the cross-sectional area of the flue system used.

Ventilation

Ventilation requirements for range cookers will be different from those for a more traditional cooker, which of course is flueless. Generally a range would be treated the same as any open flued appliance, as described in Ventilation Sizing in Part 7, page 246.

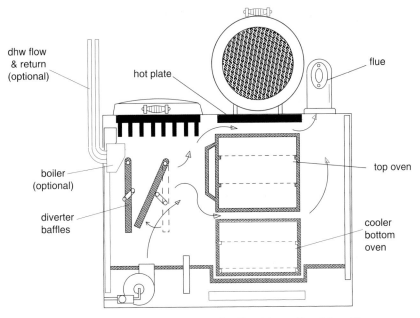

dhw flow
& return
(optional)

hot plate

flue

boiler
(optional)

top oven

diverter
baffles

cooler
bottom
oven

Section Through a Domestic Cast Iron Cooking Range

Domestic Cast Iron Cooking Range

Commissioning and Servicing Cookers

Work Record for Cookers

The Work Record opposite is purely given as a guide to the many tasks to be undertaken when servicing/maintaining and commissioning gas cookers. In order to complete the form you may need to refer back to Part 8, page 264, where many of the tasks were explained in further detail. However, the additional 'appliance specific' activities for cookers include some of the following:

Cleaning grill frets

When servicing a cooker, particular attention should be given to the inspection of the grill frets for damage, such as buckling or splits that might result in flame impingement, etc., leading to the production of high levels of carbon monoxide (CO). This is often overlooked by the inexperienced gas engineer but it has been responsible for a number of poisoning incidents.

Checking the oven by-pass

Allow the oven temperature to rise for 10–15 minutes at, say, gas mark 5. The temperature should eventually be satisfied and the thermostat allows the flame to drop to the by-pass rate. You could turn the temperature down to its lowest setting, causing the thermostat to close. If the flame goes out, the by-pass is blocked and the screw to the side of the thermostat will need to be removed to dislodge any excess grease, etc.

Checking the simmer settings on the hob

Turn the gas tap to its lowest setting. If the flame goes out the smallest hole in the valve is blocked, possibly with grease.

Checking the appliance for gas leaks

All exposed pipework and fittings should be sprayed with a leak detection solution when the gas is flowing to the burners. Particular attention should be paid to the pipe leading up to the grill, where a compression joint is often found. When a gas tap has been reassembled following a repair, the valve should also be sprayed.

Lid safety cut-off device

To check the correct operation of this safety device, with the hot plate flames burning, lower the glass lid and check that the flame is extinguished.

Checking the door seals

This is a simple test carried out by trapping a piece of paper (0.25 mm thick) in the top and both sides of the door and pulling it out. A resistance should be felt. Where there is little resistance, heat could possibly escape, so cooling the oven and, invariably, drying out the control taps. The base of the door does not need to be checked as this is where the air is drawn into the oven.

Certificate/Record of Space Heater Service/Commission

Gas Installer Details	Client Details	Appliance Date Badge Details	
Name :	Name :	Model/Serial No:	
CORGI Reg. N° :	Address :	Gas Type: Natural ☐ LPG ☐	
Address :		Heat Input: max. . . kW min. . . kW	
		Burner Pressure Range: . . .–. . . mbar	
		Gas Council N°/ CE N°:.	
Date:	**Appliance Location:**. . . . . **Install/Commission ☐ Service/Commission ☐**		
Preliminary System Checks Compliance with manufacturer's instructions			PASS ☐ FAIL ☐
General visual inspection of pipework			PASS ☐ FAIL ☐
Clearance from combustible materials			PASS ☐ FAIL ☐
Gas connection e.g. bayonet and cooker hose in sound condition			PASS ☐ FAIL ☐ N/A ☐
Stability bracket/chain effective			PASS ☐ FAIL ☐ N/A ☐
Appliance level and secure			PASS ☐ FAIL ☐
Electrical connections			PASS ☐ FAIL ☐ N/A ☐
Bonding maintained			PASS ☐ FAIL ☐ N/A ☐
Fuse rating:. . . . . .amps			PASS ☐ FAIL ☐ N/A ☐
System tightness test; to include let by			PASS ☐ FAIL ☐ N/A ☐
Standing pressure of system			PASS ☐ FAIL ☐ N/A ☐
Appliance/system is purged of air			PASS ☐
Service/Commission Checks Clean primary air ports			PASS ☐ FAIL ☐ N/A ☐
Clean injectors; burners; burner rings and grill frets			PASS ☐ FAIL ☐ N/A ☐
Check, ease and grease, if necessary, control taps			PASS ☐ FAIL ☐ N/A ☐
Ignition devices effective e.g. condition of electrodes, leads and battery			PASS ☐ FAIL ☐ N/A ☐
Pilot flame correct			PASS ☐ FAIL ☐ N/A ☐
Burner pressure. . . . . .mbar			PASS ☐ FAIL ☐
Total heat input. . . . . .kW			PASS ☐ FAIL ☐ N/A ☐
Flame picture good			PASS ☐ FAIL ☐ N/A ☐
Simmer settings to all burners and oven by-pass			PASS ☐ FAIL ☐
Flame supervision device operational			PASS ☐ FAIL ☐ N/A ☐
Oven flueway is clear			PASS ☐ FAIL ☐ N/A ☐
Appliance tightness check			PASS ☐ FAIL ☐ N/A ☐
Lid safety cut off device effective			PASS ☐ FAIL ☐ N/A ☐
Oven and door seals effective			PASS ☐ FAIL ☐ N/A ☐
Oven thermostat correct			PASS ☐ FAIL ☐ N/A ☐
Ancillary equipment (timers; oven lights and fans etc)			PASS ☐ FAIL ☐ N/A ☐
Openable window or equivalent			PASS ☐ FAIL ☐
Additional ventilation grill, where necessary. . . . . .cm^2			PASS ☐ FAIL ☐ N/A ☐
Post System Checks Meter working pressure (. . . . . .mbar)			PASS ☐ FAIL ☐ N/A ☐
Working pressure drop across system (max. . . . . .mbar)			PASS ☐ FAIL ☐ N/A ☐
Safe operation of appliance explained to customer			YES ☐ NO ☐
Recommendations and/or Urgent Notification			
Appliance Safe to Use YES ☐ NO ☐ **Next Service Due:**.			
Installer's Signature:		**Customer's Signature:**	

Checking the oven thermostat

This is achieved using an oven thermometer to compare gas settings against those of the manufacturer. *Note*: It is generally not possible to re-calibrate the thermostat and therefore it may need replacing. The following may be used as a general guide to oven cooking temperatures.

Typical oven thermostat temperature settings

Gas mark	Approx. °C	Oven heat
S/E	105–120	Very cool
1	135–140	Cool
2	150	Cool
3	160	Warm
4	175	Moderate
5	190	Fairly hot
6	202	Hot
7	220	Hot
8	230	Very hot
9	250	Very hot

Typical Oven Thermometer

Fault Finding

Many of the faults to be found with cookers are the result of spillage of food, blocking burner ports, injectors and ignition electrodes. Faults with hobs and grills are usually self-evident. Oven defects such as uneven cooking are more difficult to diagnose and may be the result of misplaced linings or distorted shelves or the cooker itself being out of level. The chart opposite lists some of the many faults that may be encountered.

Sometimes the only way to diagnose a fault is to undertake a test bake to ensure the correct customer operation. A typical test is to bake a light Victoria sandwich sponge as per the recipe from a typical cookbook. The cake is then inspected against the following possible faults:

- *Uneven browning*: Poor door seals.
- *Cake cooked thinner on one side*: Oven not level.
- *Undercooked in the specified time allowed*: Faulty thermostat or FSD.
- *Overcooked in the specified time allowed*: Faulty thermostat.
- *Cake burnt at the back*: Sponge possibly too close to burner? Check that oven shelf is not upside down.
- *Uneven cooking in parts of the oven*: Oven lining incorrectly positioned.

Fault diagnosis chart

Fault	Possible cause
Incorrect flame picture or insufficient heat	• Incorrect gas pressure • Blocked injector or wrong size • Blocked or damaged burner head or frets • Aeration port blocked or setting incorrect • Insufficient ventilation to room • Poor door seals • Faulty oven thermostat • Combustion outlet blocked
No or poor ignition	• Damaged, dirty or wrongly positioned electrodes • Faulty batteries or electric ignition system • Incorrect gas pressure • Aeration port blocked or setting incorrect • Damaged flash tube
Uneven cooking	• Appliance not level • Blocked or damaged burner head or grill frets • Blocked injector or wrong size • Poor door seals • Faulty oven thermostat or incorrectly positioned probe • Faulty or misplaced oven linings • Combustion outlet blocked • Faulty flame supervision device • Distorted shelves • Recipe not followed correctly and fully
Hot control taps	• Poor oven door seals

Instantaneous Gas Water Heaters

Relevant ACS Qualification
WAT1

Relevant Industry Documents
BS 5546 and BS 6700

There are two types of instantaneous water heater: single point and multipoint heaters. The combination boiler is a form of multipoint, but it also functions as part of the central heating system, and will therefore be dealt with later in the section on boilers. The general principle of the instantaneous heater is that when cold water passes into the unit it flows up around the combustion chamber and through the heat exchanger where it is rapidly heated to the required temperature as at the hot draw off point. The water flow rate will therefore determine the outlet temperature. As the water flows through the unit the movement will be detected by some form of differential valve. This in turn brings on the gas supply to the main burners, warming the water. When the supply is turned off, the static no-flow state is identified and the gas supply is cut off. The remaining cold water in the heater now cools the unit and so prevents the water boiling. The water supply may be either mains fed or fed via a feed cistern. Early designs of water heaters used a bi-metallic strip as the flame supervision device, however, apart from in older models, a thermal-electric flame supervision device that uses a thermocouple is usually used. One of the biggest problems with these heaters is the problem of scale build up in hard water districts, a condition that manifests itself by a low rate of water flow through the heater, squealing and kettling noises, and damage to the heat exchanger or combustion chamber, where the heat has been restricted from escaping from the unit.

The Single Point

These heaters are used in close proximity to their point of use and are usually used to serve only one or two sanitary appliances. They may have a swivel spout or be plumbed in fully to a system of pipework, running to the various outlets. A typical heater has less than 11 kW net input, with a water flow rate of around 2.5 litres per minute, allowing a temperature rise of about 50°C.

This type of unit may be either flued or flueless. If a flueless heater is used, a warning sticker needs to be fixed in a prominent position on the front of the heater warning of a maximum running period of five minutes; this is designed as a safeguard against raising the levels of combustion products, in particular carbon monoxide, within the room. A flueless heater must also be installed in the room where it is to be used and not in an adjoining room, where it would need to be flued.

The Multipoint

This heater is designed to serve several outlets and typical heat inputs of around 30 kW net, giving water flow rates of around 6.5 litres/minute and a 50°C temperature rise can be expected. Today all models are of the room sealed type. A combination boiler serves effectively as a multipoint, yet has the additional function of providing central heating. This boiler is described on page 338.

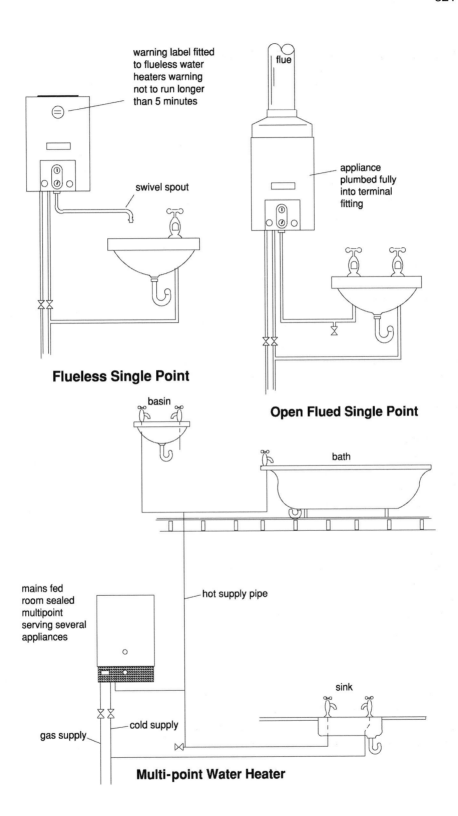

warning label fitted to flueless water heaters warning not to run longer than 5 minutes

flue

swivel spout

appliance plumbed fully into terminal fitting

Flueless Single Point

basin

Open Flued Single Point

bath

mains fed room sealed multipoint serving several appliances

hot supply pipe

sink

gas supply

cold supply

Multi-point Water Heater

Instantaneous Water Heater Operation

For the instantaneous water heater to function, the water and gas controls work in conjunction with each other and are usually specifically designed for the appliance. There are many variations for individual heaters, however the components and operation identified here are of a typical design. The diagram opposite shows the appliance in the off position, with no water flowing. Over the page the appliance operating with the water and gas flowing through the appliance can be seen.

Components and Their Operation

Differential pressure control valve

This valve has the function of automatically opening/closing the gas supply to the main burner when water is flowing through the heater. It also ensures that sufficient volume flows before bringing on the gas to prevent the water temperature rising above 55°C, which would lead to scaling up and overheating. The valve works by passing the water through a venturi, which has the effect of reducing the pressure from above a diaphragm, causing it to lift in response to the negative pressure.

The venturi works on the theory that reducing the bore of a pipe causes the water flow to increase in velocity, just like putting your finger over the end of a hosepipe. The increased velocity is an energy force and the energy is obtained by giving up some other form of energy, in this case pressure.

As the diaphragm lifts, it forces the gas valve open, allowing gas to the injector manifold to be ignited by a previously established pilot flame. When there is limited flow or no water flow through the valve the pressure differential between each side of the diaphragm is minimal and the spring will cause the gas valve to close.

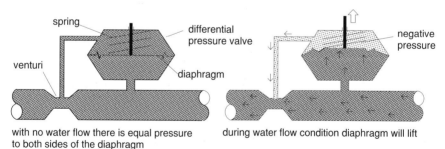

with no water flow there is equal pressure to both sides of the diaphragm

during water flow condition diaphragm will lift

Valve Lift Due to the Venturi

Water governor and throttle

For cistern fed water heaters where the pressure is low, adjustment of the water flow is usually by means of a *water throttle*, which is no more than a screw-in restrictor. For mains fed appliances, however, variable water flows are experienced and the use of a water governor is required. The governor usually forms part of the appliance and consists of a spring loaded valve acting on the underside of the diaphragm. As

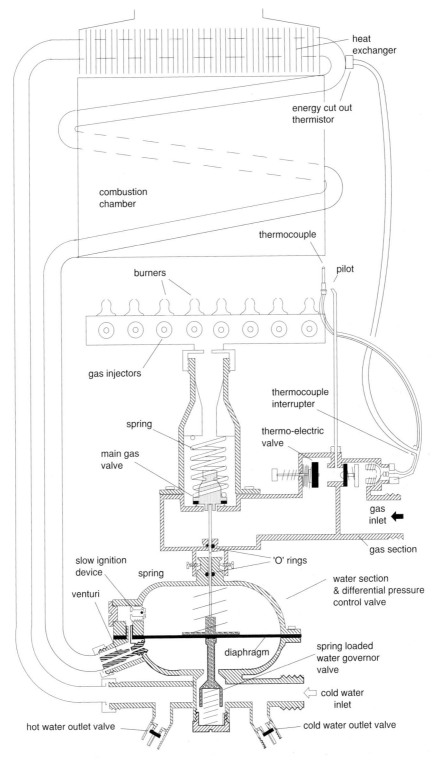

heat exchanger

energy cut out thermistor

combustion chamber

thermocouple

pilot

burners

gas injectors

thermocouple interrupter

thermo-electric valve

spring

main gas valve

gas inlet

'O' rings

gas section

slow ignition device

spring

venturi

water section & differential pressure control valve

diaphragm

spring loaded water governor valve

cold water inlet

hot water outlet valve

cold water outlet valve

**Instantaneous Water Heater
Shown With No Water Flowing Through Unit**

water flows into the heater it passes through and around this valve and on through the venturi, which allows the diaphragm to lift. If the water pressure is too high, the governor valve assumes its highest position, reducing the inlet flow through the seating. However, where a low pressure is experienced the strong spring acting on the gas valve tries to re-close the diaphragm and in so doing allows the governor valve to open further, enabling increased water flow.

Temperature control

The temperature of the water flowing out at the draw-off points is determined by the volume and speed at which the water flows through the unit. With some units this adjustment is made by a temperature selector located on the front, thus altering the volume of water flow through the selector and enabling the user to select variable temperatures for the outlet point. Some models have a water throttle in line, which restricts the water flow. As the appliance heats up, a thermostat operates and a volatile fluid is forced into the bellows chamber. This forces open the throttle, allowing a greater volume of water to flow and so the temperature is maintained. To prevent the unit overheating an energy cut off device is incorporated with most modern units. This is simply a thermistor, which is located on the top section of the heating unit and works in conjunction with a thermocouple interrupter (see page 108).

Slow ignition device

This is a device that controls the speed at which the diaphragm lifts and, consequently, the speed at which the gas flows into the burner. If the diaphragm lifts too fast, the gas will flow into the burner and be ignited with something of an explosive force. When it closes, the device responds rapidly to prevent the appliance overheating. There are several design of slow ignition device. The one described here is positioned in the low-pressure duct from the venturi. As the water flows from the space above the diaphragm, due to the action of the venturi, its velocity is restricted by a small ball bearing, blocking a central port hole, so the valve can only lift slowly as the water is drawn out. Conversely, when the cold supply is closed, the water flowing back to this space above the diaphragm can pass quite rapidly through the central hole as well as through the restricted space. The amount/rate of diaphragm lift can be adjusted by screwing the device further in or out of the housing until smooth ignition is achieved.

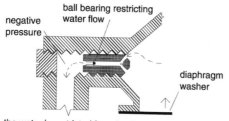

the water is restricted from flowing due to the position of the ball bearing, thus allowing for a slow lifting of the diaphragm and consequently a slow opening of the gas valve

action as water flows through the appliance

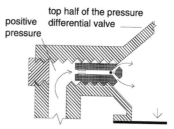

water can flow back quickly as the ball bearing no longer restricts the through way, thus the gas valve quickly closes

action as water stops flowing into the appliance

Operation of the Slow Ignition Device

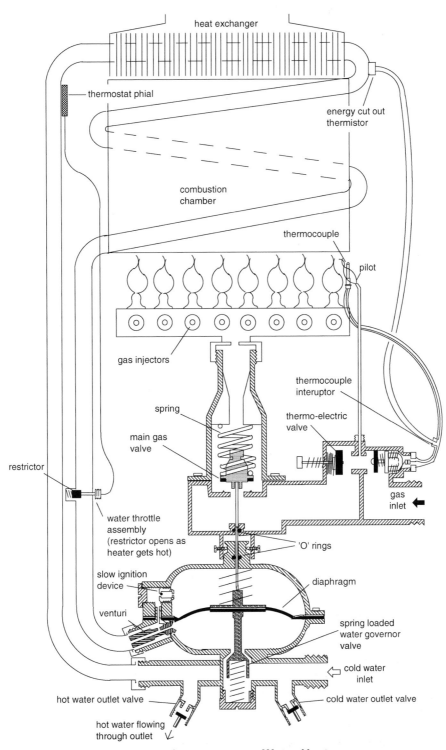

Instantaneous Water Heater
Shown with Water Flowing Through Unit Plus Additional Thermostat

9 Domestic Appliances

Gas Storage Water Heaters

There are several designs of gas storage water heater. The basic model consists of a cylinder in which a large volume of water is contained. At the base is an atmospheric gas burner. The gas products exit up through a central flue, passing right through the water chamber and are expelled outside. To assist the transference of heat, twisting baffles are positioned to direct the hot flue products on to the flue–water surface. The cylinder itself is well lagged to conserve heat. The gas storage heater may be either open flued or room sealed. However, apart from the fluing arrangement, the operating principle is the same for both. One major disadvantage of these units is their tendency to produce large volumes of condensation, because the flue passes through the comparatively cold water, cooling the flue products to the dew point of water. This condensate drops back down the central flue and causes major corrosion problems to the burner and base of the unit. The heat input to these appliances depends on the design and model chosen, however domestic appliances range from 5.5 kW to 25 kW, with water capacities of 75–115 litres.

Commercial Storage Water Heaters

The heat input of larger commercial appliances ranges to over 90 kW for units with atmospheric burners and they can have inputs of over 200 kW where blown gas is used. The larger appliances also generally have a multi-flue arrangement. Water storage capacities also cover a wide range to over 250 litres. High efficiency models are available with the burner positioned at the top forcing the products down and up through an extended flue way/heat exchanger, thereby extracting a larger amount of the latent heat from the flue products. Clearly, with this design an additional condensate pipe would be needed at the lowest point to remove the large volumes of condensation that are generated.

Water Supplies

The water supply to these appliances may be taken directly from the water supply main or alternatively a low-pressure system can be installed where the water is fed via a cistern. Mains pressure systems are referred to as an unvented supply and, where these are to be installed, the operative will need to hold the appropriate competency card before attempting to work on the appliance. A description of these systems and the controls used associated with the unvented system is outside the scope of this book, and therefore additional reading/research will be required to understand their operation fully.

Water Temperature

With the domestic sized unit no electrical supply is required and the temperature control is via a rod type thermostat connected directly into the multifunctional gas control block. For the unvented model an additional high limit thermostat would also be required.

The latest design of open flued appliance includes the addition of an atmospheric sensing device (ASD) within the draught diverter. This is designed to interrupt the thermocouple operation, closing down the gas supply in the event of excessive spillage.

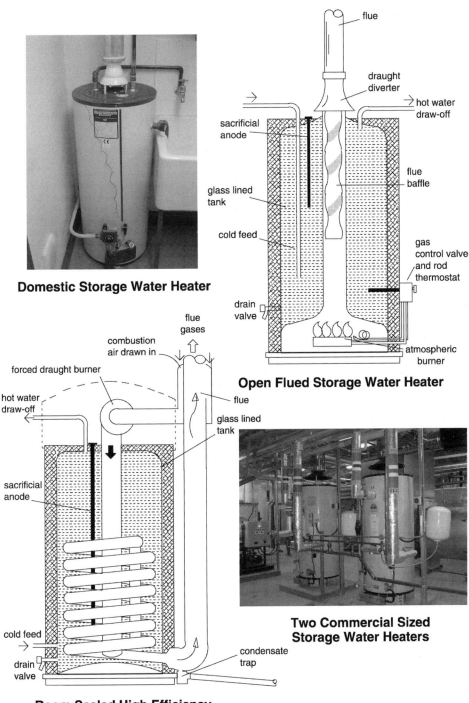

Domestic Storage Water Heater

flue

draught
diverter

hot water
draw-off

sacrificial
anode

flue
baffle

glass lined
tank

cold feed

gas
control valve
and rod
thermostat

drain
valve

atmospheric
burner

Open Flued Storage Water Heater

flue
gases

combustion
air drawn in

forced draught burner

flue

hot water
draw-off

glass lined
tank

sacrificial
anode

cold feed

drain
valve

condensate
trap

**Room Sealed High Efficiency
Storage Water Heater**

**Two Commercial Sized
Storage Water Heaters**

Installation of Water Heaters

Relevant ACS Qualification
WAT1

Relevant Industry Documents
BS 5546 and BS 6700

Location

If the appliance is to be room sealed, it may be installed in any location within reason, with the exception of an LPG appliance, which must not be located below ground or in a low lying area. Care needs to be observed to ensure that any compartment is prevented from getting too hot and clearly the terminal would have to be located in accordance with the manufacturer's instructions. If the appliance is fan assisted, there are special requirements in bath and shower rooms to ensure that all electrical components are inaccessible to anyone in the bath; the connection must also be via a fixed fused spur located outside the bathroom itself.

For open flued and flueless appliances greater care needs to be observed and Regulation 30 of the Gas Regulations in addition to the appropriate BS and manufacturer's instructions should be referred to. For example, it would not be permissible to install an open flued or flueless appliance in any of the following locations:

- bath or shower room;
- bedroom or bed sitting room where the appliance is more than 14 kW gross input; where it is less than 14 kW it would need to incorporate some form of atmospheric sensing device;
- for LPG – below ground level or basement type areas.

Rooms in which open or flueless water heaters are installed will also require some form of air vent, the size of which has already been discussed in Part 7, Ventilation.

Gas and Water Supplies

The positioning of a water heater depends on several factors, not least being the distance from the sanitary appliances that it is to serve. Should the distance be too great undue water cooling may be a problem. In general, the appliances should be within the distances from the heater listed below or they should be thermally insulated.

Table 7 Recommended maximum lengths of un-insulated hot water pipework

Maximum outside diameter	Maximum length in metres
12–22 mm	12
22–28 mm	8
>28 mm	3

Care should be taken, when installing a single point using an outlet spout or when fitting a tap, to ensure compliance with the Water Regulations, in particular the prevention of backflow or back-siphonage of water into the supply main. This is accomplished by maintaining an air gap of at least 20 mm as shown.

The gas supply to any water heater needs to be of fixed pipework and installed in accordance with the required 'installation practices'. Where a water heater is a new installation, supplementary bonding between the gas and water pipework may be required and should be undertaken by a competent operative. Where this work cannot be completed by the gas engineer, under the Gas Regulations the responsible person for the property should be informed, usually in writing, that it should be checked out by another competent person for electrical safety; the fact that no electrical connections have been made to the appliance is irrelevant.

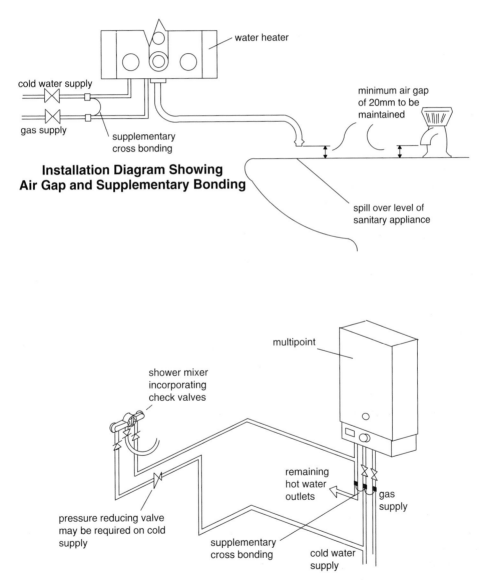

Installation Diagram Showing Air Gap and Supplementary Bonding

Installation Diagram for Shower Systems

Commissioning and Servicing of Water Heaters

Commissioning and Servicing Work Record

The Work Record opposite is purely given as a guide to the many tasks to be undertaken when servicing/maintaining and commissioning a water heater. The form can be used for both instantaneous and storage water heaters. In order to complete the form you may need to refer back to Part 8, page 264, where many of the tasks are explained in more detail. However additional 'appliance specific' checks on these appliances include some of the following activities:

Clean and Free Movement of Water Governor and Slow Ignition Device (Instantaneous Water Heaters Only)

This is simply undertaken when the water is isolated by removing the appropriate housing screw in the water section. In general, the working parts should move freely and, where appropriate, silicon based grease can be used. Other working parts in the water section should also be checked, including the free movement of the diaphragm and gas valve spindles. Care should be taken when replacing the slow ignition device to ensure that the ignition sequence is a smooth quiet operation, adjusting in or out as necessary. *Note*: With some models certain components are not removable.

Checking the flow rate and temperature rise

The flow rate is simply undertaken using a flow cup as shown over the page. The water volume is read from the scale as the water flows through the cup. The temperature rise should be taken first with the water running cold, then again at its allocated temperature, any adjustment in temperature rise being made in accordance with the manufacturer's instructions.

Checking the operation of an atmosphere sensing device

The manufacturer of the appliance may give instructions as to the method of testing the ASD, if one is fitted to an appliance. It is possible to position a metal plate on top of the heat exchanger of an instantaneous water heater or temporarily block the flue outlet to check the operation of a heat-sensing device mounted in the heat exchanger body or check the draught diverter to confirm that the thermocouple drops out within a specified period, usually 90 seconds.

Scale build-up

Scale can be a major problem with all direct water heaters, in particular instantaneous heaters, where the waterways can become blocked or reduced in size. Scale is the result of calcium carbonate deposits that are carried in suspension in hard water districts. Should the water be heated to a temperature in excess of 60°C the lime-scale, as it is called, is given up and deposited in the vessel. It is possible to de-scale the heat exchanger of an instantaneous water heater by slowly passing a proprietary descalent solution through the heater unit.

Certificate/Record of Space Heater Service/Commission		
Gas Installer Details	**Client Details**	**Appliance Date Badge Details**
Name :	Name :	Model/Serial No:
CORGI Reg. Nº :	Address :	Gas Type: Natural ☐ LPG ☐
Address :		Heat Input: max. . . kW min. . . kW
		Burner Pressure Range: . . .–. . . mbar
		Gas Council Nº/ CE Nº:. . . . .
Date:	**Appliance Location:**. . . . . **Install/Commission** ☐ **Service/Commission** ☐	

Preliminary System Checks Compliance with manufacturer's instructions	PASS ☐ FAIL ☐
General visual inspection of pipework	PASS ☐ FAIL ☐
Clearance from combustible materials	PASS ☐ FAIL ☐
Flue notice plate located and correctly filled in	PASS ☐ FAIL ☐ N/A ☐
Visual inspection of flue and flue flow performance test	PASS ☐ FAIL ☐ N/A ☐
Appliance level and secure	PASS ☐ FAIL ☐
Electrical connections	PASS ☐ FAIL ☐ N/A ☐
Bonding maintained	PASS ☐ FAIL ☐ N/A ☐
Fuse rating:. . . . . .amps	PASS ☐ FAIL ☐ N/A ☐
System tightness test; to include let by	PASS ☐ FAIL ☐ N/A ☐
Standing pressure of system	PASS ☐ FAIL ☐ N/A ☐
Appliance/system is purged of air	PASS ☐
Service/Commission Checks Clean primary air ports and lint arrestor	PASS ☐ FAIL ☐ N/A ☐
Clean/Check condition of injectors; burners	PASS ☐ FAIL ☐ N/A ☐
Condition of heat exchanger for scale build up and sacrificial anode	PASS ☐ FAIL ☐
Clean and free movement of water governor, slow ignition device & venturi	PASS ☐ FAIL ☐ N/A ☐
Check, ease and grease, if necessary, control taps	PASS ☐ FAIL ☐ N/A ☐
Ignition devices effective including condition of electrodes, leads	PASS ☐ FAIL ☐ N/A ☐
Clean and check operation of fans	PASS ☐ FAIL ☐ N/A ☐
Pilot flame correct	PASS ☐ FAIL ☐ N/A ☐
Burner pressure. . . . . .mbar	PASS ☐ FAIL ☐
Total heat input. . . . . .kW	PASS ☐ FAIL ☐ N/A ☐
Flame picture good	PASS ☐ FAIL ☐ N/A ☐
Flame supervision device operational	PASS ☐ FAIL ☐ N/A ☐
Atmosphere sensing device operational	PASS ☐ FAIL ☐ N/A ☐
Flow rate: . . . . . . .l/s Temperature rise:. . . . . . .°C	PASS ☐ FAIL ☐
Condition of frame and combustion seals effective	PASS ☐ FAIL ☐ N/A ☐
Appliance tightness check	PASS ☐ FAIL ☐ N/A ☐
Spillage tests	PASS ☐ FAIL ☐ N/A ☐
Operating thermostat correct	PASS ☐ FAIL ☐ N/A ☐
Flue guard fitted to low level terminals	PASS ☐ FAIL ☐ N/A ☐
Additional direct ventilation grill required:. . . . . .cm^2	PASS ☐ FAIL ☐ N/A ☐
High level compartment ventilation where required:. . . . . .cm^2	PASS ☐ FAIL ☐ N/A ☐
Low level compartment ventilation where required:. . . . . .cm^2	PASS ☐ FAIL ☐ N/A ☐
Post System Checks Meter working pressure (. . . . . mbar)	PASS ☐ FAIL ☐ N/A ☐
Working pressure drop across system (max. . . . . . .mbar)	PASS ☐ FAIL ☐ N/A ☐
Safe operation of appliance explained to customer	YES ☐ NO ☐

Recommendations and/or Urgent Notification

Appliance Safe to Use   YES ☐ NO ☐   **Next Service Due:**. . . . .

Installer's Signature: **Customer's Signature:**

Alternatively the solution can be made up of 10 parts of water to 1 part of hydrochloric acid. Descaling is carried out as follows:

1. Drain the heater and remove the heating unit.
2. Turn the heater upside down and connect to an acid resisting tank with suitable rubber hose, as shown.
3. Fill the container with the acidic descalent solution. Ideally the water should be hot. Add the acid to the water and NOT the water to the acid.
4. Open the control cock to allow the liquid to flow slowly through the unit into another collecting tank, continue until the solution stops bubbling; it may be necessary to replenish the solution.
5. Thoroughly flush out the unit before reinstating it to the water heater.

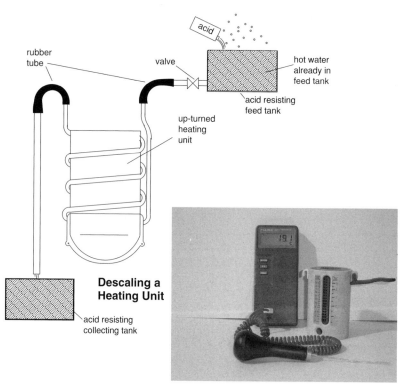

Descaling a Heating Unit

Flow Cup and Digital Thermometer

Sacrificial anode check

The sacrificial anode is simply a rod of magnesium positioned in the top of a storage water heater. It will corrode as the result of electrolysis before any other metal in the system. Electrolysis is the destruction of one metal due to the chemical reaction of another. The anode's condition can be simply checked by withdrawing it from the vessel after isolating the water supply. If it is extensively corroded it is simple to replace.

Instantaneous water heater fault diagnosis chart

Fault	Possible cause
Pilot will not light	• Gas supply turned off • Air in pipe • Pilot injector blocked or incorrect size • Incorrect spark gap • No spark or lead not connected properly
Poor pilot flame and will not stay alight	• Thermocouple connection loose • Energy cut-out connection loose • Pilot flame too small or pilot tube blocked • Faulty thermocouple to thermo-electric valve • Faulty thermal switch or energy cut out device • Inadequate ventilation
Main Burner will not light	• Low water flow rate, e.g. blocked filter • Low gas pressure • Faulty diaphragm • Gas valve push rod jammed • Slow ignition device incorrectly set or stuck
Poor water flow rate	• Blocked filter • Scaled heat exchanger • Poor inlet water supply • Water governor sticking
High water flow rate	• Faulty diaphragm or water governor sticking • Gas valve push rod sticking
Low water temperature	• Gas pressure too low • Faulty diaphragm or water governor sticking • Gas valve push rod sticking • Slow ignition device incorrectly set
Noisy heater	• Scaled heat exchanger • Incorrectly set slow ignition device • Burner ports blocked
Smells	• Faulty case or flue seals • Flueless heater and failure to open window • Newness of appliance

9 Domestic Appliances

Domestic Gas Boilers

Relevant ACS Qualification
CEN1

Relevant Industry Document
BS 6798

There are two types of hot water boiler: the open flued and the room sealed appliance. Since the introduction of the Boiler Efficiency Regulations of 1995 and its subsequent amendments the natural draught open flued boiler fails to meet the requirements as laid down and, apart from a few replacement back boilers, these are no longer installed. Open and room sealed appliances are discussed in more detail in Part 6, Flues.

SEDBUK Efficiency Rating

SEDBUK is an acronym for Seasonal Efficiency of Domestic Boilers in the UK. The SEDBUK rating is the average annual efficiency obtained in a typical domestic situation for any particular boiler. The efficiency of an individual boiler is classified by the letters A to G.

Band	SEDBUK rating
A	90–94%
B	86–90%
C	82–86%
D	78–82%
E	74–78%
F	70–74%
G	<70%

Typical seasonal efficiencies

Old boiler (heavyweight)	55%
Old boiler (lightweight)	65%
New boiler (non-condensing)	75%
New boiler (condensing)	88%

By 2005 nearly all new installations and replacement boilers, with only a few exceptions, will need to satisfy the requirements of band 'A' or 'B' and to meet this they will need to be of the high efficiency or condensing types.

SAP Rating

SAP is an acronym for Standard Assessment Procedure. This is the Government's standard methodology for home energy rating. The SAP rating currently runs from 1 to 120 (1 being the least efficient and 120 being extremely energy efficient). Scores above 80 are generally to be sought for an energy efficient home. All new dwellings and conversions need to be SAP assessed and, as mentioned above, require a high efficiency boiler.

Boiler Type and System Design

There are many central heating and hot water system designs, many more than this book can adequately describe. For more information further reading is required. The system illustrated opposite shows an example of a modern design. The water in the system may be open to the atmosphere and fed via a feed and expansion vessel, called an open system, or the boiler may be fed directly from a mains supply, via a temporary filling loop and incorporate a sealed expansion vessel, in which case the system is called a closed, or sealed system.

The boiler serving the system can be one of several designs:

- a conventional boiler;
- a combination boiler;
- a condensing boiler.

When upgrading/replacing an existing boiler in order to comply with the Building Regulations the system design must be fully pumped, with thermostatic radiator valves (TRVs) fitted to all radiators except those located within the same room as the room thermostat. It may also be necessary to replace the domestic hot water cylinder with one that is more efficient. The system must be fully interlocked so that the boiler will not fire unless heat is called for.

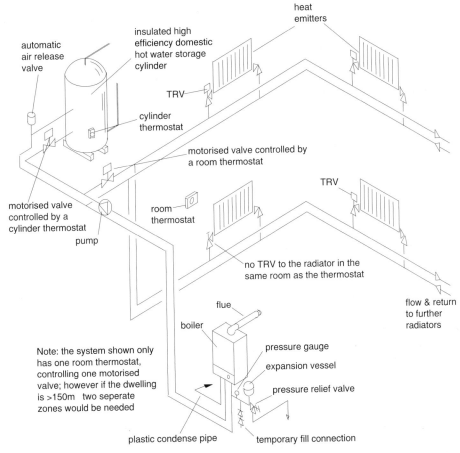

Fan Flued Condensing Boiler to a Sealed Heating System

Conventional or Regular Gas Boilers

The traditional gas boiler consists of a multifunctional gas valve, a burner/fire box, a heat exchanger and the combustion products collection point/connection to the flue system. Some models incorporate a fan to assist in the expulsion of the combustion products and the unit may be either open flued or room sealed. The boiler may be free standing, wall mounted or incorporated with a gas fire and installed within the sitting room, located within the chimney catchment space, hidden behind the fire. Sometimes very small heaters, designed just to heat the domestic hot water, are installed; they are referred to as circulators. They may be independent but are often incorporated with a warm air unit.

The Heat Exchanger
The heat exchanger consists of a chamber through which the hot combustion products pass on their way to the flue system. It consists of a series of water-ways. As the heat of the gases pass through the small spaces between the walls, the water is heated as the metal warms up. The heat exchanger is the heart of any boiler and the key to its efficiency. The larger the amount of heat extracted from the combustion products, as they pass through the heat exchanger, the more efficient is the boiler. Some heat exchangers consist of heavy cast iron chambers. These have excellent qualities in terms of life expectancy, however they tend to hold the heat and therefore limit the amount of heat transfer. More modern alternatives use thinner-walled materials such as stainless steel or aluminium that transfer the heat rapidly through the walls. These thinner materials also allow for greater wetted surface areas to be exposed to the hot flue products. When initially commissioning a boiler, the return water temperature needs to be such that it is not particularly cool because if water temperatures of below 55°C are encountered (typical dew point) condensation will form on the outside of the combustion chamber walls giving rise to excessive corrosion problems. These basic boilers have no provision for this water accumulation. Condensing boilers, as seen on page 340, do not have this water accumulation problem due to their design.

Heat Input/Heat Output
The amount of heat that is put into the appliance due to the combustion of fuel is not the amount of heat that is available for use by the system it serves. Clearly some heat will be lost through the flue system and from the appliance itself. When commissioning an appliance the gas engineer is primarily concerned with the heat input. It is the designer or central heating installer who is concerned with the heat output, i.e. what heat is actually available for use. Thus when 'selecting' a boiler for a particular purpose it is the heat output available that should be important. Conversely when 'commissioning' an appliance the gas installer confirms the heat input by calculating the gas consumption used over a period of time. See also 'Gross Rates and Heat Input' on page 56.

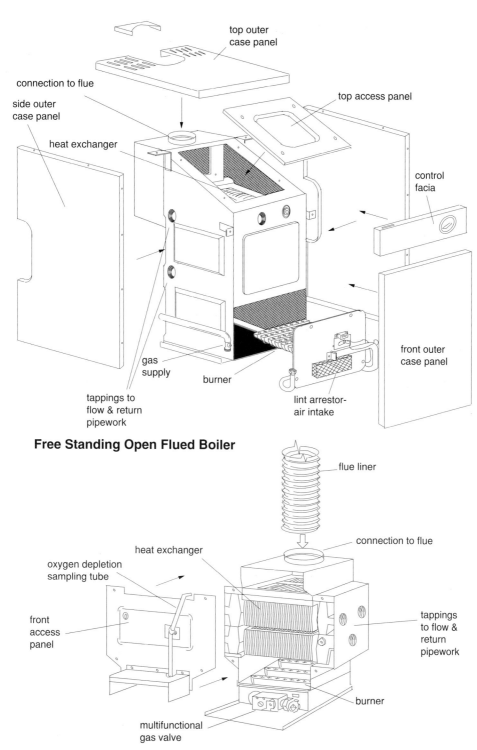

top outer
case panel

connection to flue

side outer
case panel

heat exchanger

top access panel

control
facia

gas
supply

burner

tappings to
flow & return
pipework

lint arrestor-
air intake

front outer
case panel

Free Standing Open Flued Boiler

flue liner

connection to flue

heat exchanger

oxygen depletion
sampling tube

front
access
panel

tappings
to flow &
return
pipework

burner

multifunctional
gas valve

Back Boiler Unit

The Combination Boiler

Unlike the conventional/traditional gas boiler, the combination boiler is usually sold as a complete unit, incorporating all the components required to operate and control the domestic hot water (d.h.w.) and central heating (c.h.) systems. The heart of the combination boiler lies in its ability to warm the d.h.w. water instantly. It does this by diverting the hot water flow from the central heating system and temporarily passes it through a heat exchanger that, in turn, rapidly warms the d.h.w. to the desired temperature. Because the water is heated instantaneously, a saving can be made in that no stored domestic hot water is required. However, in relation to property size, the following need to be considered:

1. the size of the incoming water supply, i.e. is it large enough to feed all the cold and hot water outlets within the dwelling, in terms of pressure and flow, and
2. it must be remembered that while the domestic hot water is being heated, no warming of the central heating will take place.

Backflow Of Domestic Hot Water Into The Water Companies Service Main
In order to prevent the domestic hot water that expanded on heating from passing back into the water authority supply, a non-return or check valve is fitted to the supply pipe prior to the appliance. Any expansion may also cause additional forces within the pipe that may lead to possible damage to the pipework or the appliance itself. Therefore it may be necessary to install a small expansion vessel in the pipe; this chamber is approximately 80 mm in diameter and about 60 mm deep. Sometimes the manufacturer incorporates this in the design, however a check needs to be made.

Operation of the Combination Boiler
A schematic section through a combination boiler is shown opposite. It operates as follows.

Central heating mode
If the time clock and thermostat call for heat, the pump runs. This allows the water to flow through primary pipework within the boiler and eventually out into the central heating circuit. As the water flows past the central heating flow switch within the boiler it allows the exhaust fan to operate, which initiates the gas flow. This in-turn generates the flame within the combustion chamber. The boiler continues to run until the thermostat is satisfied.

Domestic hot water mode
When water is drawn off from a tap, water flows through the boiler and secondary heat exchanger and in so doing activates the d.h.w flow switch. This brings on the pump and operates the three-way control to allow water to divert and flow through the heat exchanger. With water flowing round the boiler within the primary circuit the c.h. flow switch allows the boiler to fire up, as above. Heat flows round within the boiler and, as it passes over the secondary heat exchanger, warms the water instantaneously. Note, as stated above, the d.h.w has precedence over the c.h. and, as a consequence, if the c.h. and d.h.w. were activated at the same time, the c.h. would temporarily cease.

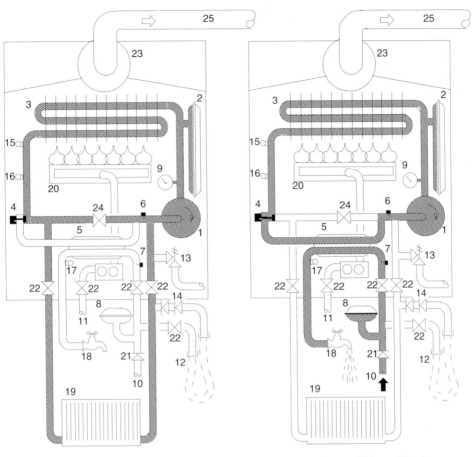

Central Heating Mode

Domestic Hot Water Mode

1. pump
2. primary expansion vessel
3. primary heat exchanger
4. three way control valve
5. secondary heat exchanger
6. c.h. flow switch
7. d.h.w. flow switch
8. secondary expansion vessel
9. pressure gauge
10. cold supply main
11. gas supply
12. temporary connection
13. pressure relief valve
14. double check-valve
15. high limit thermostat
16. c.h. thermostat
17. d.h.w. thermostat
18. hot draw-off point
19. heat emitter
20. burner
21. non-return valve
22. isolation valve
23. exhaust fan
24. by-pass
25. flue

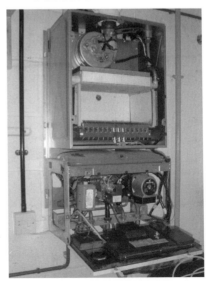

**Combination Boiler
Exposing Burner and Heat Exchanger**

The Condensing Boiler

The condensing boiler is a highly efficient appliance that extracts as much of the heat from the flue gases as possible, rather than allow it to be dispersed to the external environment and waste fuel. In order to gain the maximum efficiency from a condensing appliance, it is essential that the system is designed to operate at lower water temperatures. Systems such as those relying on cooler flow and return temperatures prove to be the most effective, such as those using radiant heating, with pipes embedded within the walls and floors. Where radiators are used, these should be larger than those traditionally fitted in the 1970s and 1980s, working on cooler temperatures. Alternatively the return to the boiler should have a greater temperature differential, as much as 17°–20° lower than the flow. *Note*: A forced draught burner is invariably used to give improved efficiency.

Principle of Operation
When gas is burnt, water vapour (H_2O) is produced as a result of the combustion process (see page 28). In the more traditional/conventional appliance this water vapour is dispelled from the appliance along with the other combustion products. However, due to the design of the combustion chamber in a condensing boiler, consisting of a very tight network of closely fitted waterways and baffles or in some models two individual heat exchangers, the flue gases are cooled extensively down to temperatures, typically of around 35°–50°C, often allowing for a plastic flue. Water vapour condenses to its liquid form at a temperature of about 55°C, therefore water forms within the appliance and runs down the inside, collecting in a condensate trap in the base of the unit. Condensate will also form within the flue system and, as a consequence, the flue will need to be routed in an uphill direction so that this water can also drain back to the condensate trap. Clearly, for such an effective heat exchanger to work, the flue products will need to be expelled from the appliance by the use of a fan draught flue system.

Disposal of the Condensate
The condensate collected in the trap at the base of the appliance discharges to a drain or soakaway when sufficient volume has collected. Discharging as a volume rather than a continued dripping effect prevents the water freezing within the discharge pipe to the drain. Because of the nature of the condensate (slightly acidic), copper pipe is not recommended for the condensate drain and plastic materials are used. The drain should have a minimum fall of 2.5°.

Pluming Effects
Because the water content of the flue gases have condensed or are close to the point where condensation occurs, a white vapour is often seen discharging from the flue terminal; this is referred to as pluming. It can be seen at all times of the year, however it tends to be more pronounced when the weather is cooler. Because of the nuisance factor involved, particularly with neighbours, the site of a flue terminal should be carefully chosen.

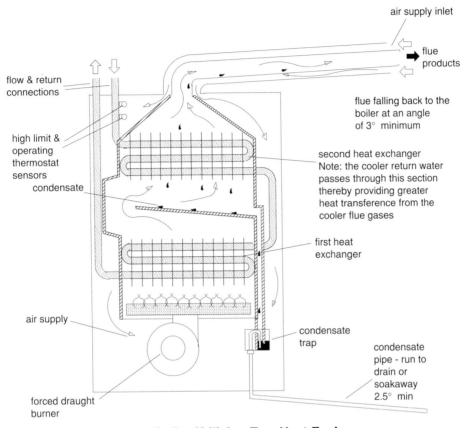

air supply inlet

flue products

flow & return connections

flue falling back to the boiler at an angle of 3° minimum

high limit & operating thermostat sensors

condensate

second heat exchanger
Note: the cooler return water passes through this section thereby providing greater heat transference from the cooler flue gases

first heat exchanger

air supply

condensate trap

condensate pipe - run to drain or soakaway 2.5° min

forced draught burner

Condensing Boiler Utilising Two Heat Exchangers

Condensing Boiler Utilising One Heat Exchanger

Installation of Domestic Gas Boilers

Relevant ACS Qualification   Relevant Industry Document
            CEN1   BS 6798

Restricted Locations

Open flued appliances are totally restricted from showers and bathrooms, and for bedrooms and bed-sits no open flued appliance over 12.7 kW net input (14 kW gross) is allowed; this includes adjoining cupboards. Therefore, these are not places where the boiler should be sited. If the appliance is room sealed, then no such ruling exists and the appliance has no prohibited locations. LPG boilers, however, are not allowed to be located in basements and similar positions where any possible gas escapes could allow gas to build up. If a room sealed boiler is located in a bathroom, the electrical switch controls must be mounted outside.

Compartments

A very common place to locate the boiler is within a compartment. A compartment is not a specific size, it may be small or large, depending on the size of the appliance. In general it is a purpose-built rigid structure in which nothing else is used or stored. Sometimes an airing cupboard is used, however the boiler needs to be separated from the clothes by a perforated partition. This may consist of expanded metal with perforations no larger than 13 mm. Where a flue passes up through the clothes section, this too must be protected, with a minimum 25 mm air space between the pipe flue and contents of the cupboard. The compartment should not be located under the stairways of buildings higher than two storeys. Some of the main requirements, in the absence of manufacturer's specific instructions, for a compartment include the following:

- Internal surfaces need to be 75 mm away from the boiler, unless non-combustible.
- Adequate ventilation must be provided and open flued appliances must not communicate with a bathroom or bedroom, see above.
- Adequate space must be provided to service the boiler and permit its removal if necessary.
- A notice should be located at a suitable position warning against storage.

Roof Spaces

Where a boiler is to be located in a roof void, specific additional installation points need to be considered, including the following:

- A permanent means of easy access (e.g. a ladder) must be provided and the area around the loft opening should have a guard fitted.
- A suitably floored route to the boiler must be provided, with sufficient area for servicing, as necessary.
- Items stored in the roof void must not be allowed to be in contact with the boiler.
- Within the loft fixed lighting must be installed and a means of isolating the boiler from inside and from outside the roof space provided.

Flue Termination

Most domestic gas boilers installed today are of the room sealed type, therefore the flue requirements are generally quite specific as laid down by the manufacturer of the appliance. However, where a fan assisted open flued appliance is installed or one wants to confirm the correct siting of an existing installation terminal, the points discussed in Part 6 should be referred to.

Ventilation

The ventilation requirements for a boiler depends on its type and location, if it is installed in a compartment, see the note opposite. Where it is room sealed, no ventilation is necessary and if it is open flued it requires 5 cm² for every kilowatt in excess of 7 kW (see page 248).

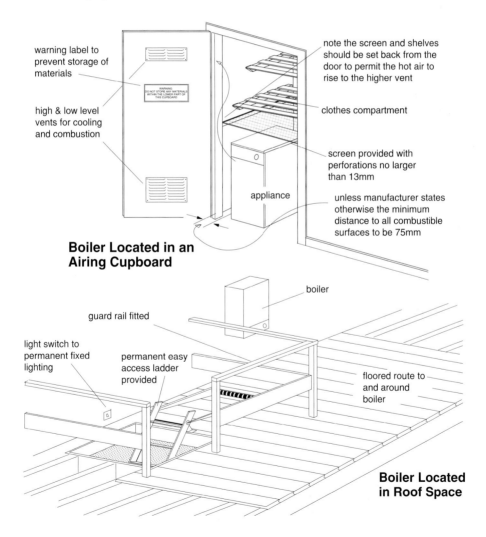

warning label to prevent storage of materials

high & low level vents for cooling and combustion

WARNING
DO NOT STORE ANY MATERIALS
WITHIN THE LOWER PART OF
THIS CUPBOARD

note the screen and shelves should be set back from the door to permit the hot air to rise to the higher vent

clothes compartment

screen provided with perforations no larger than 13mm

appliance

unless manufacturer states otherwise the minimum distance to all combustible surfaces to be 75mm

Boiler Located in an Airing Cupboard

boiler

guard rail fitted

light switch to permanent fixed lighting

permanent easy access ladder provided

floored route to and around boiler

Boiler Located in Roof Space

9 Domestic Appliances

Commissioning and Servicing Gas Boilers

Commissioning and Servicing Work Record

The Work Record opposite is given purely as a guide to the many tasks to be undertaken when servicing/maintaining and commissioning a boiler. In order to complete the form you may need to refer back to Part 8, page 264, where many of the tasks were explained in more detail. However, the additional 'appliance specific' checks on boilers include some of the following activities.

Notice plate located and correctly filled in See page 197.

Check condition of the heat exchanger and sweep through On many of the modern heat exchangers that use low water content heat exchangers, greater care is required not to damage the delicate fins, which may require no more than a gentle brushing. With the older more robust cast iron heat exchangers, a selection of stiff brushes can be passed through the waterways in the hope of removing any loose rust, etc.

Gas/Air pressure differential Some high efficiency boilers require the setting of the gas pressure to be made in conjunction with the pressure generated by the air-flow generated by the fan. Where such a pressure reading is required, the manufacturer of the appliance will give clear instructions how to obtain it, explaining where and how to make any adjustments if necessary.

Atmosphere sensing device operational There is no laid down procedure for checking the correct operation of an atmosphere-sensing device located in an open flued boiler. Some systems rely on oxygen depletion, whereas others rely on sensing the temperature at the draught diverter (see page 108). However, as an installer/service engineer you can check the accumulation of dust, etc. around the sensing points and check that the components are correctly and securely in place.

Condensate trap cleaned and condensate pipe effective For the condensating type of boiler, the condensate trap may need to be removed in order to give it a thorough clean; debris may have accumulated inside, which may lead to a blockage. The discharge pipe leading to the outside should also be checked to ensure that it is not damaged and still operates effectively.

Condition of frame and combustion seals effective Generally this requires no more than a good visual inspection of all joints for corrosion or poor sealing/gasket material. However, where the appliance case is subject to positive fan pressure, a lighted match or taper should be positioned close to all the seals with the appliance fan running to check that there is no leakage of air or combustion products.

Pressure relief valve effective For sealed systems the pressure relief valve test lever should be operated to confirm that the valve opens and discharges the water safely into the discharge pipe and closes effectively on completion. The discharge pipework should also be inspected for damage and correct location.

Certificate/Record of Hot Water Boiler Service/Commission		

Gas Installer Details	Client Details	Appliance Date Badge Details
Name :	Name :	Model/Serial No:
CORGI Reg. N° :	Address :	Gas Type: Natural ☐ LPG ☐
Address :		Heat Input: max.....kW min.....kW
		Burner Pressure Range: ... –....mbar
		Gas Council N°/ CE N°:..................

Date:	**Appliance Location**:.. **Install/Commission** ☐ **Service/Commission** ☐	

Preliminary System Checks Compliance with manufacturer's instructions	PASS ☐ FAIL ☐
General visual inspection of pipework	PASS ☐ FAIL ☐
Clearance from combustible materials	PASS ☐ FAIL ☐
Visual inspection of flue	PASS ☐ FAIL ☐ N/A ☐
Flue notice plate located and correctly filled in	PASS ☐ FAIL ☐ N/A ☐
Flue flow performance test	PASS ☐ FAIL ☐ N/A ☐
Electrical connections	PASS ☐ FAIL ☐ N/A ☐
Bonding maintained	PASS ☐ FAIL ☐ N/A ☐
Fuse rating:...........amps	PASS ☐ FAIL ☐ N/A ☐
System tightness test; to include let by	PASS ☐ FAIL ☐ N/A ☐
Standing pressure of system	PASS ☐ FAIL ☐ N/A ☐
Appliance/System is Purged of Air	PASS ☐ FAIL ☐ N/A ☐
Service/Commission Checks Clean primary air ports and lint arrestor	PASS ☐ FAIL ☐ N/A ☐
Clean/check condition of injectors; burners	PASS ☐ FAIL ☐ N/A ☐
Check condition of heat exchanger and sweep through	PASS ☐ FAIL ☐
Ignition devices effective including condition of electrodes, leads	PASS ☐ FAIL ☐ N/A ☐
Clean and check operation of fans	PASS ☐ FAIL ☐ N/A ☐
Pilot and main burner flame correct	PASS ☐ FAIL ☐ N/A ☐
Burner pressure.........mbar	PASS ☐ FAIL ☐ N/A ☐
Gas/Air pressure differential.........mbar	PASS ☐ FAIL ☐ N/A ☐
Maximum heat input.............kW	PASS ☐ FAIL ☐ N/A ☐
Flame supervision device operational	PASS ☐ FAIL ☐ N/A ☐
Atmosphere sensing device operational	PASS ☐ FAIL ☐ N/A ☐
Condensate trap cleaned and condensate pipe effective	PASS ☐ FAIL ☐ N/A ☐
Combustion seals effective	PASS ☐ FAIL ☐ N/A ☐
Appliance tightness check	PASS ☐ FAIL ☐ N/A ☐
Spillage tests	PASS ☐ FAIL ☐ N/A ☐
Flue gas analysis with good combustion (printout attached Yes ☐ No ☐)	PASS ☐ FAIL ☐ N/A ☐
Operating thermostat correct	PASS ☐ FAIL ☐ N/A ☐
Flue guard fitted to low level terminals	PASS ☐ FAIL ☐ N/A ☐
Pressure relief valve effective	PASS ☐ FAIL ☐ N/A ☐
Additional direct ventilation grill required:.............cm²	PASS ☐ FAIL ☐ N/A ☐
High level compartment ventilation where required:.............cm²	PASS ☐ FAIL ☐ N/A ☐
Low level compartment ventilation where required:.............cm²	PASS ☐ FAIL ☐ N/A ☐
Post System Checks Meter working pressure (.............mbar)	PASS ☐ FAIL ☐ N/A ☐
Working pressure drop across system (max.............mbar)	PASS ☐ FAIL ☐ N/A ☐
Safe operation of appliance explained to customer	YES ☐ NO ☐
Central Heating Checklist Completed	YES ☐ NO ☐

Recommendations and/or Urgent Notification Benchmark Logbook completed YES ☐ NO ☐	
Appliance Safe to Use YES ☐ NO ☐	**Next Service Due:**.............
Installer's Signature:	**Customer's Signature:**

Central Heating Checklist Completed In addition to the boiler service or installation it is always good practice to use this opportunity to inspect the condition of the central heating system and pipework. As with all tasks a checklist such as that shown below could be completed.

Inspection Record of Hot Water Central Heating System		
Gas Installer Details	**Client Details**	**System Details**
Name :	Name :	Open System ☐ Closed System ☐
CORGI Reg. N° :	Address :	Fully Pumped ☐ Gravity Primaries ☐
Address :		Conventional Boiler ☐
		Combination Boiler ☐
		Condensing Boiler ☐
Date:	This inspection completed in conjunction with a check of the gas installation of the boiler: **Yes** ☐ **No** ☐ Boiler location:..	

Components Inspected	Notes
Boiler: location acceptable	PASS ☐ FAIL ☐
Pipework: no leaks, secure and in accordance with good practices	PASS ☐ FAIL ☐
Pump: Speed correct: (Flow and return temperature differential........°C)	PASS ☐ FAIL ☐ N/A ☐
TRVs fitted to all radiators except rooms with the thermostats	Yes ☐ No ☐
System balanced	PASS ☐ FAIL ☐ N/A ☐
No pumping over air drawing in air from F & E	PASS ☐ FAIL ☐ N/A ☐
Pump noise minimal	PASS ☐ FAIL ☐ N/A ☐
Motorised valve operational	PASS ☐ FAIL ☐ N/A ☐
Radiator/Heat emitter valves operational	PASS ☐ FAIL ☐ N/A ☐
Sealed system at correct pressure (sealed/closed system)	PASS ☐ FAIL ☐ N/A ☐
Temporary filling loop disconnected (sealed/closed system)	PASS ☐ FAIL ☐ N/A ☐
Expansion vessel at correct pressure (sealed/closed system)	PASS ☐ FAIL ☐ N/A ☐
Pressure relief operational at designed pressure (sealed/closed system)	PASS ☐ FAIL ☐ N/A ☐
Discharge from pressure relief at safe location (sealed/closed system)	PASS ☐ FAIL ☐ N/A ☐
Automatic air admittance valves working freely	PASS ☐ FAIL ☐ N/A ☐
F and E cistern adequately insulated	PASS ☐ FAIL ☐ N/A ☐
F & E water level adjusted and float operated valve functioning	PASS ☐ FAIL ☐ N/A ☐
Overflow pipe secure & at visible location	PASS ☐ FAIL ☐ N/A ☐
Condition of Water within system showing no major corrosion	PASS ☐ FAIL ☐ N/A ☐
Removed air from system	Yes ☐ No ☐
System flushed	Yes ☐ No ☐
Corrosion inhibitor added	Yes ☐ No ☐
Programmer and time clock correctly adjusted	Yes ☐ No ☐
Room thermostats, cylinder and frost thermostats operational	Yes ☐ No ☐
All other external controls operational	Yes ☐ No ☐
Condensating pipe correctly installed	PASS ☐ FAIL ☐ N/A ☐

Recommendations and/or Urgent Notification

Installer's Signature:	**Customer's Signature:**

Fault Diagnosis

Many of the faults are associated with the modern fan-assisted appliances using printed circuit boards (PCBs). To assist fault diagnosis quite extensive fault finding charts will be found at the back of the manufacturer's installation instructions, taking you through the various options available in diagnosing why a boiler will not work. Some manufacturers have even gone that bit further and have incorporated a point to enable the connection of a computer that tells you exactly what is required for the operation of the boiler. The following chart covers a few simple common faults.

Basic fault diagnosis chart

Fault	Possible cause
No flame will establish within the boiler, e.g. pilot flame or main burner in the case of electronic ignition	• Gas or electrical supply turned off • Faulty fuse • Air in pipe • Pilot injector blocked or incorrect size • Incorrect spark gap • No spark or spark lead not connected properly • Faulty fan or tubes to pressure switch • Faulty pressure switch or poor connections
Poor pilot flame and will not stay alight	• Thermocouple connection loose • Pilot flame too small or pilot tube blocked • Faulty thermocouple of thermo-electric valve • Inadequate ventilation • Defective atmosphere sensing device or oxygen depletion tube
Poor heat output and lack of heat from the appliance	• Inadequate gas pressure • Blocked gas injectors • Incorrectly fitted gas injectors • Faulty thermostat
Poor flame picture	• Incorrect gas pressure • Insufficient air intake or lint arrestor blocked • Blocked heat exchanger • Poor or blocked flue system • Incorrect injectors • Damaged burner
Noisy Boiler	• Gas pressure too high • Pump speed incorrect • Flames impinging onto heat exchanger • Scale build up within boiler • Bypass insufficiently open • Loose screws in boiler casing or incorrectly fitted case

9 Domestic Appliances

Domestic Ducted Warm Air Heaters

Relevant ACS Qualification
DAH1

Relevant Industry Document
BS 5864

A warm air unit consists of an enclosed burner, around which air is passed through a heat exchanger, assisted by the draught created by a fan. The warmed air is then circulated around the building, through a system of ductwork to be discharged into the various rooms via register grilles. The cooler air returns to the warm air heater simply by being sucked from the room via a second grille in the wall leading back via the passageway or hall. Some older systems worked by allowing circulation by convection currents, but these systems are now quite antiquated. There are three types of fan-assisted warm air unit, each illustrated opposite, these include:

• the down-flow unit;
• the up-flow unit;
• the horizontal unit.

Some warm air units, referred to as 'Modairflow' or Even Temperature (ET) are designed to give variable heat outputs, thus saving fuel. Basically, used in conjunction with a thermistor type room thermostat, they will bring on the firing of the warm air unit only if needed and will turn off the burner intermittently, simply circulating the warm air at a reduced fan flow rate.

Some units incorporate a circulator to provide a hot water supply system, or to be used as additional background heating. Where this is the case, the circulator operates as an individual appliance within the unit, however it shares the same flue system. If a circulator is incorporated with a warm air unit, it is essential that sufficient ventilation is provided to serve both appliances.

Warm Air Duct System

Various ducting systems are outside the scope of this book. However, as a brief introduction, four typical design layouts are illustrated opposite: the stub duct, the radial duct, the extended plenum and the stepped duct.

As warm air leaves the heat exchanger of the appliance, it collects inside a box-shaped plenum chamber. This plenum is designed to equalise the air pressure inside and distribute it to the various supply ducts. For the down-flow unit the plenum chamber needs to be strong enough to support the weight, as it stands on the floor on which the heater is placed.

Return Air Ducting

The return air is circulated back to the hall, as stated above, and passes back into the heater, via a filter, for reheating. Where the heater is located in a compartment it is essential that the return air is suitably ducted from outside the compartment to the return air inlet to ensure that the operation of the fan, which creates negative pressure, does not adversely affect the safe operation of the flue.

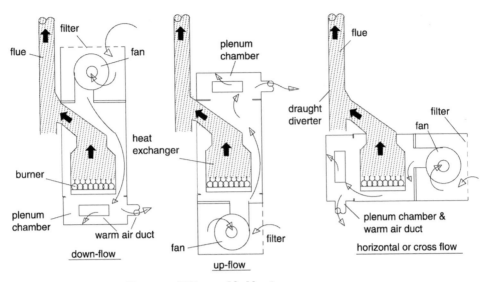

Types of Warm Air Heater

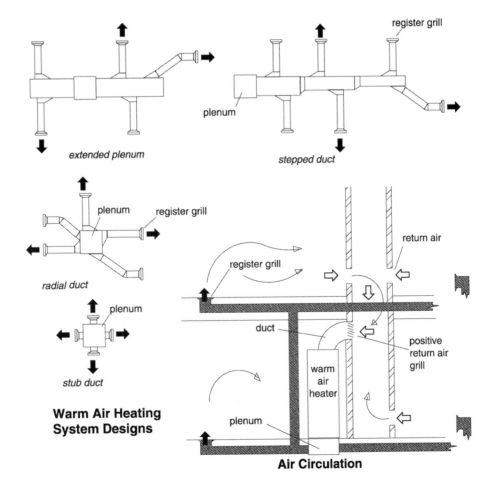

Warm Air Heating System Designs

Air Circulation

Installations of Warm Air Heaters

Warm air units are installed as either freestanding, slot-fix or within a compartment; the latter tends to be the most common. The freestanding model is often made to suit the height of the room by using a top closure set and is designed to be located back to the wall or, ideally in a corner. Slot-fix models are specially designed to be located within a purposely-designed area between two surfaces that protect the sides. Only appliances that are designed for slot-fix applications can be fitted in this way.

Restricted Locations

The restricted locations for warm air heaters as the same as those for boilers as described on page 342.

Compartments

Where the warm air unit is installed within a compartment the following points need to be considered:

- The internal surfaces should be 75 mm from the heater, unless non-combustible.
- Adequate ventilation must be provided, both for cooling and combustion air (see page 246) and open flued appliances must not communicate with a bathroom or bedroom, see Restricted Locations for Boilers, page 342.
- Adequate space must be provided to service the heater and permit its removal if necessary.
- Return air grilles must be permanently connected via a suitable duct to the return air inlet sited outside the compartment.
- A notice should be located at a suitable position warning against storage.

Sometimes an airing cupboard is used, however the heater needs to be separated from the clothes by a suitable perforated partition. This may be of expanded metal with perforations no larger than 13 mm. Where a flue passes up through the clothes section, this too must be protected with a minimum 25 mm air space between the pipe flue and contents of the cupboard.

Locating a compartment under a stairway of a building of more than two storeys should be avoided and, where chosen, all surfaces including the floor must be lined with a material that has a minimum fire resistance of 0.5 h.

Specific Notes on the Installation of Return Air Grilles

- Grills should not be located more than 450 mm above the floor to prevent the spread of smoke in the event of a fire.
- Communication between bedrooms should be avoided for the sake of privacy.
- No return air should be taken from kitchens, bath/shower rooms and toilets to avoid smell and moisture transference.

Noise Transmission from the Return Air Ducting

Because of the speed of the fan, noise is sometimes transmitted through the return air duct to the habitable area adjacent to the compartment. This can be avoided by incorporating a bend or two in the return air duct, thus increasing its length or, alternatively, lagging the duct may help.

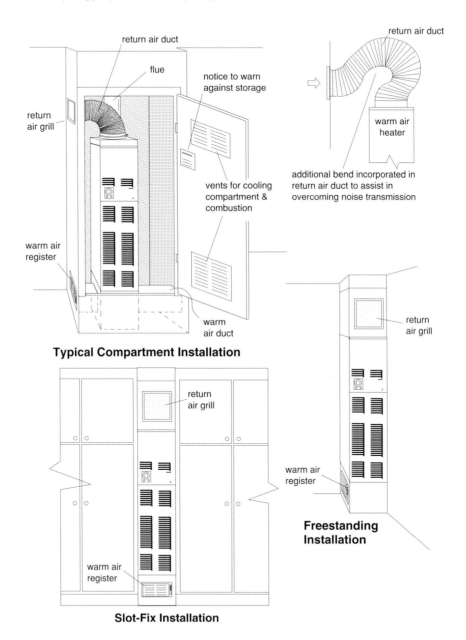

return air duct

flue

notice to warn against storage

return air grill

vents for cooling compartment & combustion

warm air register

warm air duct

Typical Compartment Installation

return air duct

warm air heater

additional bend incorporated in return air duct to assist in overcoming noise transmission

return air grill

return air grill

warm air register

Freestanding Installation

return air grill

warm air register

Slot-Fix Installation

9 Domestic Appliances

Commissioning and Servicing Warm Air Heaters

Commissioning and Servicing Work Record

The Work Record opposite is given purely as a guide to the many tasks to be undertaken when servicing/maintaining and commissioning a warm air unit. In order to complete the form you may need to refer back to Part 8, page 264, where many of the tasks are explained in further detail. However the additional 'appliance specific' checks on these units include some of the following.

Notice plate located and correctly filled in See page 197.

Check condition of heat exchanger

Distortion often occurs, resulting in cracking at the welded seams due to metal fatigue. This is due to the method of manufacturing the heat exchanger from pressed steel, and the continual expansion and contraction of the metal. Therefore it is essential that a good close visual inspection is undertaken because if a crack should develop the circulation air would be blown into the heat exchanger causing flame turbulence and possible spillage. Less likely, but also possible, is that combustion products may get drawn into the air duct system and be discharged around the whole dwelling. The heat exchanger should be checked by shining a powerful torch inside to allow close inspection of the welded joints. A good indicator is that the flame picture will be disrupted when the fan kicks in to blow the warm air through the building. When in doubt, the heat exchanger could be tested as follows:

- Light the appliance and allow it to heat up for some 5–10 minutes.
- Switch off the appliance and insert a smoke pellet, placed on a non-combustible surface, into the heat exchanger, towards the rear.
- Whilst it is burning, close all warm air register grilles, except the one nearest the heater.
- Finally, switch on the circulation fan and observe the open register grille for traces of smoke.

Where necessary the heat exchanger should be removed, this may need to be undertaken anyway where difficulty is experienced in cleaning the flue ways.

Positive return air path

Because the operation of convection currents within an open flue system caused by the air flow fan can be affected, the return air must be suitably ducted back to the air inlet within the heater compartment. Where this is not the case in an existing installation, the appliance must be regarded as At Risk.

9 Domestic Appliances

Certificate/Record of Warm Air Unit Service/Commission

Gas Installer Details	Client Details	Appliance Date Badge Details
Name :	Name :	Model/Serial No:
CORGI Reg. N° :	Address :	Gas Type: Natural ☐ LPG ☐
Address :		Heat Input: max... kW min... kW
		Burner Pressure Range: ...–...mbar
		Gas Council N°/ CE N°:................

Date: **Appliance Location:**...................................
                   Install/Commission ☐   Service/Commission ☐

Preliminary System Checks Compliance with manufacturer's instructions	PASS ☐ FAIL ☐
General visual inspection of pipework	PASS ☐ FAIL ☐
Clearance from combustible materials and compartment warning notice up	PASS ☐ FAIL ☐
Visual inspection of flue	PASS ☐ FAIL ☐ N/A ☐
Flue notice plate located and correctly filled in	PASS ☐ FAIL ☐ N/A ☐
Flue flow performance test	PASS ☐ FAIL ☐ N/A ☐
Appliance level and secure	PASS ☐ FAIL ☐
Electrical connections	PASS ☐ FAIL ☐ N/A ☐
Bonding maintained	PASS ☐ FAIL ☐ N/A ☐
Fuse rating:.........amps	PASS ☐ FAIL ☐ N/A ☐
System tightness test, to include let by	PASS ☐ FAIL ☐ N/A ☐
Standing pressure of system	PASS ☐ FAIL ☐ N/A ☐
Appliance/system is purged of air	PASS ☐
Service/Commission Checks Clean primary air ports and lint arrestor	PASS ☐ FAIL ☐ N/A ☐
Clean/Check condition of injectors, burners	PASS ☐ FAIL ☐ N/A ☐
Check condition of heat exchanger	PASS ☐ FAIL ☐
Check, ease and grease, if necessary, control taps	PASS ☐ FAIL ☐ N/A ☐
Ignition devices effective including condition of electrodes, leads	PASS ☐ FAIL ☐ N/A ☐
Clean and check operation of fans and filters	PASS ☐ FAIL ☐ N/A ☐
Pilot flame correct	PASS ☐ FAIL ☐ N/A ☐
Burner pressure...........mbar	PASS ☐ FAIL ☐
Total heat input............kW	PASS ☐ FAIL ☐ N/A ☐
Flame picture good	PASS ☐ FAIL ☐ N/A ☐
Flame supervision device operational	PASS ☐ FAIL ☐ N/A ☐
Combustion seals and plenum seals effective	PASS ☐ FAIL ☐ N/A ☐
Positive return air path	PASS ☐ FAIL ☐ N/A ☐
Appliance tightness check	PASS ☐ FAIL ☐ N/A ☐
Operating thermostats correct, including high limit stat and fan switch	PASS ☐ FAIL ☐ N/A ☐
Temperature differential through unit:........°C	PASS ☐ FAIL ☐
Spillage tests	PASS ☐ FAIL ☐ N/A ☐
Distribution grills and dampers	PASS ☐ FAIL ☐ N/A ☐
Additional direct ventilation grille required:...........cm^2	PASS ☐ FAIL ☐ N/A ☐
High level compartment ventilation where required:...........cm^2	PASS ☐ FAIL ☐ N/A ☐
Low level compartment ventilation where required:...........cm^2	PASS ☐ FAIL ☐ N/A ☐
Post System Checks Meter working pressure (........ mbar)	PASS ☐ FAIL ☐ N/A ☐
Working pressure drop across system (max ....mbar)	PASS ☐ FAIL ☐ N/A ☐
Safe operation of appliance explained to customer	YES ☐ NO ☐

Recommendations and/or Urgent Notification

Appliance Safe to Use   YES ☐ NO ☐   **Next Service Due:**...........

Installer's Signature:	**Customer's Signature:**

Operating thermostats correct, including high limit stat and fan switch The high limit or overheat thermostat should be checked as per manufacturer's instructions. If the air does not circulate through the warm air heater, the heat exchanger will quickly become overheated. One of two controls may overcome this problem: a limit thermostat set at 95°C and an overheat thermostat set at 110°C. Either one or both may be included.

The *limit stat* turns off the gas supply at the upper limit of 95°C and re-lights the appliance when the temperature falls to about 80°C. This automatic resetting causes the heater to cycle on and off every few minutes.

The *overheat stat* differs in that if the temperature reaches the upper limit of 110°C, the appliance will shut down with no automatic reset facility. The overheat stat is often an additional control in down-flow heaters because of the high heat rise that can be experienced when the fan is slow to switch off.

To test this control it may be possible to block the filter inlet with a piece of card or dust sheet or, alternatively, run the appliance with the heater alight, but the fan disconnected. The heater should shut down within 2–3 minutes.

The fan switch In order to prevent cold air blowing into the rooms before the heat exchanger has had time to warm up, a thermally operated switch is incorporated; this only allows the fan to operate when a pre-determined temperature is reached. Operating temperatures are: fan on 58°C, fan off 38°C.

Summer/Winter switch For summer operation the user simply switches the control to the 'summer' position to allow cold air to be blown to the rooms.

Temperature differential through the basic unit (not Modairflow) This is a check of the temperature rise through the appliance. It is simply carried out by taking a temperature reading at the air intake to the appliance and at the first warm air diffuser grille. The rise should be as indicated by the manufacturer, usually in the region of 50°C ± 5°, which may mean balancing the system or adjusting the fan speed if necessary.

Spillage Test This is usually carried out as described on page 223. However, in some instances the draught diverter is totally inaccessible. Therefore an alternative method is employed as follows:

1. Pre-heat the appliance for some 5–10 minutes.
2. Turn off the appliance and insert a small smoke pellet on a non-combustible surface inside the combustion chamber, replacing the cover plate. The pellet selected should not be too large, otherwise the volume of smoke emitted would give an unrealistic test.
3. Look for the presence of smoke in the general area of the draught diverter.

Distribution grilles and dampers Check to see that the register grilles located in all the rooms are obtaining sufficient heat, adjusting the damper located at the rear of the distribution grille as necessary. A check should also be made to ensure that the diffusers open and close freely.

Basic fault diagnosis chart

Fault	Possible cause
Poor pilot flame and will not stay alight	• Gas supply turned off or air in pipe • Thermocouple connection loose • Pilot flame too small or pilot tube blocked • Faulty thermocouple or thermo-electric valve • Inadequate ventilation • Defective atmosphere sensing device or oxygen depletion tube
Pilot established but main burner will not ignite	• Electricity supply turned off or faulty fuse • Controls not calling for heat or set to 'summer' setting • Loose electrical connections • Faulty gas solenoid valve • Faulty thermostat (room or limit stat)
Main burner lights but fan fails to operate after preheat period.	• Faulty fan assembly, e.g. electrical connection loose, defective switch or fan belt (if fitted) • Fan setting incorrect • Operating gas pressure too low
Main burner lights intermittently with fan running	• Operating gas pressure too high • Fan speed incorrect • Air filter or return air path restricted • Most outlet diffusers closed
Fan running intermittently with main burner on	• Fan switch setting incorrect • Operating gas pressure too low
Fan runs for long time after main burner shuts off, or intermittently	• Fan switch setting incorrect Poor flame picture • Split or blocked heat exchanger • Insufficient air intake or lint arrestor blocked • Poor or blocked flue system • Incorrect or blocked injectors
Noisy operation	• Gas pressure too high • Noisy fan motor • Fan speed too high
Main burner does not switch off	• Faulty multifunction valve

9 Domestic Appliances

Domestic Tumble Dryers

Relevant ACS Qualification	Relevant Industry Document
LAU1	BS 7624

The domestic gas tumble dryer looks to all intents and purposes just like the electrical version, which is far more common. It consists of the same components, including an electronic drive belt, drum and control panel for the user. The only real difference is the method employed to heat the air that is warmed to pass through the tumbling clothes. This is achieved by using a gas burner located at the base of the unit that rapidly warms the air drawn in for combustion. A fan pulls the products through the drum and eventually expels them from the rear of the appliance. An exhaust vent, supplied with the unit, is to disperse the products to outside. The dryer should not operate with the door in the open position; when the door is closed it operates a timer switch allowing the appliance to operate.

Component Parts

Burner

This is usually of pressed steel construction and is fitted inside a metal tube, with the gas flame burning horizontally in the direction of air flow to the back of the appliance. Because the draught is induced through the combustion chamber, flame detection is usually by means of flame rectification. To provide an added level of safety, two safety shut off valves are usually incorporated as shown opposite.

Drum

This is the heart of the tumble dryer, it consists of a stainless steel cylinder of around 115 litre capacity and turns at approximately 50–60 revolutions/minute, driven by a belt fixed to the outer circumference and motor. It usually rotates in a clockwise direction but it stops and turns anticlockwise for short periods to enable the clothes to untangle so that they are free to move.

Thermostat control

Two operating thermostats are generally incorporated to provide drying temperatures of around 50 and 60°C. These are located in the exhaust duct just after the fan. In order to prevent overheating and damage to the unit or clothes, a thermostat operating at around 110°C is also incorporated to the top rear of the drum housing. This thermostat will cut off the gas supply but allow its re-ignition as the drum cools. A final high limit overheat thermostat, operating at 120°C, is also located at this point and will shut down the appliance if the 110°C thermostat fails.

Lint filter

This is located just inside the front door. This collects the vast amount of fluff, etc. generated from the clothes during tumble drying and it is essential that the customer is instructed how to remove and clean out the filter regularly, failure to do this will result in the appliance short cycling.

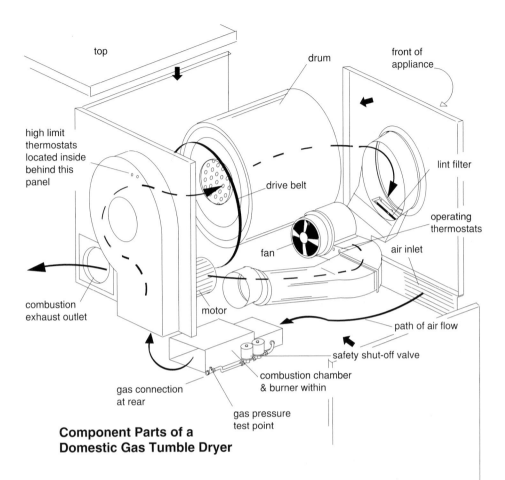

top

drum

front of appliance

high limit thermostats located inside behind this panel

lint filter

drive belt

operating thermostats

combustion exhaust outlet

fan

air inlet

motor

path of air flow

safety shut-off valve

gas connection at rear

combustion chamber & burner within

gas pressure test point

Component Parts of a Domestic Gas Tumble Dryer

Domestic Gas Tumble Dryer

Installation of Domestic Tumble Dryers

Restricted Locations

Tumble dryers must not be installed in bath/shower rooms and can only be installed within a bedroom/bed sitting room or garage where the manufacturer permits and the room volume is at least 7 m^3/kW appliance input. The positioning in a protected stairway, such as in flats over two storeys high, is also restricted. LPG appliances in basements, etc. are also not allowed.

The exhaust vent should not discharge into the confines of a covered alleyway, such as between two adjoining properties, where the combustion products may accumulate. These products may contain CO, which may eventually find its way into an inhabited area.

Location and Clearances

The tumble dryer will normally fit in a space 600 mm × 600 mm. Where the dryer is to be positioned under a worktop a 15 mm minimum space between the top of the dryer and worktop should be maintained to allow for ventilation. A free space to the front of the appliance should be allowed so that air can be drawn into the appliance and to allow the appliance to be pulled right out for maintenance and servicing purposes. It is possible to stack a tumble dryer directly on top of a compatibly sized washing machine, providing the correct recommended stacking kit is used. In such a case, a restraining device must be used.

Gas Connection

The gas supply is connected via a flexible hose and bayonet connection. This allows for limited movement and, where frequent movement is to be expected, a restraining device should be used to prevent undue damage.

Ventilation

The domestic gas tumble dryer has an heat input up to 6kW and is categorised as a flueless appliance The minimum requirement is to have a window or similar opening directly to outside. In addition, if the room is less than 3.7 m^3/kW a permanent air grille, direct to outside, of 100 mm^2 is required. Any air grille to the outside should be at least 300 mm from the exhaust vent termination.

Exhaust Venting

In all cases exhaust venting should be made to outside. Prior to the current British Standard this venting was permitted directly into the room, however amongst other things it gave rise to high levels of moisture-laden air and condensation problems where extensive condensation would cause rotting to the base units and worktop. The exhaust vent may simply be hung out of an opened window where the input is <3 kW or a more permanent arrangement can be made where the hose is connected to a purposely provided wall or window terminal grille. This grille should be designed to

9 Domestic Appliances

give the required airflow and be up to the same standards as those used for fixed free ventilation. The terminal should be at least 300 mm above ground level to prevent its blockage by leaves or snow. Where the hose is to pass through a cavity wall it should be sleeved. It is also a good idea to run the sleeve slightly downwards to the external face to prevent the entry of rain, etc. The number of 90° bends should also be restricted to a maximum of three.

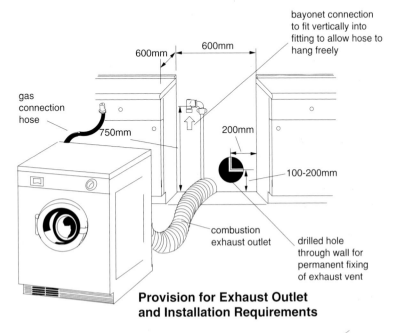

Provision for Exhaust Outlet and Installation Requirements

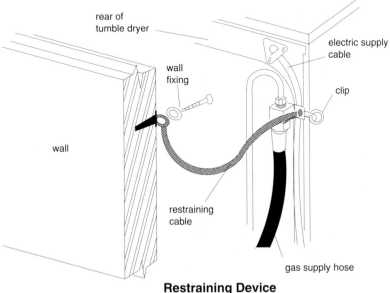

Restraining Device

9 Domestic Appliances

Domestic Gas Refrigerators

Relevant ACS Qualification
REFLP2

The gas fridge often operates as a multi-fuel appliance, particularly when used in caravans or touring vehicles, using either gas or electricity to warm an ammonia–water mix. Gas refrigerators are of the absorption type and work as follows:

1. A mix of ammonia and water is heated within the heater compartment of a sealed unit by a small gas flame and, as the liquid heats up, ammonia is given off as a gas which rises to the condenser.
2. Air surrounding the condenser allows the ammonia to cool and it condenses back into its liquid form and drains by gravity into the evaporator which is located in the cooling compartment of the refrigerator.
3. Hydrogen gas within the evaporator lowers the pressure and, as a result, the ammonia evaporates. For evaporation to occur the ammonia needs to extract heat from its surroundings, in this case from the air inside the fridge.
4. The hydrogen–ammonia mix then falls, due to its density, to the absorber. The absorber is a series of small tubes fed with a trickle of water from the heater compartment. This water rapidly dissolves the ammonia, leaving the hydrogen, which is now free to rise back to the evaporator.
5. The newly replenished water–ammonia mix now flows to the lowest point in the system where there is a coil located around the combustion chamber. As the liquid heats up, bubbles form in the boiling liquid and rise up into the top of the heater, whereupon the ammonia is driven off and so another cycle begins.

Most of the cooling unit is mounted behind the back of the refrigerator cabinet, with only the evaporator located inside at the top. The liquids contained within the unit, referred to as the refrigerants, cannot be replaced on site and where damaged or faulty the unit will need to be replaced. The actual cabinet usually consists of an inner lining, usually of plastic or aluminium and an outer skin of mild steel, with the space between filled with polyurethane foam or, for insulation near the boiler, glass fibre.

The burner

Located at the base of the boiler will be the burner, which consists of an injector and small burner head that generates a very small flame. As heat rises in the combustion chamber it is deflected by a twisted stainless steel baffle that is suspended from the top of the heater and is designed to radiate and deflect the heat towards the boiler coil. The heat then passes up through a small central flue to be discharged either into the room or run externally. The amount of heat generated is controlled by a liquid expansion thermostat with a variable setting to compensate for hot and cold weather.

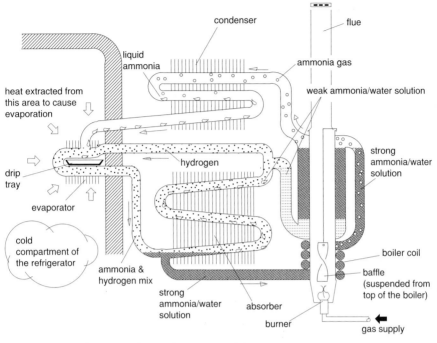

Section through a gas refrigerator

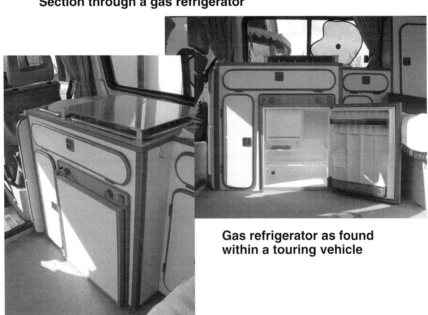

Gas refrigerator as found within a touring vehicle

Leisure Appliances

Relevant ACS Qualification   Relevant Industry Documents
LEI1   BS EN 498 and BS 2977

The term leisure appliances includes appliances that are associated with outdoor activities and may be situated around a terrace or patio. They include the following.

Barbecue

The gas barbecue may be a free standing on a flat surface or built into a non-combustible structure. One of the most common designs is that used with LPG as a mobile unit. The barbecue consists of a burner located beneath a grill cooking plate and with long-lasting ceramic briquettes spread out to provide the heat distribution medium. A barbeque should be sited away from all combustible materials and over-hanging shrubbery. The gas connection to a barbecue is made via a flexible connection from gas bottles sited underneath or, where the gas is supplied from a piped supply, from a locally fixed micro-point.

Patio heater

Free-standing patio heaters operate typically with an heat output from 6 to 14 kW, which will enable an area of 25 m^2 to be warmed. They stand at a height of around 2.3 m. Basically, they consist of a burner with a round heat emitter positioned at the top of a pedestal, with a reflector above to direct the heat downwards. A whole range of heaters is available, including a tabletop version, operating at around 2.8–6 kW. The heater may be fixed and secured firmly to the ground and connected to the gas supply by a permanent rigid pipe, or it may form part of a mobile unit on wheels and be supplied with an LPG gas bottle. The siting/positioning of these heaters needs some thought and generally a minimum distance of 1.5 m should be maintained between the heater and any combustible surface. The heater is ignited with either piezo or electronic ignition.

Gas light and flambeaux

There are several designs of gas-light. The traditional gas lamp uses a mantle on which the flame burns, and a bright light is emitted. This design was very common in older style caravans, but today they are more commonly found as a single light, located on a pedestal or wall in a courtyard or similar venue using a gas mantle. Another form of gas light does not use a mantle but has a traditional fishtail burner to give an old-world effect with a flickering or vestal flame. The last type of gas light to be described, known as the gas flambeaux, is basically a decorative gas-burning torch. These come in a whole range of shapes and designs including wall units, bowls and mounted on pedestals. Their sizes range from a small gas flame of around 6 kW to a more dramatic flame, outside the scope of leisure appliances, of up to 60 kW input. As with the other appliances listed here, these appliances can run on LPG or natural gas. The modern gas light switches on automatically, possibly using a photo-electric cell to open the gas solenoid to the main burner as the light fades.

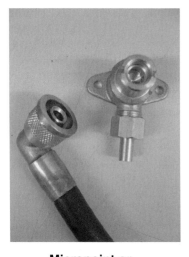

**Micropoint or
Leisure Gas Connection**

Gas Barbecue

Gas Lamps

Gas Street Lamp

Patio Heater

Gas Flambeaux

9 Domestic Appliances

Greenhouse Heaters and Gas Pokers

Relevant ACS Qualification   Relevant Industry Document
LEI1   BS 3929

Greenhouse Heater

Gas greenhouse heaters can be supplied to operate on either LPG or natural gas, usually operating between 2 and 4 kW. A 2 kW heater is large enough to provide frost protection over an area of 11 m^2. Larger heaters of 4 kW would be sufficient to provide frost protection over an area of 32 m^2. The heater may be free standing or it may be permanently installed on a flat non-combustible surface or wall-mounted, depending on the appliance manufacturer's instructions. Where the heater is installed in a greenhouse that forms part of a dwelling, a rigid gas connection must be made. If the greenhouse is independent of the house a flexible connection is used. The heater incorporates a self-extinguishing device in case it is knocked over. The single greenhouse heater is a flueless appliance and as such it is essential that the greenhouse is adequately ventilated. One feature of a greenhouse heater is that it produces a vast quantity of carbon dioxide (CO_2), which many plants thrive on. One of the biggest problems with these heaters is spiders' nests that are built in them during the summer months when the heater is not being used. These cause all sorts of problems, particularly with the pilot assembly where the injector is almost impossible to clean and replacement is often needed. The best solution is to disconnect and store the heater away during the summer months.

Ventilation

For greenhouses attached to a dwelling the minimum ventilation required is an openable window. However, additional ventilation may be required, depending on the manufacturer's instructions. It must also be noted that the size of a greenhouse heater is restricted to 90 W/m^3 of its volume where there is an opening into a dwelling. Where the greenhouse stands alone and has no openings into any buildings and the appliance net input rating does not exceed 2.7 kW, no additional ventilation is required. However, for appliances with inputs greater than this, two fixed free air grilles will be required, one low level and one at high level, providing a minimum effective area of 39 cm^2 for every kilowatt in excess of 2.7 kW.

Gas Poker

The gas poker, although rarely seen today, consists of a handle and flattened tube through which gas can pass at various positions. This is attached to a specially designed reinforced flex, suitable for ambient temperatures up to 70°C and 95°C touch temperatures. Basically the gas poker can be used to establish a coal fire without the need to use paper and wood by bringing the fuel up to ignition temperature.

When using a gas poker it is essential to ensure that all the holes through which the gas can pass are alight, otherwise un-burnt gas can pass to the fire bed and subsequently up the chimney, resulting in a 'boom'.

Greenhouse Heater

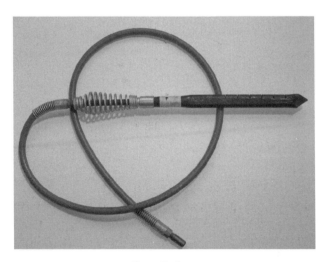

Gas Poker

Part 10
Commercial Appliances

Commercial Boilers

Relevant ACS Qualification   Relevant Industry Documents
            CIGA1   BS 6644 and IGE/UP/10

Boilers over 70 kW net heat input fall within the category of commercial appliances, but in terms of commercial boiler operation it would not be uncommon for appliances to operate at inputs of 500 kW to over 5000 kW. The range is quite vast. Today, with improved boiler designs and the need for weight reduction, boilers are much smaller than their predecessors. Sometimes a series of boilers installed in a row and connected together via an arrangement of headers, referred to as modules, are used to make installation and servicing far easier. Boiler design tends to be of one of the following types.

Sectional Boiler

This appliance consists of a number of individual cast iron sections that contain the waterways. Each section is joined to the adjacent section by the flow header at the top and usually two return connections at the bottom. An appropriate gasket material is used between each section and the whole system is clamped together via long bolts. The number of sections joined together denotes the total heat output that the boiler can provide. The burner is specifically selected to suit the final size of the boiler.

Shell Boiler

These boilers can generally operate at higher pressures than sectional boilers, and are often used where high temperature hot water or steam generation is required. There are several designs of shell boiler but they all work on the principle that the flue products are forced through a series of fire tubes that are surrounded by water, on their way to the flue system, as shown. In order to keep down the size of the boiler, the flue products are often turned through 180° to make a pass through another heat exchanger, thereby extracting more heat from the combustion products.

Modular Boilers

The modular boiler is basically a series of between two and six smaller boilers installed together and connected by a series of headers. As a result of advances in technology, these boilers are lighter in weight and are suitable for rooftop installation. Mounting the boiler at the rooftop also cuts down on the total weight of the boiler, as it no longer needs to support the additional weight of water in a large system. Each module is usually small enough to pass through doorways and can be transported in a passenger lift. Above all, they are more easily installed and serviced. It is possible to service an individual boiler without shutting down the whole system and, where a fault condition occurs, the supply is not totally lost. Modular boilers also have the added advantage of being able to cope with variable heating loads.

10 Commercial Appliances

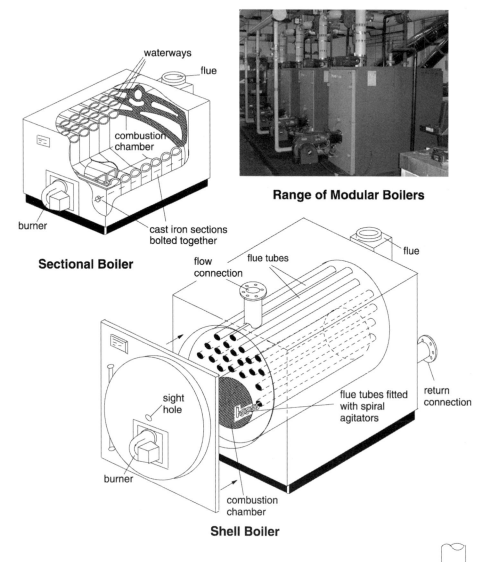

waterways

flue

combustion chamber

burner

cast iron sections bolted together

Sectional Boiler

Range of Modular Boilers

flow connection

flue tubes

flue

sight hole

flue tubes fitted with spiral agitators

return connection

burner

combustion chamber

Shell Boiler

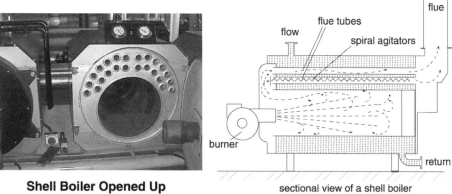

Shell Boiler Opened Up

flue

flow

flue tubes

spiral agitators

burner

return

sectional view of a shell boiler

10 Commercial Appliances

Commercial Boiler Gas Control Systems

Relevant ACS Qualification	Relevant Industry Documents
CIGA1	BS 5885 and BS EN 676

The gas supply to a large commercial boiler may consist of a single multifunctional gas valve or it may consist of a 'gas train' comprising of separate gas controls such as a regulator and safety shut off valves.

Burner Operating Sequence

The control unit fitted to the burner runs through a set sequence of checks before allowing the main volume of gas to flow to the burner, thus preventing an explosive ignition. This sequence is often shown in the form of a graph in the manufacturer's instructions as illustrated opposite, each stage being displayed as a shaded time period. The general sequence is as follows:

1. When the thermostat contacts are made, the air fan operates for a set period to pre-purge the appliance of any combustion products.
2. This is followed by a set period to establish a pilot flame.
3. With the pilot flame confirmed, the main gas flame commences.
4. On completion of the run period and the satisfaction of the thermostat, the boiler will switch off. *Note*: Some burners also have a post purge, to ensure that all products are removed from the appliance and flue system.

Safety shut-off valves (SSOV)

For operation see page 76.

These valves must be installed in series to the burner as shown in the following table.

Appliance/System input	Safety shut-off valve requirements
60–600 kW	One class 1 and one class 2 valve
600–1000 kW	Two class 1 valves
1000–3000 kW	Two class 1 valves with a system check
>3000 kW	Two class 1 valves with pressure proving system

Pressure Proving System

Pressure proving systems check the correct operation of the safety shut-off valves before the burner can begin its operating cycle. Where a problem is encountered at any stage of proving, the system will go into lock-out and prevent ignition. There are several proving systems including the following, which relies on *sequential proving* (see diagram opposite):

1. With all SSOVs closed, valve 'A' opens for 2–3 seconds to release the pressure from the enclosed section of pipework and re-closes.
2. The pressure is now monitored for a set period by pressure switch 1 to check that the pressure does not rise above 5 mbar, thus indicating that no let-by is occurring from the valves, in either direction.
3. Assuming no pressure rise, the inlet SSOV opens for 2–3 seconds to charge the section between the valves and re-closes.
4. The pressure is now monitored for a set period by pressure switch 2 to check that the pressure does not drop, indicating a leak through the downstream valves. If proved sound, the boiler ignition cycle is initiated.

Operation	pre-purge	pilot/low fire ignition	pilot/low fire proving	main flame established	burner run period	post-purge
Boiler calling for heat	/////	/////	/////	/////	/////	
Air fan 'on'	/////	/////	/////	/////	/////	/////
Spark ignition		/////				
Pilot gas 'on'		/////	/////	/////		
Main gas 'on'				/////	/////	
Time in seconds	30	2 - 5	5 minimum	2 - 5		30

Typical Burner Control Sequence Diagram

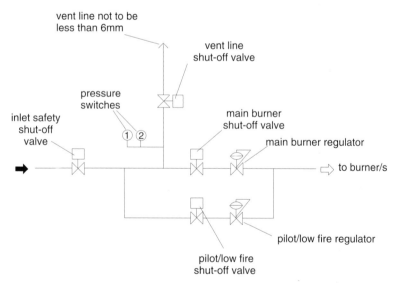

Gas Train showing Location of Safety Shut-Off Valves
(for sequential proving, see text opposite)

Commercial Warm Air Heaters

Relevant ACS Qualifications
CIGA1 and CDGA1

Relevant Industry Document
BS 6230

Warm air heating falls into two categories:

1. direct fired air heaters and
2. indirect fired air heaters.

Both designs of air heater may be floor mounted, secured to a wall or installed suspended from the roof trusses of the structure. These high level heaters are commonly referred to as unit air heaters. They prove popular where floor space is at a premium. Where high level mounting is to be considered, the fixing height should be approximately 2.5 m from the floor to the underside of the heater and generally a 1 m distance is required between the heater and ceiling. Manufacturer's instructions will provide greater details and, as always, must be followed.

The air heater is fitted with similar controls to those used for domestic warm air heaters as found on page 348 and include a temperature limiting thermostat and fan control to ensure that the air temperature blown into the room is at an acceptable level. They also have the facility to blow cool draughts in summer. The heaters shown opposite are independant serving the area where they are located, however units can be found utilising a series of ductwork, serving several outlets.

Ducted Air Distribution and Return Air

Generally any ductwork needs to be as short as possible, giving the experience of some warmth without the need for an excessive positive draught from the warm air outlet grill. Materials used for ductwork must not be a fire risk and they need to be of adequate strength and of sufficient durability to withstand any internal and external temperatures and loads under normal operating conditions. If the air heater is installed in a plant room, the return air intake and warm air outlet must be fully ducted to prevent any interference with the safe operation of the flue. The openings into the plant room must also be suitably fire stopped. Return air ducts and inlet points must not be located in areas where smells, dust or fumes could be drawn into the appliance, affecting its combustion performance and distributing the smells and odours around the entire building.

Hazardous Areas

Air heaters should not be used to supply warmed air to hazardous areas unless:

- all incoming air is from outside and
- the outlets from register grilles are at least 1.8 m above the floor.

Pipework

Gas pipework is usually made using a rigid pipe connection, however for suspended appliances it may be possible to use a semi-rigid flexible connector or metallic gas flex, which the manufacturer's instructions will specify.

Indirect Warm Air Heater

Direct Warm Air Heater

Indirect Warm Air Heater
(flue not shown)

Commercial Direct Fired Air Heaters

Relevant ACS Qualification   Relevant Industry Document
      CDGA1   BS 6230

A direct fired air heater is one in which the products of combustion mix freely with the heated air and are passed out, through distribution outlet grilles, into the space to be heated. Generally the air required for combustion is taken directly from outside, however the appliance may be positioned where it is dependant on natural ventilation from within the room. Because these heaters discharge their products into the heated environment it is essential that during commissioning environmental analysis checks are carried out in the room to ensure that the CO and CO_2 levels remain acceptable. The maximum exposure limits in any space should not exceed 0.001% CO (10 p.p.m.) and 0.28% CO_2 (2800 p.p.m.). Some units have fitted, as part of their design, a CO_2 limiting control to monitor the environment continually, shutting down the appliance should the levels rise excessively. Where such a control is incorporated, it needs to be regularly calibrated. The heater may be independent and a permanent fixture, incorporating various outlet diffusers and connected to a system of ductwork or a mobile air heater may be found. Mobile or transportable heaters must not be controlled by time switches or other remote controls.

The temperature of the air discharged from these heaters should not exceed 60°C; to maintain such temperatures invariably dilution of the heated air with fresh or room air may be necessary. The efficiency of a direct fired air heater is very high, typically over 90%, as there is no heat exchanger or fuel wasted into a flue system. A well-designed direct fired heater installation tends to slightly pressurise the room in which the heated air passes. This has the effect of preventing cold draughts from entering the room and, as such, proves particularly suitable for areas such as a swimming pool. Areas such as restaurants that are adjacent to kitchen areas are also suitable for slight room pressurisation as the pressure tends to keep the smells from travelling into the eating area.

Ventilation

Where a direct fired heater is located in a large open space no additional ventilation is generally required, provided that the maximum exposure limits, identified above, are not exceeded. Where the heater is installed in a plant room, passing the warmed air into the room to be heated by a system of ductwork, ventilation will be required to keep the relatively small compartment cool. The sizing of such a ventilation grille was discussed page 246, in Part 7, Ventilation. Where a direct fired heater is installed in a environment that relies on a closable ventilation system or an extract system, a full system of safety interlock must be provided to shut down the appliance if the air movement is interrupted.

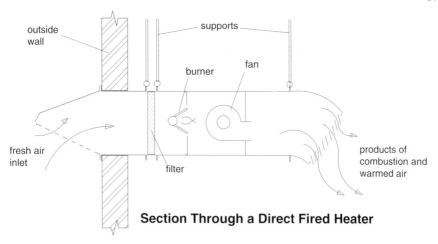

outside wall

supports

burner

fan

fresh air inlet

filter

products of combustion and warmed air

Section Through a Direct Fired Heater

Direct Fired Heater

Side Panel Removed Exposing the Burner

CO$_2$ Analyser

Burner Removed from Heater

Portable Direct Fired Heater

Commercial Indirect Fired Air Heaters

Relevant ACS Qualification   Relevant Industry Documents
         CIGA1   BS 6230 and IGE/UP/10

Unlike the direct-fired air heater, previously described, the products of combustion are kept separate from the heated air that is discharged into the space to be heated. Indirect air heaters have a heat exchanger and flue system through which the combustion products pass.

Indirect heaters are manufactured in a range of sizes from 10 to 140 kW if suspended; this increases to 440 kW for the free standing model. A fan is located in the heater to draw cool air into the heat exchanger from the room and force it across the heat exchanger where it is rapidly heated and discharged through the system of ductwork or, as is often the case, an outlet louvre grille forming part of the unit.

Air heaters up to 100 kW may have either an atmospheric or forced draught burner installed. However, as the output increases appliances generally use forced draught burners only.

Ventilation

The standard for ventilation air should follow the same general guidelines as for all open flued appliances installed in rooms or plant rooms, as described in Part 7, Ventilation. However the following specific points also need to be considered: Air should not be taken from areas where it is likely to be contaminated, for example by odours and exhaust fumes from vehicles, etc., as these would subsequently be blown out through the warm air distribution system. Particular attention needs to be paid to the warm air carrying high levels of water vapour, which is released as condensation when the air touches the cold surfaces of steel structures and windows or through natural cooling in poorly ventilated roof spaces. In determining the ventilation requirements, an assessment of the maximum heat input rate for all appliances needs to be considered, to include that required for combustion, all process and other equipment, and other products of combustion in the heated space, which may include those from motor vehicles.

Fluing

The standard for fluing to be maintained follows the same general guidelines as for all open or room sealed appliances as described in Part 6, Flues. In general, wherever possible, flue systems should be designed to operate freely under natural flue draughts and systems requiring an additional fan should be avoided.

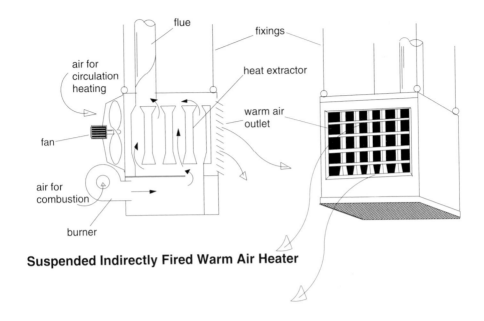

Suspended Indirectly Fired Warm Air Heater

Freestanding Indirect Fired Warm Air Heater

(Flue not shown)

Overhead Radiant Heaters

Relevant ACS Qualification	Relevant Industry Documents
CORT1	BS EN 416 and BS 6896

Radiant Heating

Before one can begin to grasp the concept of overhead radiant heating, one must understand the principles on which this form of heating works. Radiation is the method of direct heat transference of infra-red energy from the source to the point at which the heat rays land, and this is where the heat is absorbed. Anything blocking the path of the heat ray will block the heat transference. The best analogy is the feeling of the sun's rays directly on your skin on a hot day; the rays cannot reach you when you are in the shade. When heat is emitted, the greatest intensity is on a surface parallel to the heater panel with the heat ray travelling at right angles to it. As the angle at which the heat strikes the surface falls below 90°, so does the heat intensity, falling gradually to the point where no radiation is experienced. This means that a single heater might not be sufficient to heat an area: several heaters could be required, spaced as evenly as possible around the area to be heated. Each point in the room should be heated from at least two directions, more if possible, to prevent shadowing.

Heater Design

There are two basic designs of radiant heaters:

1. luminous heaters – ceramic plaque heaters;
2. non-luminous heaters – radiant tube heaters.

The Ceramic Plaque Heater

This is a non-flued appliance and the combustion products are discharged directly into the premises. Sometimes the water vapour that is produced condenses onto cold walls and steelwork. The closer these heaters are fitted to the working area, the more the radiant heat intensity and the less the area that is heated. Generally one should aim to provide approximately 150–160 W/m^2 at working level. This can be achieved by positioning a typical heater some 4–5 m off the ground, but the manufacturer's instructions will give more information as to the ultimate fixing height based on the heat output. Ceramic heaters may, if required, be installed at an angle of up to 60° from the horizontal, thus directing the heat into the room.

The Radiant Tube Heater

These heaters can be flued or un-flued and can be individually installed or put in as a multi-tube installation. The heater consists of a burner located at one end of a steel tube, typically 65 mm in diameter, and a fan at the end to suck the combustion products through the tube. The fan suction also draws the air/gas mixture into the

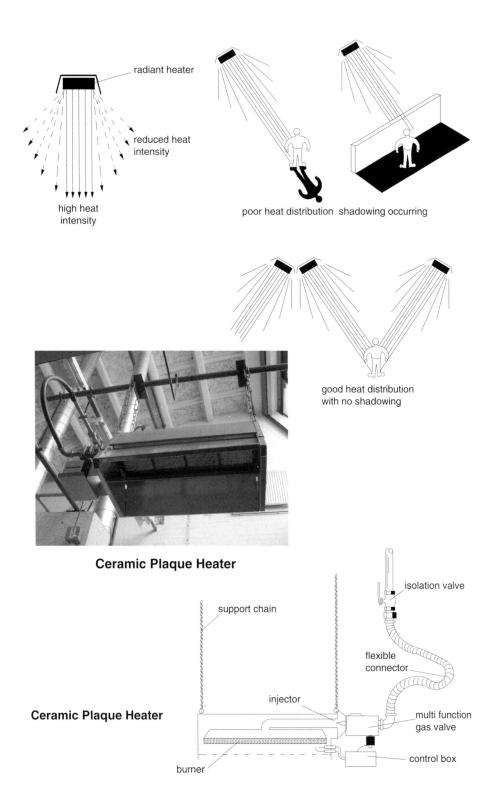

radiant heater

reduced heat intensity

high heat intensity

poor heat distribution shadowing occurring

good heat distribution with no shadowing

Ceramic Plaque Heater

Ceramic Plaque Heater

support chain

isolation valve

flexible connector

injector

multi function gas valve

control box

burner

10 Commercial Appliances

burner. The length of the tube and its diameter vary according to the heat output required and the manufacturer's design. Typical individual heaters are approximately 7 m long, with the tube turning back on itself, in a U shape, some 3.5 m from the burner. Above the tube is positioned a polished aluminium or stainless steel reflector panel, designed to direct the infra-red heat down into the room. The fixing height is similar to that for the ceramic plaque.

With the multi-tube installation, a large centrifugal vacuum fan is located at the point of discharge from the building to provide the necessary suction pressure to pull all the products through the entire system. In order to achieve maximum efficiency and sufficient draught through all the tubes, this design of system would use a series of dampers in the tubes to control the flow. To maintain a high enough temperature in the tube, several burners are positioned throughout its length.

The Black Bulb Sensor

Because radiant heating systems do not heat the air in the room in which they are situated, it would be pointless to use the usual type of room thermostat to control the temperature. Instead a sensor, consisting of a bi-metallic strip or electronic thermostat, enclosed in a bulb-shaped hemisphere coloured matt black, is used. The black finish readily absorbs heat. The heated surface of the bulb warms the trapped air within, which acts on the thermostat making or breaking the electrical supply as necessary, so controlling the operation of the heaters.

Installation Requirements

The heaters can be fixed high up on a sidewall or, alternatively, can be suspended from a bracket or chain, an arrangement that invariably uses a flexible gas connection. The flue pipe, where installed, is generally not subject to high temperatures due to the cooling effect of the long tube, therefore it may be run in combustible surfaces of 25 mm, unlike the 50 mm that is usually required for flues. The ventilation requirements for overhead heaters can be found on page 246; they should be referred to for further information. However, it should be noted that the vents should, where possible, be on two sides of the building at high and low levels at distances greater than 3 m apart at high and low level.

Work Record and Additional Maintenance and Servicing

When working on overhead heaters the tasks include the generic gas installation commissioning tasks listed on page 265, plus the following appliance-specific work:

- cleaning and polishing the reflectors;
- removing dust build up from the face of the ceramic plaques; this may be accomplished with a fine jet of compressed air or lightly brushing;
- replacing damaged or badly linted plaques;
- brushing through the tubes and removing loose material from the radiant tubes;
- confirming the suitability of the supports and ensuring their sound construction.

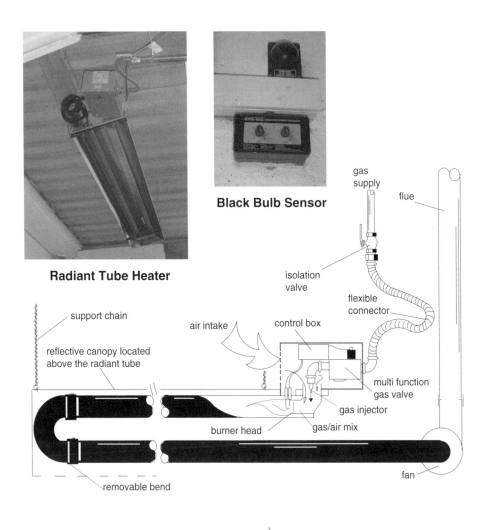

Radiant Tube Heater

Black Bulb Sensor

gas supply

flue

isolation valve

flexible connector

support chain

air intake

control box

reflective canopy located above the radiant tube

multi function gas valve

gas injector

burner head

gas/air mix

removable bend

fan

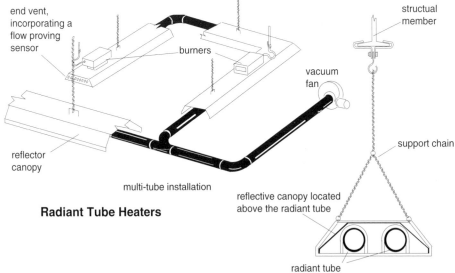

end vent, incorporating a flow proving sensor

burners

vacuum fan

structual member

reflector canopy

support chain

multi-tube installation

reflective canopy located above the radiant tube

Radiant Tube Heaters

radiant tube

Commercial Laundry Equipment

Relevant ACS Qualifications   Relevant Industry Document
     CCLNG1 and CLE1   BS 8446

Gas tumble dryers, washing machines and rotary gas ironers are available for use in commercial laundries. An important safety feature to check when commissioning these appliances is that the gas supply to the burner is extinguished if the door is opened during operation.

Location of Appliances
The appliances should be positioned on a firm level surface that is capable of taking the weight of the appliance when fully loaded. There must be adequate room to enable the appliance to be loaded or worked at without causing an obstruction. Appliances should only be stacked where the manufacturer's instructions permit. Care also needs to be taken to avoid combustible materials coming into close proximity with the appliance and any exhaust system.

Gas Supply and Appliance Connection
Typical kilowatt ratings vary depending on the appliance, but the ironers operate at around 4.5 kW, whereas at the other end of the scale are washing machines that operate at up to 30 kW. By definition these appliances are non-domestic, however the pipe size to any an individual appliance would not normally exceed 28 mm in diameter. The final connection to the appliance is usually made by means of a flexible connection to BS 699 with an individual isolation valve included adjacent to each appliance. A restraint device is also required to prevent excessive movement of the appliance away from the supply connection. In addition to fitting the isolation valve, a manually operated emergency control valve should be located by the exit door. If it is not possible to fit a manually operated valve at this point, then an automatic valve should be installed, operated via a stop button, along with the appropriate proving system to ensure downstream pilots, etc. are turned off before the gas is reinstated.

Tumble Dryer and Rotary Ironer
Both these appliances are regarded as flueless and, as a consequence, one needs to take into consideration the removal of the products of combustion and water vapour. This is usually achieved using an exhaust duct. The exhaust ducts from tumble dryers and ironing machines must be kept separate.

Exhaust and Ducting Systems
The exhaust system simply acts as an extractor fan to remove the vast amount of water vapour and combustion products to the external environment. Owing to the substantial air movement through the building caused by the exhaust venting, any open flued appliances may be subjected to increased spillage problems and therefore additional spillage checks will be required. These should be carried out with all

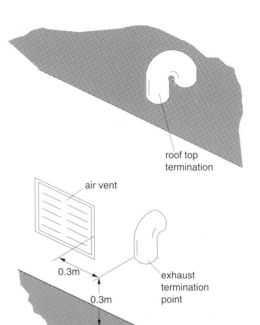

roof top termination

air vent

0.3m

0.3m

exhaust termination point

Location of Exhaust Terminal

Commercial Tumble Dryer

Rotary Ironer

10 Commercial Appliances

the appliances running. The exhaust duct will need to be sized to comply with the manufacturer's requirements. The number of bends should be kept to a minimum in order to limit the resistance to air flow. The outlet duct should terminate facing down towards the ground to prevent rain entering. In order to prevent any re-entry or products into the building, the duct should terminate at least 0.3 m (preferably 2 m) from any openings such as ventilation air inlets. Owing to the vast amount of lint produced, no grilles should be incorporated in the duct, as they rapidly become blocked. For the same reason it is essential that adequate access is made to the duct for maintenance purposes. Where a single duct is used as a header, serving several appliances, it needs to be sized with an overall cross-sectional area of at least the sum total of all appliances serving the duct. These branch connections are often made using a flexible duct not exceeding 2 m in length. In order to prevent the accumulation of condensation in the ductwork, the duct should be run on an incline to the outside, with a condensate trap at the lowest point.

Ventilation

Ventilation is required for combustion and drying purposes and is achieved by either natural or mechanical means. Should mechanical ventilation be selected, the system needs to be fully interlocked with the gas supply, ensuring that no gas is supplied should the ventilation system fail (see Gas Interlock on page 392). The ventilator size is as recommended by the manufacturer, however, for general guidance the following table may be used.

Typical load	Minimum natural ventilation	Minimum mechanical ventilation
14 kg (30 lb)	1000 cm^2	0.17–0.34 m^3/s
23 kg (50 lb)	1500 cm^2	0.34–0.38 m^3/s
34 kg (75 lb)	2250 cm^2	0.38–0.55 m^3/s

Gas Washing machine

This appliance is an open flued appliance and, as such, special care needs to be taken over the vast amount of extraction used where a tumble dryer or rotary ironer are employed in the same vicinity, as spillage may result. The gas connections are as identified on the previous page. The size of flue system selected will vary depending on the input of the appliance. In no case would this be less than 125 mm diameter, or 100 mm × 100 mm square section, but sizes up to 250 mm may be required. A draught diverter is required. Generally each appliance will have its own individual flue system, however should a common flue be used, fan assisted appliances must not be installed with appliances operating on natural draught. Should a fan assisted flue be employed, it is essential to incorporate a suitable proving system to isolate the appliance in the event of fan failure. Gas heated appliances must not be installed in the same environment as rooms containing dry cleaning machines using perchlorethylene or solvents containing CFCs. Combustion products mixing with the escaping vapours would condense to form hydrochloric acid, resulting in damage to the machine and laundry.

385

Commercial Gas Washing Machine

Boosters and Compressors

Relevant ACS Qualification   Relevant Industry Document
CBHP1   IGE/UP/2

A gas booster or compressor may be incorporated in the installation pipework to increase the gas pressure or to assist in maintaining the pressure where the pipework is undersized. Generally a gas booster is used in installations of up to 7 bar and a compressor is used above this.

A *booster* consists of a centrifugal fan driven by an electric motor. The impellers of the fan are in direct contact with the gas, drawing it in through a central point and forcing it out by centrifugal force with increased pressure.

A *compressor*, on the other hand, increases the pressure by forcing the gas through a rotary screw or reciprocating pump.

A *rotary screw* is effectively two large Archimedes' screws turning tightly together, one turning clockwise and one counter clockwise, forcing the gas between them, reducing its volume, and thereby increasing the pressure.

A *reciprocating compressor* operates by allowing gas to be drawn into a chamber using a piston. With the inlet valve closed, the piston forces down on the gas, compressing it, again reducing its volume and finally allowing it out through another valve with increased pressure.

Pressure Increase and the effect of Boyle's law

Although the pressure is increased in the installation, it must be remembered that there is no increase in gas volume. In fact the opposite will occur, as shown above and demonstrated below, applying Boyle's Law (see page 60).

Example If the supply pressure of 1 m^3 of gas is increased from 21 mbar to 60 mbar, the resultant volume will be:

$$Pressure\ 1 \times Volume\ 1 \div Pressure\ 2 = Volume\ 2$$
$$\therefore (1013 + 21)\,\text{mbar} \times 1\,\text{m}^3 \div (1013 + 60)\,\text{mbar} = \underline{0.96\,\text{m}^3}$$

Note: The 1013 above is atmospheric pressure, which needs to be included with the calculation.

Where the pressure in not to be increased but just maintained, such as in an undersized gas supply, no such loss in volume occurs.

Installation Requirements

As stated above, the booster or compressor does not increase the volume of gas required and therefore it is essential to ensure that there is sufficient gas supply to deliver the required gas load. The booster or compressor should ideally be located as close as possible to the point where the elevated pressure is needed. Where the entire installation needs to be boosted for a specific requirement, it is recommended that the booster/compressor is not located in the meter compartment. The unit needs to be located on a firm flat horizontal surface. To reduce noise levels transmitting through

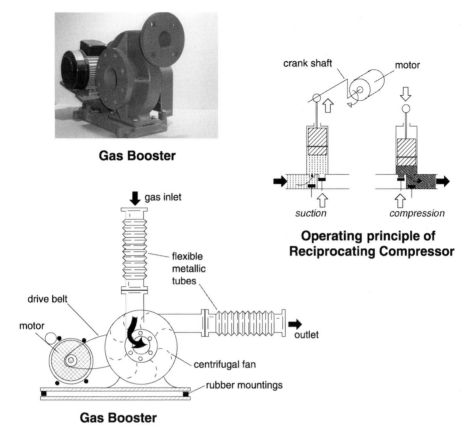

Gas Booster

Operating principle of Reciprocating Compressor

Gas Booster

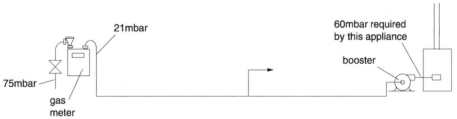

Gas Booster used to elevate pressure to a single appliance

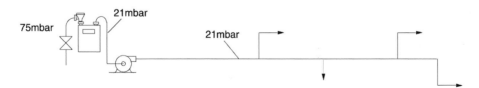

Gas Booster used to maintain pressure within an undersized system

10 Commercial Appliances

the pipework it may be desirable to install the unit on anti-vibration mountings, with flexible metallic tubes to the inlet and outlet connections. A booster or compressor should be located in a dry, well-ventilated area. The size of the ventilation will depend on the size of the room, but it should be large enough to ensure that the ambient temperature does not exceed 45°C; for this it is usual to provide high and low level ventilation. Where mechanical ventilation is provided, this needs to be interlocked with the booster/compressor.

Pressure Fluctuation in the Pipework

To prevent pressure fluctuations up and down stream of the installation, the booster/compressor is sandwiched between a *low-pressure cut-off* and *non-return valve*, as shown.

The *low pressure cut-off* (LPCO) is a pressure switch that senses the pressure in the pipeline preceding the unit; where the gas pressure falls to a predetermined level it shuts down the electrical supply to the motor thereby preventing a dangerous situation occurring, such as the ingress of air into the system. Where the pressure switch is activated there should be no automatic restart and manual intervention will be required. It is the responsibility of the gas engineer to ensure that the LPCO operates correctly and the setting should be as high as possible to prevent nuisance operation.

A *non-return valve* is required where:

- the boosted pressure is to exceed 70 mbar or a compressor is installed;
- two boosters are connected in parallel and gas re-circulation may occur;
- reverse rotation is possible;
- reverse pressure surges occur, due to large volume outlet pipework.

The non-return valve prevents air from entering the system in the event of reverse flow and protects against increased back-pressure supplies. Where excessive high pressures may be experienced, such as from a compressor, a pressure relief valve must be incorporated on the outlet. The valve should be set to open at a pressure sufficient to protect any equipment with the discharge made via a full capacity valve, venting to outside.

Should the gas supply to the booster require a reduction in pipe size, it is necessary to use concentric reducers or tapers in order to prevent unnecessary turbulence. Where the inlet supply pressure to a booster or compressor has a tendency to suffer the continued effects of reduced pressure problems, it is possible to install a volume accumulator or reservoir prior to the unit. This in effect stores a supply of gas and thus prevents the supply drastically fluctuating or dropping as gas is drawn from the pipe. Alternatively a modulating speed motor may be used. A high pressure receiver may also be employed to hold a reserve supply of compressed gas.

Warning Notices

Where a booster or compressor is installed the following notices must be permanently displayed:

- a line diagram showing the location of the isolating valves;
- operating instructions and emergency procedures;
- notification that the meter valve needs to be fully open;
- notification at the meter of the booster/compressor installed.

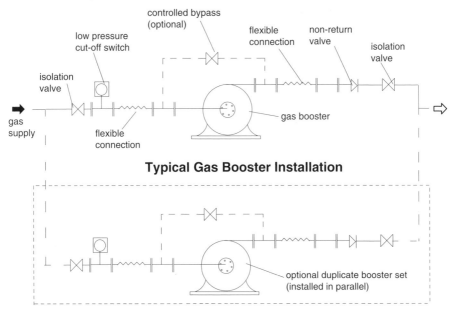

Typical Gas Booster Installation

optional duplicate booster set
(installed in parallel)

Gas Booster Serving a Boiler

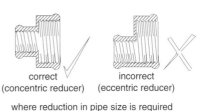

correct
(concentric reducer)

incorrect
(eccentric reducer)

where reduction in pipe size is required
concentric reducers should be used

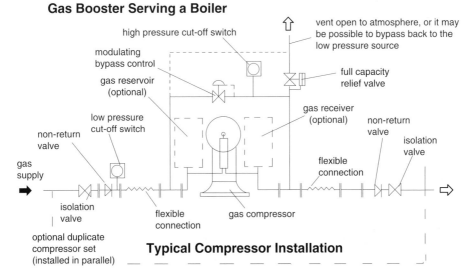

Typical Compressor Installation

optional duplicate
compressor set
(installed in parallel)

Commercial Catering

'Commercial catering' establishments include a variety of restaurants, cafes, school kitchens and pubs, etc. and factories and bakeries where food is processed on a large scale: producing biscuits, cakes and bread, etc. There are many kinds of appliances in commercial kitchens that cater for cooking, warming and cleaning. The following list provides a guide to the cooking processes that may be involved.

Cooking Processes

Baking

This involves cooking food dry in an oven, with no oil, etc. added to the process.

Roasting

This involves cooking in an oven or on a rotating spit using a little oil or fat to keep the food moist. Often the food is initially cooked at a high temperature, e.g. at a high oven setting or by boiling prior to roasting at a slower cooler temperature.

Braising

This is a method of slowly cooking the meat or vegetables in a closed container in an oven. The food is surrounded in a liquid or sauce that provides an accompanying gravy or sauce.

Blanching

This is a process in which the food is usually immersed in boiling water for 2–5 minutes and then immediately cooled in order to remove strong odours, whiten or assist in the removal of skin from a fruit e.g. tomatoes. Blanching is also carried out to limit the action of enzymes that would cause frozen food to deteriorate.

Boiling

This is a process in which the food is cooked in water at 100°C. Food, such as root vegetables, may be placed in cold water and brought to the boil. Other foods, such as green vegetables and meat, are immersed in the boiling water.

Simmering

This is a process of cooking in which the liquid is just below the boiling temperature; at this temperature the liquid shows only an occasional sign of movement. Typically soups and broths are cooked in this way.

Poaching

Here the temperature is just below boiling point, but above the temperature used for simmering, i.e. around 93–95°C. At this temperature there is a gentle occasional bubbling of the liquid. Eggs and fish are typically poached; containers are usually used to contain the food, keeping it separated from the water.

Steaming

This is a process in which the food is cooked in an enclosed container, surrounded by steam from boiling water. Often, to assist steaming, the pressure of the food container is raised; this has the effect of reducing the cooking time required.

Stewing

This is a kind of simmering process in which meat and vegetables are slowly cooked in a stock mixture that just covers the food. Where the food is cooked in the oven, it is often referred to as a hot-pot or casserole.

Deep Frying

This involves submerging the food completely in hot oil or lard/fat at temperatures usually between 180 and 195°C. Often the food is coated with a flour and egg mixture (battered) in order to give a desired taste.

Shallow Frying

Unlike deep frying, in shallow frying the food is cooked in only a small quantity of oil/fat; the food needs to be turned to ensure that all sides are adequately sealed and cooked. Chips (French fries) and other vegetables cooked in this way are known as sautéed.

Dry Frying or Griddling

With this method of cooking a very small dash, if any, of oil is put on a hot plate and all excess fat from the food is allowed to drain away down the slightly tilted plate or griddle.

Grilling

In this method of cooking, the food is cooked by radiant heat from a heated mesh or grill. Grilling can be achieved in many ways: the heat source may be above, below or horizontal to the food source to give the desired effect. For example, where the grill is located below the food (under-fired or flame grilled), as the liquid fat drops from the food it flares up to give the food a distinctive smoky burnt taste and flavour.

Toasting

This is a process, particularly common with bread, in which the food is warmed or browned to give a crisp texture.

Typical Commercial Kitchen

Catering Establishments

Relevant Industry Document
BS 6173

Catering Establishments

An array of specialist gas equipment used in the process of cooking and cleaning will be found in the commercial kitchen. The layout should be such that fryers and ranges should not be located next to areas where water is used, which could lead to a hazard unless suitable splash shields are in place.

Gas Supplies, Emergency Isolation and Gas Interlock

The gas supply to a kitchen requires a manually operable valve to be located in an accessible location, ideally outside the kitchen area, and typically installed near to the exit for emergency purposes. Often an automatic isolation system is used in conjunction with the electrical emergency stop buttons and fire protection systems. Where mechanical ventilation or extract systems are employed, automatic isolation must be provided to prevent the gas flowing to the kitchen unless the extractor fan has been proved to be operational. Sometimes these control systems have the added facility of proving that the supply pipework is gas tight, ensuring no open ends, before gas is allowed to flow through into the system. This is achieved using a low-pressure cut-off valve as shown on page 82. A suitable notice should be positioned at the emergency control or automatic gas valve reset position giving the following information:

In the Event of an Emergency

Turn off the Gas
Prior to re-establishing the supply turn off all burners and pilot valves to the appliances.

After extended shut-off: purge before restoring gas.

Pipes should be spaced 25 mm away from the wall to facilitate cleaning, and an isolation valve should be located prior to each appliance. The final connection to mobile appliances should be made using a flexible connector hose with braid, and white hygienic cover. Mobile appliances should also have a suitable restraint fitted to prevent excess movement; the wheels need to be fitted with a locking device.

Electrical Supplies

The electrical supply to specific appliances needs to be in accordance with the manufacturer's instructions and supplies are usually made via one of the following: fused double pole switch, plug and 13 amp switched or un-switched socket outlet. The flex used to connect to the appliance should be long enough for the appliance to be withdrawn for cleaning, etc.

Fire Precautions

Fire precautions need to be observed in catering establishments to ensure that there is no additional risk of fire, particularly where hot surfaces and liquids are encountered. It is essential to ensure that appliances are not sited where adjacent combustible surfaces could exceed 65°C and, where necessary, the surfaces should be protected. In general, manufacturers' instructions give clear guidance as to the separation distances needed.

Where ductwork, for pipes or ventilation passes through the buildings structure it must be designed so as to prevent the spread of fire or smoke along it. There is a special need for precaution where deep fat fryers are concerned and a notice must be clearly displayed describing what action to take in the event of a fire. Such a notice may read as follows:

Action to take in the Event of a Fire

Turn off Appliance and close lid to fryer.
Where safe to do so, use adjacent fire blanket.

DO NOT USE WATER TO EXTINGUISH THE FIRE

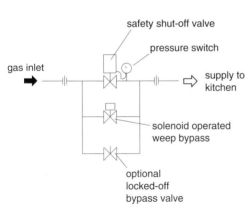

Typical Bypass and Gas Interlock Arrangement

Gas Interlock System

10 Commercial Appliances

Ventilation/Extraction in Commercial Kitchens

Relevant Industry Document
BS 6173

Ventilation Requirements

The ventilation requirements for a catering establishment should be adequate to provide sufficient clean cool air for the occupants to breath and remain comfortable and remove the excess hot air, odours, vapour and steam from the cooking and washing activities. The designer has to match the ventilation for the cooking load, the number of staff/customers and the equipment used. This is usually achieved by a combination of hoods and extractor fans and a system of fresh air supply to replace that removed, providing 20 to 40 air changes per hour. Smaller kitchens often rely on sufficient replacement air being drawn in through grilles in the walls and windows. Typically, 85% of the fresh air needed is drawn from outside with the remaining 15% taken from adjoining areas; this is achieved by maintaining a slightly negative pressure within the environment. This also assists in ensuring that cooking odours do not escape. Air taken from adjoining areas should not be taken from rooms where smoking is permitted. In addition, make-up air velocities from adjoining rooms should not exceed 0.25 m/s, otherwise draughts may be experienced. The ventilation requirements should be supplied by the appliance supplier, however the following provides a guide to that needed.

Typical ventilation requirements

Appliance	m^3/min	Appliance	m^3/min	Appliance	m^3/min
Bain-marie	12	Pastry oven	18	Under-fired grill	18
Café boiler	15	Single-range oven	18	Over-fired grill	27
Boiling pan	18	Steaming oven	18	Deep fat fryer	18

Where the ventilation requirements for the cooking appliances is unavailable, it may be possible to design the air-flow rate based on the following velocities: 0.25 m/s for light cooking; 0.4 m/s for medium and 0.5 m/s for heavy duty cooking. If no canopy is provided and ventilation is to be achieved via, for example, a ventilated ceiling, then as a guide the minimum ventilation rate should be 17.5 l/s per m^2 of floor area, providing not less than 30 air changes per hour. If the kitchen is subdivided into several rooms, a lower air change rate may be advisable to cut down on draughts.

Note that due to the nature of the negative environment no natural draught flued appliances must be installed in the kitchen, as spillage would result.

Canopy Design

It is advisable to take most of the kitchen extraction from directly above the appliances that generate the heat, smell and vapour. The plan dimension of an island canopy should extend a minimum distance of 250 mm all round and, where the canopy is wall mounted, it should overhang the front by 250 mm and the sides by 150 mm. The height should be such that it does not form an obstruction; a typical height to the lower edge would be 2 m.

The material used should be non-combustible. Materials of anodised aluminium and stainless steel prove most popular. A condensation channel should be fitted around the lower edge, sometimes fitted with a drain, to prevent condensation dripping. The canopy should be designed to enable any grease filters to be easily accessible for cleaning, which should be carried out monthly. Some grease vapour will pass through the filters, therefore some form of access must be provided to enable cleansing inside the duct. Clearly all fans need to incorporate anti-vibration mountings and the point of termination must not cause a nuisance to adjoining properties.

A spillage check should be made at the lower edge of the canopy to ensure that the combustion products are being safely taken away.

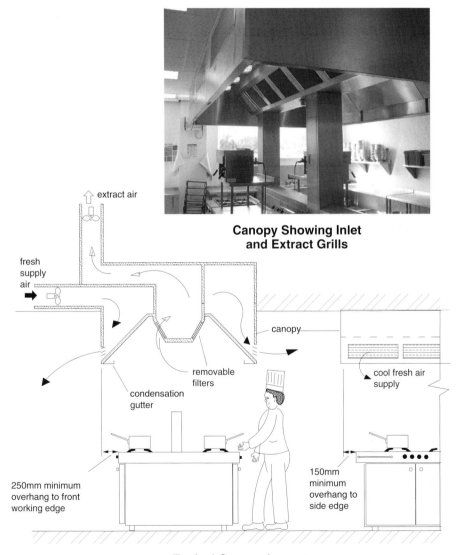

**Canopy Showing Inlet
and Extract Grills**

Typical Canopy Layout

Commercial Cookers and Ovens

Relevant ACS Qualification
COMCAT1

Relevant Industry Document
BS 6173

There is a large range of ovens and hotplates for commercial kitchens; the choice depends on the kind of cooking that is to be undertaken. The basic models are described here. The following page shows some specialist models.

General Purpose Boiling Tables, Ranges and Ovens

The boiling table is simply a hob or hotplate mounted at a working height of around 800 mm. It may consist of an open topped ring burner or it may have a solid top, with the burner located below the metal plate surface, referred to as a boiling table. A storage shelf is usually fitted below the cooking surface. A smaller version, between 450 and 600 mm high and with one or two burners, is referred to as a stockpot stove and is used to heat a large heavy container. The general purpose oven, operating between 7 and 16 kW, may be free standing or may include an open top hob and/or hotplate as part of its design; this is referred to as a range. The component parts of a commercial oven are similar to those previously described for a domestic oven, as described on page 308.

Forced Convection Ovens

The forced convection oven uses a fan to provide even heat distribution throughout, as opposed to the stratified heat (i.e. hotter at the top) in the conventional oven. A version of this oven consists of two ovens or tiers positioned on top of each other. There are two such designs:

1. **Those that are externally heated** With this design a gas burner, located outside the oven, passes the hot combustion products around the outside shell and discharges them out through the flue. A fan located in a section of the oven compartment circulates the warmed air emitted from the oven inner walls.
2. **Those that are semi-externally heated** With this design the flue products pass through the oven itself. The burner is not located inside the oven, but passes the flue products through small holes in the side panels of the oven. The combustion products are drawn into the oven by a fan that creates a negative pressure. Owing to its design, with a throat restriction as shown, some of the air is re-circulated, via a route at the rear of the oven, while some are discharged through the flue.

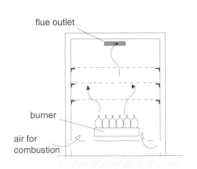

flue outlet

burner

air for combustion

internally heated oven
(combustion products enter the oven)

Conventional Oven

Boiling Table

Range

Forced Convection Double Oven

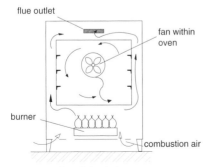

flue outlet

fan within oven

burner

combustion air

externally heated oven
(combustion products do not enter the oven)

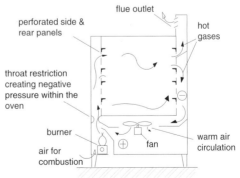

flue outlet

perforated side & rear panels

hot gases

throat restriction creating negative pressure within the oven

burner

air for combustion

fan

warm air circulation

semi-externally heated oven
(combustion products enter the oven)

Forced Convection Ovens

Specialist Ovens and Steamers

The ovens described here are found in specialist kitchens, depending on the type of cooking that is to be carried out.

Pastry Ovens

These are similar in design to the externally heated forced convection ovens, previously described, operating between 9 and 14 kW. The pastry oven is used for baking cakes and pastries and often has several levels or compartments between 125 and 300 mm in height stacked on top of each other, thereby comprising up to four ovens in a group. Each oven may have its own burner or it may be constructed with a single burner located in the base, however separate thermostats for each oven are provided. To change the conditions in an individual oven, vents in the doors are incorporated. These create a moist atmosphere when closed and a dryer heat when opened.

Proving Ovens

This is a large low-temperature oven working at between 26°C and 32°C. It is designed to provide sufficient heat for fermentation, enabling dough to rise prior to bread making, etc. A water container, fed from a feed cistern, is located in the base of this oven to ensure a moist atmosphere.

Steaming Ovens

These ovens are used to cook vegetables, fish and puddings. They usually operate at between 13 and 20 kW and simply heat water in the base to produce a constant large volume of steam. There are two designs: atmospheric and pressurised steaming ovens.

The atmospheric steaming oven

These have a cold fill cistern located adjacent to the oven. This maintains the water level in the oven for the purpose of heating. It can be seen from the diagram that the top is open to allow the steam to pass out slowly and unrestricted.

The pressurised steaming oven

A pressure relief valve located at the top of the oven will only open at around 35 mbar above atmospheric pressure, although there are models that work to much greater pressures that are used to cook frozen foods. Pressurised steaming ovens tend to cook the food much faster than the atmospheric types. The water level is maintained by a float-operated valve located in the oven compartment itself.

Specialist Baking Ovens

There are many types of specialist baking ovens, each are designed to produce a different effect on the cooking process. Many are quite large and they may include many racks, rotating reels or use a conveyor system to cook the food for a set period, according to the speed of the gears. These ovens are particularly suitable for large-scale or batch production.

Combination Oven

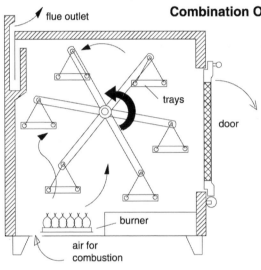

flue outlet

trays

door

burner

air for
combustion

Rotating Reel Oven

The combination oven, shown above,
has a total capability, to include
proving, steaming and convection
cooking. This includes baking and
brazing. It can be programmed to change
humidity levels and
temperatures to ensure the food is
cooked to perfection.

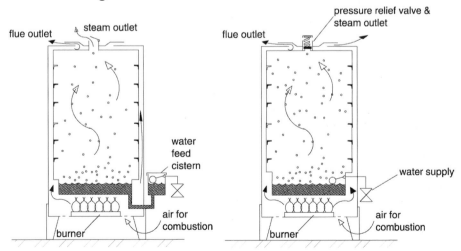

flue outlet   **steam outlet**

water
feed
cistern

air for
combustion

burner

Atmospheric Steaming Oven

flue outlet

pressure relief valve &
steam outlet

water supply

air for
combustion

burner

Pressurised Steaming Oven

Boiling Pans, Hot Cupboards and Bains-Marie

Relevant ACS Qualification
COMCAT1-2

Boiling Pan

These are large-capacity containers that are used to boil up large volumes of water, etc. for bulk cooking. They hold typically up to 180 litres in volume and operate at between 9 and 37 kW. There are several designs, including those that allow for the contents to be directly heated by a burner located beneath a single cooking pot, and those that indirectly heat the pan contents by hot water or steam surrounding the pot, this type is often referred to as a jacketed pan. This indirect method of heating reduces the amount of sticking and burning that can occur when cooking foods such as custards, toffees and jams. Boiling pans often include a large diameter draw-off tap at their base to assist in draining the liquid contents.

Hot Cupboards

These are basically insulated cabinets that are designed to keep previously cooked food warm or they may be used to warm plates. They operate between 2 and 7 kW, depending on their size and the temperature required. Typically 60°C is required to warm plates, whereas a higher temperature of around 80°C is required to keep food warm. Two methods are employed to warm the cupboard: one uses a burner located inside the base of the cabinet area so that the contents are warmed directly by the hot gases as they pass through to the flue outlet. The other design indirectly warms the cabinet by circulating a flow of hot gases or steam, generated in a water compartment below the base of the cabinet, around through channels inside the unit. Cupboards that are indirectly heated tends to be more humid and have the advantage of ensuring that the food does not dry unduly.

Bains-Marie

The bains-marie is another design of appliance designed to keep food hot prior to serving. The word 'bain' is French, meaning bath. Basically a bains-marie is a trough like unit that holds a quantity of water heated to around 80°C. Pans containing the food are placed in it. It operates at around 5–11 kW. There are several designs, including ones with a perforated shelf, below the water line, on to which various sized pots and pans can be placed. In another design, special serving containers that are supplied with the appliance rest on a lip above the water line. There is also a dry heat version in which no water is used and all the heat is generated from the hot gases of the burner.

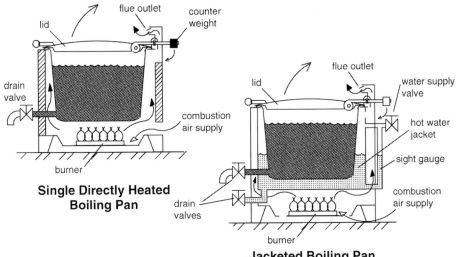

Single Directly Heated Boiling Pan

Jacketed Boiling Pan

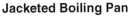

Directly Heated Hot Cupboard

Hot Cupboard

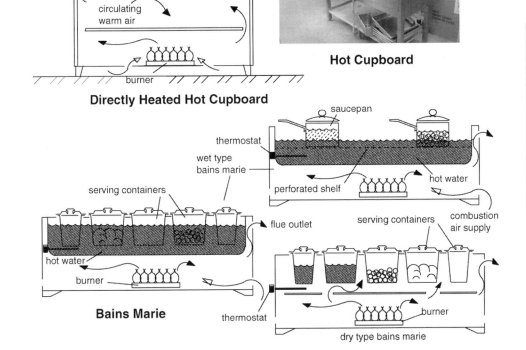

Bains Marie

Grills, Griddles and Fryers

Relevant ACS Qualification
COMCAT3-4

Relevant Industry Document
BS 6350

Grillers

Under-fired or *flare grills* give a heat source below the food, operating up to 30 kW. The grill consists of a burner firing upwards on to a bed of refractory ceramic or pumice material. The food is placed on a series of bars. As the hot fat from the food drops, it is burnt off in the form of an 'uncontrolled' flame, giving the meat a characteristic appearance and flavour.

Horizontally fired grills use surface combustion plaques, between 3 and 12 kW, and are often accompanied by a rotisserie to cook foods, such as kebabs and chicken. The fat from the food can drop to be collected in a container at the base.

Over-fired grills or *salamanders* have the heat source above the food and operate between 5 and 16 kW. They can also be used for toasting. This type of grill often has a grooved solid aluminium branding plate on to which the food is placed. This plate allows the hot fat to drain away, but at the same time assists in cooking the food on both sides due to absorbed heat that is held in the aluminium plate.

Griddles

The griddle is effectively a hotplate that is heated from below by burners to give a surface temperature of up to 300°C, operating between 5 and 15 kW. It is basically used for dry-frying, and as such it is sometimes called a dry fry plate. A channel is often incorporated to one side to enable the fats from certain foods to be drained away. As with the grill, these appliances are often installed around the counter area of some types of snack bar.

Deep Fryers

These are appliances that hold a quantity of oil or fat in which food such as chips, fish and doughnuts, etc. are cooked. There are two different designs: one that has a flat bottom with the oil heated directly from below and a second type has oil that is heated at a location above the base, so maintaining a cool zone below the flame in which old food particles can sink without carbonising and spoiling the flavour of the food load. There are two methods of heating the oil using this design of fryer. In the first method there is what is known as a V-shaped fryer, with the lower V section below the burner (see illustration). The second method requires the gas flame to be fired directly into tubes that pass through the mass of the oil. With the flat bottom fryer it is essential that, before each frying, all old particles are strained from the oil. The heat input can vary between 8 and 40 kW, depending on the size of fryer.

Tilting Fryers and Brat Pans

These are shallow cast iron flat-bottomed appliances used for shallow frying or dry frying. They have a tilting function to enable the surface to be tilted forward by the operation of a lever or hand wheel.

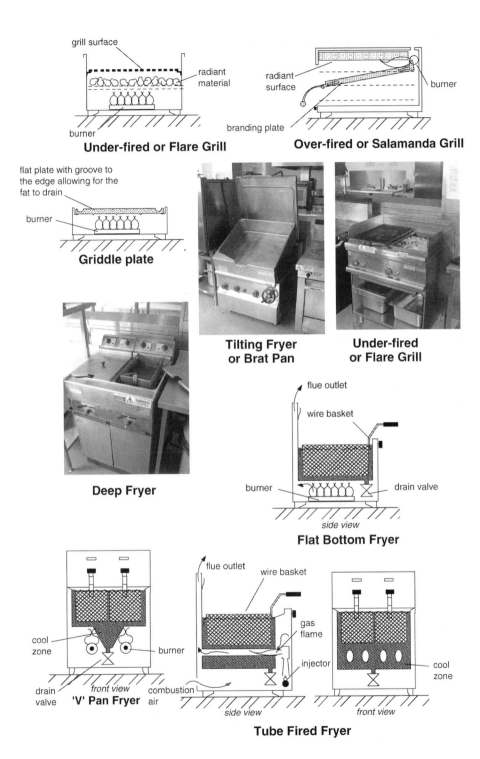

Under-fired or Flare Grill

Over-fired or Salamanda Grill

Griddle plate

Tilting Fryer or Brat Pan

Under-fired or Flare Grill

Deep Fryer

Flat Bottom Fryer

'V' Pan Fryer

Tube Fired Fryer

10 Commercial Appliances

Drinking Water Boilers

Relevant ACS Qualification
COMCAT2

Several different designs of water boiler will be found for the purpose of providing hot water for drinking purposes. These range from the simple urn to the pressure boilers found in a café. Some of these boilers are described below.

The Urn
The simple urn consists of a container with a lid into which water is poured from a jug. Beneath is a gas burner that simply heats the water. The urn has limited controls and just allows the water to boil away. A more sophisticated version has many extras, including a water fill connection with sight glass and a means of thermostatic control. With the urn, water is drawn off from the bottom of the container and so the water is only hot enough when the whole volume has been sufficiently heated. A variation on this is the jacketed urn, which consists of a container within a container. The inside container contains a liquid that easily burns and sticks when heated, such as milk. The outside container is filled with water. This overcomes the problem of hot spots.

The Expansion Boiler
Unlike the urn above, which requires all the water to be heated before it is ready for use, in the expansion boiler the draw-off point is located at a high point, above the water level of the cold water. As water is heated it expands and it is only the boiling water that would reach this draw-off tap. The cold water is fed to the boiler from a cistern with a float-operated valve located at the back of the boiler.

The Pressure Boiler (Café Steam Boiler)
This is a design of boiler that produces both steam and hot water. The boiler is often located beneath the counter top and the hot water is forced up to the draw off point by the steam pressure contained in the boiler. Cold water enters the boiler and stops when the water level reaches a set height, leaving a void at the top of the boiler. As the water is heated and boils, the steam generated cannot escape, as with the two previous boilers, but accumulates in the top void. As the pressure of the steam begins to build it acts on the pressure stat (described on page 78), which in turn closes down the gas supply. Nothing will happen if the hot draw-off tap or steam outlet is opened prior to the production of steam, however if there is steam pressure in excess of atmospheric pressure, opening the hot draw-off tap will cause the steam to force the boiling water up the discharge tube and discharge from the outlet. Conversely, if the steam outlet is opened it too would discharge from its outlet. Both a pressure relief valve and high limit stat, interlocked with a thermocouple interrupter, are incorporated with this type of boiler. Often these boilers have an adjoining side urn for heating milk and coffee percolator, named a café set. The control lever of the gas and water supplies to these boilers are connected together, referred to as a safety interlock. This prevents one supply being open without the other. The safety interlock can be seen to the far right in the photo.

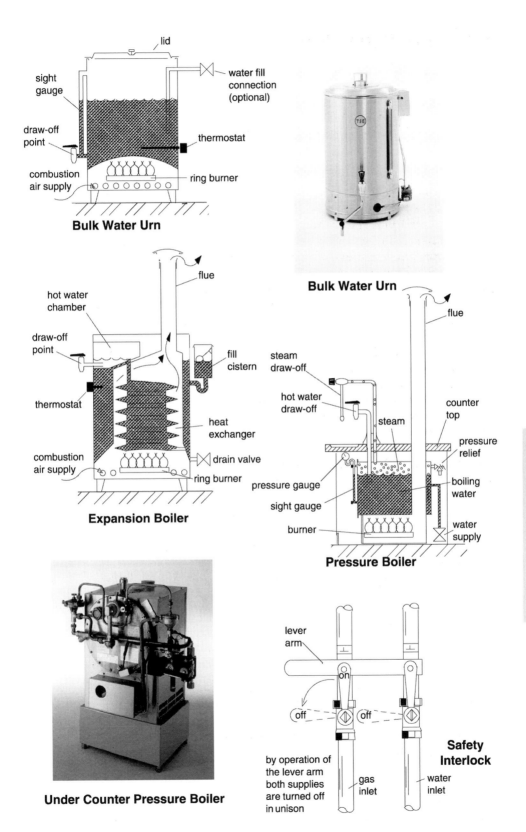

Bulk Water Urn

lid

sight gauge

water fill connection (optional)

draw-off point

thermostat

combustion air supply

ring burner

Bulk Water Urn

flue

hot water chamber

draw-off point

fill cistern

thermostat

heat exchanger

combustion air supply

drain valve

ring burner

Expansion Boiler

flue

steam draw-off

hot water draw-off

steam

counter top

pressure relief

boiling water

pressure gauge

sight gauge

burner

water supply

Pressure Boiler

Under Counter Pressure Boiler

lever arm

on

off

off

by operation of the lever arm both supplies are turned off in unison

gas inlet

water inlet

Safety Interlock

Mobile Catering Units

Relevant ACS Qualification
CMCLP1

Relevant Industry Documents
BS 6350 and LPGA CP24

The mobile catering unit could be a motor vehicle, trailer or a caravan that has been converted or it may be purposely designed and factory produced. There are many designs, including the burger or hot dog van, fish and chip vehicle and the mobile café. These vehicles and the gas appliances contained in them are covered by the Gas Regulations and all work undertaken must be completed by an approved gas fitting operative. Changing the gas cylinders is not considered as gas work.

Cylinder Carriage and Location

In transporting the gas cylinders to the location in a closed vehicle, there are specific considerations, as described on page 148. Once on site the cylinders should be located in the outside air at least 1 m from any opening into the vehicle or sources of ignition. A sign should also be positioned warning of the presence and danger of gas. It is best to site the cylinders on the nearside of the vehicle to reduce the risk of damage and explosion in the event of a road accident. The cylinders could be located in a ventilated compartment that can be accessed from outside. The compartment must be separated and gas tight from the interior of the vehicle and be of a fire resistant construction providing a minimum 30-minute fire resistance. The size of the ventilation needs to be a minimum of 1/100 of the compartment floor area, or alternatively the compartment should be totally ventilated using reinforced mesh. There must be no sources of ignition within 1 m measured horizontally and 0.3 m above the highest vent.

Pipework

Where an automatic changeover device supplies a constant gas supply to the vehicle, an emergency control valve needs to be located next to the vehicle entrance along with the appropriate warning notice identifying what to do in the event of a gas escape. For single cylinder installations, the cylinder valve is sufficient. Pipework would generally follow the same installation practices as found with a permanent dwelling, however additional protection will be necessary for pipework fixed below the vehicle to overcome the increased likelihood of damage such as by stones being 'thrown up' while the vehicle is moving. See also the notes specific to caravans on page 416.

Appliances

All appliances should be secured in position and sited so that they do not obstruct the passageway. Appliances should be installed in accordance with the manufacturers' instructions and not sited where they might create a fire hazard. It is essential that no appliance is ignited whilst the vehicle is in motion and that the gas supply is also turned off at the cylinder valve when the vehicle is to be moved. Vehicles that carry deep fryers should include a hood that incorporates a flue to the outside. Any flue needs to be capable of withstanding the vibration of the vehicle movement and the discharge needs to be in a safe place outside the vehicle at high level and away from any openings

10 Commercial Appliances

into the vehicle. The canopy should extend at least 150 mm beyond the appliance and have a minimum of 27 cm^2 of flue for every 1000 cm^2 of canopy base area.

Ventilation

The minimum ventilation openings to a mobile unit should be at least 25 cm^2 for every 1 kW heat input of all the combined appliances, or 100 cm^2, whichever is the greater. This ventilation is then divided equally between high and low levels, with the top ventilator grille being positioned as high as possible. This ventilation is in addition to that provided through the serving hatch and windows.

Visual Inspection and Maintenance

Every time the vehicle is used, the cylinders, pipework, appliances and flues should be inspected for obvious signs of damage. It is recommended that the vehicle is inspected every 6 months and in all cases the continued safety of the vehicle, in particular the tightness of the gas system, which is subjected to the continued movement brought about by travel, should be checked at intervals of less than 1 year.

**Typical Mobile
Catering Units**

Part 11

Non-Permanent Dwellings

Residential Park Homes

Relevant ACS Qualification
    CCLP1 RPH

Relevant Industry Documents
    BS 5482 pt. 2 and 3632

The residential park home, although used for permanent residence, is regarded as a *non-permanent dwelling*. The structure is produced off site, often in two halves and bolted together in situ. Basically these homes are prefabricated bungalows, often with a pitched and tiled roof. They are manufactured fully fitted to include a kitchen/dining area, sitting room, bathroom and bedrooms. Generally, everything in the home is included, including all fixtures and fittings plus carpets and soft furnishings. They are set on a concrete base and supplied with water, drainage, electricity and gas connections. These units may be supplied with natural gas or LPG. The appliances found within this type of building are similar to those of any permanent dwelling. A flue and false chimney breast accommodates a flue box and non-combustible hearth to enable a gas fire to be installed. Alternatively, a caravan holiday home type fire may be installed (see page 414). Residential park homes may also require the use of drop vents in enclosed housings to prevent the accumulation of gas in low areas, again see page 414 where this is described.

Ventilation for Habitable Areas Containing Open Flued or Flueless Gas Appliances

As stated above, because the structure is regarded as non-permanent it has different ventilation requirements to meet the needs of its structure from those of a permanent dwelling. The ventilation requirement for such a building is calculated using the following formula:

$$(2200\ G) + (440\ F) + (650\ P) + (1000\ R) = V$$

where

G is the gross kW heat input of all type 'A' appliances (flueless);
F is the gross kW heat input of all type 'B' appliances (open flued);
P is the number of people for whom the home was designed;
R is the gross kW heat input of all open flued oil-fired appliances; and
V is the minimum area of fixed ventilation in mm^2.
The ventilation is equally divided between high and low level.

Example A residential park home has an open flued radiant gas fire of 5.2 kW (gross) a gas cooker of 12.7 W (gross) and an open flued oil boiler of 17 kW (gross). The home is designed for six people.

The ventilation size would need to be:

$$(2200 \times 12.7) + (440 \times 5.2) + (650 \times 6) + (1000 \times 17) = \underline{51\,128\,mm^2}$$

Thus $\underline{25\,564\ mm^2\ (256\ cm^2)}$ high level and $\underline{25\,564\ mm^2\ (256\ cm^2)}$ low level grilles are required, positioned equally round the areas in which the appliances are located.

Ventilation in Areas that Contain No Gas Appliances Other Than Room Sealed

Although more a question of design than gas safety, ventilation is also required to overcome condensation problems, etc. in this type of room. This includes:

- *bathrooms and* WCs that require 10 cm^2 at low level and 10 cm^2 at high level and
- *bedrooms and general living rooms* that require 10 cm^2 at low level and 20 cm^2 at high level, although in the case of a bedroom this may be a single 40 cm^2 grille positioned in a high level window frame. If there are more than four occupants of the dwelling, the above set figure is not used and the following formula is applied: 6.5 cm^2/person, apportioned high and low level with the high level grille being twice the size of the low level grille.

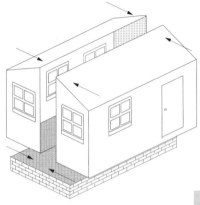

the residential park home generally consists of two or more sections bolted together

The Residential Park Home

11 Non-Permanent Dwellings

Leisure Accommodation Vehicles

Relevant ACS Qualification   Relevant Industry Documents
CCLP1 LAV   BS 5482 pt. 2 and BS EN 721, 1645-7 and 1949

Leisure accommodation vehicles are for temporary or seasonal use and fall into one of the following three categories:

1. *caravan holiday homes*, delivered to a site by a transporter and manufactured with wheels solely for manoeuvring into place. These are often grouped together, forming a holiday site;
2. *touring caravans*, towed by a vehicle;
3. *motor caravans*, which are self-propelled.

Leisure accommodation vehicles are supplied fully fitted out and, depending on their size, include kitchen/dining area, sitting area, bedrooms and often a flushing toilet and shower facility. Everything is included, even carpets and soft furnishings. An optional extra in many of the modern units is a central heating system. Connections to this type of caravan form part of the construction and include water, drainage, electricity and gas (LPG), which can be linked temporarily to the services provided by the park operator. If the unit is for hire, then it would be subject to the Landlords Gas Safety inspections and records, described on page 272, as required by the law.

Ventilation Requirements

Prior to 1999, ventilation sizing in caravans was undertaken using a different formula from that undertaken today. For the static caravan holiday home the calculation employed was the same as that currently used in boats and non-permanent dwellings. If the caravan was built before 1999 this method of ventilation should be applied; it is described on page 422. However, in order to standardise with Europe a ventilation method is now employed, based on the Table 1 opposite, which takes into account the overall plan area of the caravan itself, i.e. excluding towing bar, etc. and should be applied to all the principal habitable rooms. Where there are additional compartments or cupboards with no gas appliances other than room sealed, Table 2 is used.

The ventilation openings should:

- be located more than 300 mm from any terminal;
- be protected by an accessible grille or screen that can be easily cleaned;
- where high-level wall vented, be at least 1800 mm above the floor level and not affected by high-level bunk bed mattresses;
- where low-level wall vented, be no more that 100 mm above the floor;
- not be affected by drapes or curtains covering them;
- where ventilation is to pass through a bed box or cupboard, etc., it must not be affected by stored items and, where necessary, be ducted.

Table 1 Size of ventilation openings in the main living areas

Overall area of caravan	Low level vent (minimum)	High level vent (minimum)	
		If in roof	If in walls
≤ 5 m^2	10 cm^2	75 cm^2	150 cm^2
$>5–\leq 10$ m^2	15 cm^2	100 cm^2	200 cm^2
$>10–\leq 15$ m^2	20 cm^2	125 cm^2	250 cm^2
$>15–\leq 20$ m^2	30 cm^2	150 cm^2	300 cm^2
>20 m^2	50 cm^2	200 cm^2	400 cm^2

Table 2 Size of ventilation for compartments with no gas appliances (excluding WCs and bathrooms and those with room sealed appliances)

Low level vent (minimum)	High level vent (minimum)	
	If in roof	If in walls
5 cm^2	15 cm^2	30 cm^2

Caravan Holiday Home

Touring Caravan

Motor Caravan

Caravan Holiday Homes

Relevant ACS Qualification   Relevant Industry Documents
CCLP1 LAV   BS 5482 pt. 2 and BS EN 721, 1949 and 1647

Gas Supply

LPG is traditionally used to supply a caravan holiday home from a permanently metered supply or from cylinders. Often the manufacturer will use two sets of gas cylinders, located at different positions, to overcome the problem of an inadequate gas flow where a combination boiler has been installed. If two separate supplies feeding a caravan, adequate labelling should be provided at each emergency control, warning of the additional supply. The regulator is often mounted on a purpose-built bracket, providing stability, and the supply is taken to the van through a hose, conforming to BS3212: type 1, which has a braided stainless steel outside covering. Connections to this hose need to be crimped or worm drive proprietary fitting. All LPG hoses should be replaced every five years. If the caravan is not fully anchored down, and there is a possible risk of movement due to high winds, a quick-release self-sealing coupling or a low-pressure cut-off device should be incorporated. Installation pipework should follow the same general rules as that for permanent dwellings.

Drop Vents (Gas Dispersal Hole)

In order to prevent the accumulation of gas in base units and cupboards where un-burnt gas may have accidentally been released, holes are positioned in the floor to allow it to escape. These may also form part of the low level ventilation. It is essential that they are designed and positioned so that they do not get covered and that the gas can escape from beneath the caravan itself.

Specific Appliances

In general the appliances used in a caravan are of the same design as those found in the permanent home and for this reason the same ACS assessments and appliance designs that have been previously described apply. However, there is one specific design of gas fire, shown opposite, that incorporates a continuous 63 mm flexible flue pipe that joins the fire to the terminal. This is located within a concealed part of the van. It is essential that this void is adequately ventilated at top and bottom with grilles that each provide a minimum of 96 cm^2 free air. Access to the void also needs to be allowed for and during safety checks this void must be inspected. The flue needs to be securely fixed and supported, ensuring that it remains at least 50 mm from all combustible surfaces and passes continually upward to the terminal with no low point. The terminal itself also forms part of the supplied appliance and, in effect, seems to disregard BS specifications in terms of its discharge height, terminating just above the roof line. However, provided that manufacturers' instructions are observed fully for installation and testing this is the specified location. The ACS qualification to this design of fire is HTRLP2.

The ventilation requirements have previously been described on page 412.

**Typical 47kg Propane Gas
Cylinder Supplying Caravan**

**Typical Gas Fire
with its Supplied Flue and Terminal**

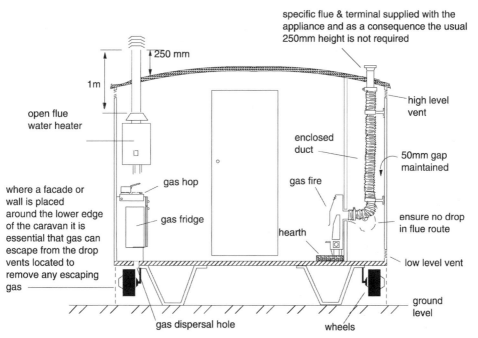

specific flue & terminal supplied with the
appliance and as a consequence the usual
250mm height is not required

250 mm

1m

open flue
water heater

high level
vent

enclosed
duct

50mm gap
maintained

where a facade or
wall is placed
around the lower edge
of the caravan it is
essential that gas can
escape from the drop
vents located to
remove any escaping
gas

gas hop

gas fridge

gas fire

hearth

ensure no drop
in flue route

low level vent

gas dispersal hole

wheels

ground
level

Caravan Holiday Home

11 Non-Permanent Dwellings

Touring and Motorised Caravans

Relevant ACS Qualification   Relevant Industry Documents
CCLP1 LAV   BS 5482 pt. 2 and BS EN 721, 1949 and 1645-6

Gas Supply

The gas supply to this type of vehicle is from small LPG gas cylinders carried on board. A supply regulator must be fitted with an overpressure shut-off device, capable of ensuring that a pressure of 150 mbar is not supplied to any appliance. The cylinders are either positioned externally within a frame or contained within an enclosed housing.

The *compartment* must:
- provide a securing device to hold the cylinders in an upright position;
- allow sufficient access to connections and for replacement of cylinders;
- ensure hot surfaces, e.g. exhaust pipes, are at least 250 mm from the side-wall and 300 mm beneath unless a thermal shield is included, as shown opposite, with a 25 mm air gap;
- housings are to have a minimum fire resistance of 20 minutes and have ventilation direct to the outside. This may be at low level only, to a minimum size of 2% of the floor area of the enclosure, with a minimum of 100 cm^2 *or* high and low level ventilation can be provided of not less than 1% of the floor area with a minimum of 50 cm^2.

Internal housings must:
- contain no more than two 16 kg sized cylinders;
- have a sealed access, with its bottom edge at least 50 mm above the floor.

Note: The ventilation size may be reduced to a 20 mm duct in the base, falling continuously to outside with a maximum length of 200 mm, if the maximum cylinder capacity is reduced to 7 kg.

External housings must:
- only be accessible from outside the vehicle;
- be sealed totally from the internal area.

Specific Appliances

The appliances used within a motorised or touring caravan are quite specialist and different from the appliances found in all other types of non-permanent and permanent dwellings. Some smaller boats make use of some of these appliances. Space heaters, water heaters and gas refrigerators must be of a room sealed design. Cookers and gas lights will be flueless. Many of the larger caravan holiday homes, described on the previous page, no longer have gas lights. However, for touring vehicles this is not the case, and many new vans have this method of lighting in addition to electrical lamps. The flue arrangement must carefully follow the manufacturer's instructions. Where fluing below the floor of a vehicle, care needs to be taken to ensure that the outlet is not within an area, such as within the chassis, where air is drawn and the products could be pulled back into the vehicle. Also terminals should not be positioned within

500 mm of the refuelling point. The warm air heaters shown below are typical. A concentric flue/ventilation duct to the outside is made to the connection at the top (see photograph: the warm air distribution duct is on the left of the unit). The return air is also taken in at this point. The units shown are also suitable for installing in a boat and can be mounted in almost any position as the flue does not necessarily need to come from the top. A warm air and hot water combination unit that may also be installed in this type of caravan is shown on page 423.

The ventilation requirements have previously been described on page 412.

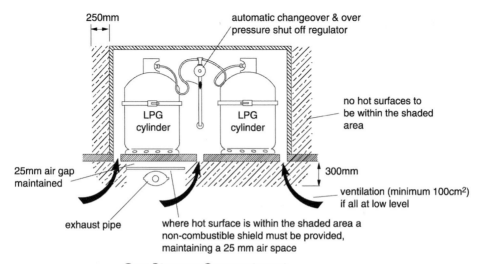

Gas Storage Compartment

Warm Air Unit
(flue and vent connection shown at the top)

Gas Supply on Boats

Relevant ACS Qualification   Relevant Industry Documents
CCLP1 B   BS5482 pt. 3 and BS EN 10239

Shore Fed Gas Supply

This type of supply would only be considered for a permanently moored boat. The boat would need to be adequately secured to prevent any movement apart from that expected due to tidal variations, flooding conditions and possible drainage of canals or due to drought. The method of gas connection can be achieved by one of the following two methods:

1. by using a loop of adequate length, allowing for the maximum movement, between the fixed rigid pipes of the boat deck and bank; precautions need to be taken to ensure that the loop cannot be fouled up due to the actions of the movement;
2. by using two flexible connectors and a fixed rigid section secured to a gang plank that is hinged at one end with a slide way at the other.

For an LPG installation the type of hose that can be used for the flexible connections is PVC coated braided armoured hose with flexible tubing manufactured to BS 3212. If the supply is natural gas, the flexible connectors need to be PVC coated, braided, armoured hose with manufactured heavy duty stainless steel to BS 6501.

If there is a possibility that the boat could break from its mooring, allowing the possibility of a gas escape, then a quick release, self sealing coupling should be incorporated. For LPG supplies this safety arrangement may be achieved by the use of a low-pressure cut-off device.

At the point of supply there should be a warning notice advising of the need to ensure safe isolation.

Onboard LPG Supply

Ideally the gas cylinders should be stored on an open deck or within a housing that is suitably ventilated so that any leakage of LPG will fall and discharge overboard. They also need to be sited at least 1 m from any openings/hatches below deck and from possible sources of ignition. It is possible to store the cylinders in a compartment below deck, or within a recess in the deck. From the base of this compartment there should be a low level drain vent made of a material suitable for the passage of LPG and of no less than 19 mm internal diameter, that discharges, with a fall to outside, at least 75 mm above the water line, even when the boat is fully laden.

Points to observe where onboard cylinder storage within a housing is used

- Provide a 30 minute minimum fire resistance to the compartment.
- The compartment should not be used for general storage.
- The compartment needs to be sealed (vapour tight) from the hull.
- Access should be gained via a lid or cover at the top
- In addition to low-level ventilation, high-level ventilation should be provided above the level of the cylinder.
- The high-pressure stage and non-adjustable regulator should be at the top of the compartment. The high-pressure hoses should be less than 1 m in length.
- An over-pressure device should be included and non-return valves incorporated in the high pressure stage connections where two cylinders are connected together.

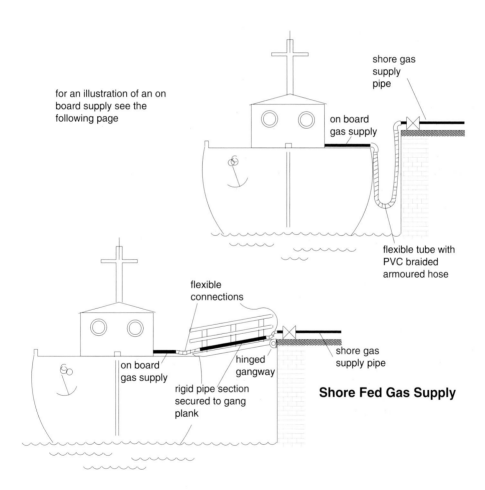

for an illustration of an on board supply see the following page

shore gas supply pipe

on board gas supply

flexible tube with PVC braided armoured hose

flexible connections

on board gas supply

hinged gangway

rigid pipe section secured to gang plank

shore gas supply pipe

Shore Fed Gas Supply

Pipework Installation on Boats

Pipework

Special precautions need to be observed when running pipework on boats as the pipework will be subjected to very harsh conditions in terms of the vast amount of water and vapour that can be expected, which may contain salt, oil and possibly petrol. The materials to be used should be restricted to the following:

- seamless copper tube to BS EN 1057;
- stainless steel, suitable for use with LPG in a marine environment;
- copper nickel alloy, suitable for use with LPG in a marine environment.

All joints to the pipework should either be hard soldered (using a solder with a melting point above 450°C) or have compression or screwed fittings. Compression fittings should use copper or stainless steel olives as appropriate. Soft soldered joints should not be used. The method of pipe sizing given on page 154 can be successfully used for boat installations.

All pipework should be supported at intervals no greater than 500 mm and, where it penetrates bulkheads, the pipe needs to be adequately protected by a suitable sealed sleeve or bulkhead fitting/grommet.

The pipe should not be installed within the following locations:

- where it is inaccessible;
- in areas dedicated to electrical equipment, including where batteries are used, or petrol engines unless run through a pipe capable of containing the gas in the event of a leak – there should be no joints in the gas pipe;
- within 100 mm of engine exhaust pipes;
- in ducts used for ventilation, electricity or telecommunications;
- below the bilge water level;
- within 30 mm of electrical cables.

Leak Detectors

Special leak detectors are often installed within the cylinder housing. With all the appliances turned off, they are designed to provide an instant visual check for leaks in the low-pressure pipework. There are two types of detector.

The bubble leak tester

When the top is fully depressed, if there is any leak on the supply pipework it will be indicated by bubbles, which can be seen in the liquid through the viewing window. When installed, a pressure gauge should also be fitted in the high-pressure side of the system.

The 'gasflow' indicator

This is fitted to the regulator. If there are no leaks the indicator stays green, but if there is a leak the indicator changes to red.

Tightness Testing

BS EN 10239 is primarily concerned with the manufacture of new craft and therefore fails to identify the requirements of testing with gas. It makes the suggestion that before gas is put into the pipework for the first time, the pipework should be subjected to a test of three times its working pressure, but no greater than 150 mbar. This pressure should, after a 5 minute stabilisation period, hold constant to ±5 mbar during a following 5 minute period. To complete any further air tests and tightness tests with gas, one should follow BS 5482, pt 3, carrying it out according to the instructions in Part 5, Tightness Testing, with the exceptions that the test period is 5 minutes and not the 2 minutes as stated and no leaks are permitted during the test time.

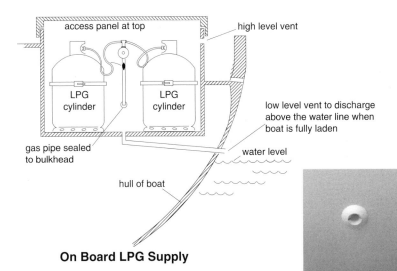

On Board LPG Supply

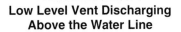

Low Level Vent Discharging Above the Water Line

Bubble Leak Tester

the plastic tube shown contains the liquid (propylene glycol) to be filled into the bubble tester, through which the bubbles rise

Looking Down Into a Small Storage Housing

a *Gasflow* leak indicator is fitted to the regulator (note the low level drop vent at the base of the housing going through the hull of the boat)

Appliances and Ventilation for Boats

Relevant ACS Qualification
CCLP1 B

Relevant Industry Documents
BS5482 pt. 3 and BS EN 10239

Appliances

In larger boats, there are all the typical gas appliances associated with the domestic dwellings, including boilers, gas fires, cookers and refrigerators. Older installations may include flueless and open flued appliances (i.e. types 'A' and 'B'.) However, for all new installations and replacement of existing appliances, apart from gas cookers, room sealed appliances (type 'C') must be installed. They need to be recommended by the manufacturer for use in a marine environment. All appliances need to incorporate a flame failure device and be installed with particular care if they are near to any combustible materials. A typical warm air/hot water combination unit is shown opposite. A second design that is also suitable for installation within a boat can be found on page 417.

Ventilation Requirements

Ventilation on board a boat is found by applying the following formula:

$$(2200\ U) + (440\ F) + (650\ P) = V$$

where

U is the gross kW heat input of all type 'A' appliances (flueless);
F is the gross kW heat input of all type 'B' appliances (open flued);
P is the number of people for whom the vessel was designed;
V is the minimum area of fixed ventilation in mm^2.

The ventilation is divided equally between high and low level, low level ventilation being achieved by the use of suitable duct work.

The size of the ventilation opening depends on the age of the craft. Since 1999, as stated above, all gas appliances installed, apart from the cooker, need to be type 'C', i.e. room sealed. However flued appliances may be found in older boats. So where there are no open flued appliances on board, the $440\ F$ is simply dropped from the calculation.

Example A four-berth boat has the following gas appliances installed: a cooker of 12.6 kW (gross); a room sealed water heater of 14.8 kW (gross) and room sealed warm air heater of 10 kW (gross). The ventilation required would be:

$$(2200\ U) + (650\ P) = V$$
$$\therefore (2200 \times 12.6) + (650 \times 4) = \underline{30\,320\ \text{mm}^2\ (303.2\ \text{cm}^2)}$$

If the water heater and warm air heater had been open flued, as was sometimes fitted in older craft, then the ventilation needed would be much larger:

$$(2200\ U) + (440\ F) + (650\ P) = V$$
$$\therefore (2200 \times 12.6) + (440 \times 24.8) + (650 \times 4) = \underline{41\,232\ \text{mm}^2\ (412.3\ \text{cm}^2)}$$

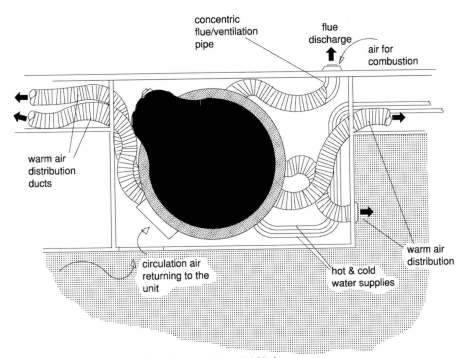

Plan View of Combi Unit

Combination Warm Air and Hot Water

Part 12
Electrical Work

Electricity

This book is not intended to make the reader competent in the skills of an electrician, it merely aims to identify the basic needs associated with electrical installation work.

What is electricity?

Electricity is basically the flow of infinitely small particles, called free electrons, along a conductor in a circuit. If there is no circuit, there can be no flow of electricity. As with water in a stream, if it is flowing it is referred to as a current. Electrical current is measured in amperes (amps). To make the current flow a force is required, referred to as an electromotive force (e.m.f.). This can be obtained from several sources, including batteries, thermocouples and generators. The e.m.f. is measured in volts. Basically, where two conductors have different voltages (i.e. there is a potential difference between them) a current will flow between them. The current needs to flow through/along a material that has free electrons within its structure, such as a metal, and the resistance to the flow, due to the conductor's size and length, will decrease or increase the speed of the current, causing the conductor to warm up due to friction. The resistance to flow is measured in ohms (Ω) and the warming effect, being the power generated, is measured in watts.

The current, e.m.f., resistance and power relate to each other and their relationship is given by Ohm's law:

$$\text{Volt} \div \text{Amps} = \text{Ohms} \quad \text{and} \quad \text{Volts} \times \text{Amps} = \text{Watts}$$

Direct Current and Alternating Current

There are two forms of electricity: one in which the current flows continually in the same direction (direct current – d.c.) and the other, where it alternates between one direction and the other (alternating current – a.c.). The way that the electricity is produced determines how it initially flows. Batteries and thermocouples, for instance, use d.c., whereas an alternating generator produces a.c.. The domestic power supply to the home is a.c. and the current switches direction between phase (+) and neutral (−) 50 times per second (50 Hz).

Magnetism

One final point in this quick introduction to electricity is that a force is generated by the flow of electricity. As the current flows, it creates a small flux or flow of energy from the conductor. Where a length of conductor wire is wound to form a coil, this flux is then magnified to produce a magnetic field. This phenomena is used in many electrical systems, operating motors and solenoids (magnetic valves) and is also used to induce a flow of electricity in another conductor, such as occurs in a transformer.

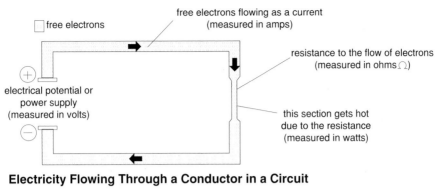

Electricity Flowing Through a Conductor in a Circuit

free electrons flowing as a current
(measured in amps)

free electrons

resistance to the flow of electrons
(measured in ohms Ω)

electrical potential or
power supply
(measured in volts)

this section gets hot
due to the resistance
(measured in watts)

ohms law

potential
volts

current
amps

resistance
ohms

power
watts

current
amps

potential
volts

With calculation triangles it is possible
to find the value of one unknown
quantity simply by covering that value
and completing a calculation of the
remaining quantities as shown.

volts
ohms
∴ volts ÷ ohms = amps

amps ohms
∴ amps x ohms = volts

volts
amps
∴ volts ÷ amps = ohms

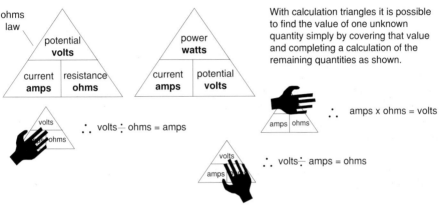

current flow in one
direction only

lamp

battery

Direct Current

a simple ac
alternator

slip rings with carbon brushes

As the loop rotates it cuts through the magnetic flux
generated between the two poles of the permanent magnet
and in so doing induces the electrons to flow

S   N

lamp

the current flows
alternately one way
then the other (ac)

Alternating Current

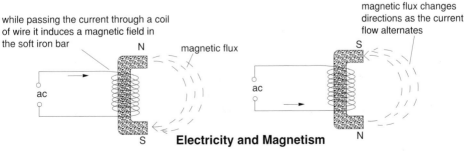

while passing the current through a coil
of wire it induces a magnetic field in
the soft iron bar

N   magnetic flux

ac

S

magnetic flux changes
directions as the current
flow alternates

S

ac

N

Electricity and Magnetism

12 Electrical Work

Electrical Safety

Relevant Industry Document
BS 7671

Using gas can involve hazards: death can be caused instantly by a gas explosion or by the slower, hidden, but equally deadly effect of carbon monoxide poisoning. One is not necessarily aware of the presence of this gas at the time of installation. Electricity can also kill by electric shock or cause a fire as a result of some poor design, which again may not have been apparent at the time of installation.

Electrical Shock
Every time a gas installer works on an appliance that requires an electrical supply, they are working on something that has the potential to kill. The size of electric current that can kill a human is as small as 80 milliamp (0.08 amps). This is 'very' low and any fuse fitted in an appliance or supply would be insufficient to protect the user from harm. Therefore it is imperative that operatives do not remove any covers to electrical components unless they are fully competent in the work that they are about to undertake. In fact, the Electricity at Work Regulations, like the Gas Safety Regulations, state that no one should undertake work unless *competent* to do so.

Electrical Fires
Electrical shock is not the only way in which electricity can kill; electricity can also cause fires. A fire caused by an electrical fault is often the result of a poor earthing arrangement. When a live wire touches an earth conductor the current should flow so rapidly that it exceeds the ampere rating of the fuse, causing the fuse to trip and break the circuit.

Assume 230 V touches the earth wire, which is in good condition, and assume that the resistance of the cable used is 0.2 Ω. Therefore, applying Ohm's law, Volts $\div$ Resistance = Amps:

$$230 \div 0.2 = \underline{1150\,\text{amps}}$$

This would rapidly break, say, a 32 amp fuse in the consumer unit.

However if there were a poor earth, with a bad connection, the resistance might increase to, say, 10 Ω then, again applying Ohm's law, Volts $\div$ Resistance = Amps

$$230 \div 10 = \underline{23\,\text{amps}}$$

This would not break the circuit where a 32 amp fuse is used! The 23 amps would continue to flow, causing the wire to get hot, and possibly leading to a fire.

When a gas installer completes the installation of a new boiler or cooker, etc., the work often involves the final connection to the electrical supply. The gas installer often approaches this with an "I can do that" attitude and proceeds to make the final connection, possibly from a local spur outlet point. There is no problem with this, providing the operative is competent to do so. However, it must be fully understood that by making this connection the operative is taking responsibility for the electrical connections where it is to be left operational.

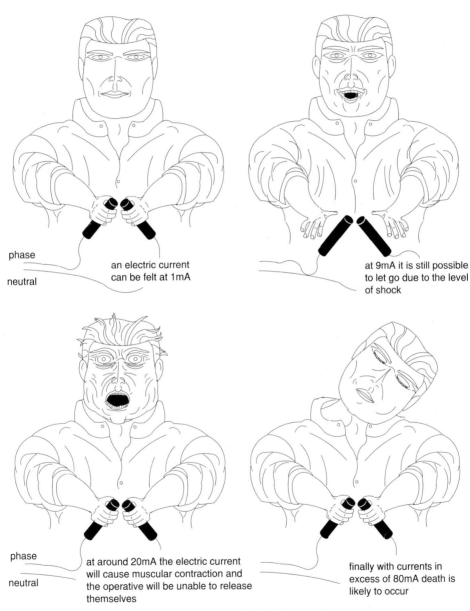

phase

neutral

an electric current can be felt at 1mA

at 9mA it is still possible to let go due to the level of shock

phase

neutral

at around 20mA the electric current will cause muscular contraction and the operative will be unable to release themselves

finally with currents in excess of 80mA death is likely to occur

Electrical Current and the Human Body

Bonding

Relevant Industry Document
BS 7671

The term bonding here relates to the joining of all metal work within a building by means of cables if necessary, called supplementary bonding wires, so that everything is at the same potential. At specific points, such as where the gas or water supply pipes enter the building a 10 mm^2 cable, known as the equipotential bonding wire, is taken back to the main earthing conductor which is situated where the electrical supply enters the property. Thus, should any live phase conductor touch any piece of metalwork within the building, instead of the metal also becoming live it allows a current to flow to earth, causing the fuse or circuit breaker to trip. If the metal was not bonded to earth, this would not happen and it would become live, causing a risk of electrocution, should anyone touch it.

Stray Currents and the Temporary Bonding Wire

For the fuse to trip, a current larger than the fuse rating needs to flow through it and if a phase conductor touches a piece of bonded metalwork this will certainly occur (see diagram (A) opposite). However, if the neutral conductor touches a section of metalwork, although it is live, it will not necessarily cause the trip to blow! This is because within the modern electrical supply system [called the Protective Multiple Earthing system (PME or TNCS system)], the neutral conductor is used as the earth. In diagram (B) the neutral wire is broken and is touching the earth and therefore the metalwork, but the light is still working. This is because the current can still flow back and forth through the series of conductors between the local supply transformer and the light without incurring a current fault, thus the fuse will not trip. When the unsuspecting gas operative removes a section of pipe it might be that is what is occurring. As the two sections of pipe are pulled apart it is like breaking the switch, so a spark may jump between the two sections as they are separated. It may also be that the operative is holding one section of the pipe in each hand, in which case the current would continue to flow through the operative, resulting is electrocution. There is no way of knowing whether this fault exists without carrying out extensive electrical tests and therefore, for reasons of safety, one must assume that it does. So as a precaution one should place a temporary bonding wire across the section of pipe when breaking into the run of pipe. Apart from reasons of safety, as just described, it is also a legal requirement to do this. Any temporary bond must be left in place until a positive bond as been remade.

Absence of Bonding

Where an operative installs new pipework or discovers that there is no bonding to an installation it is not their responsibility to correct this and likewise they should not unless they are competent to do so. However it is their responsibility to notify the responsible person for the property that bonding may be required.

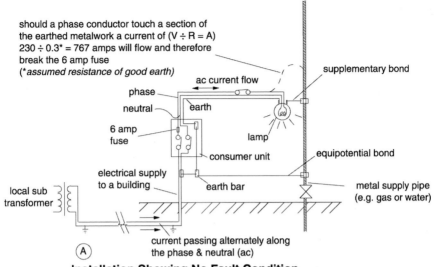

should a phase conductor touch a section of
the earthed metalwork a current of (V ÷ R = A)
230 ÷ 0.3* = 767 amps will flow and therefore
break the 6 amp fuse
(*assumed resistance of good earth)

ac current flow

phase

neutral   earth

6 amp
fuse

lamp

consumer unit

electrical supply
to a building

local sub
transformer

earth bar

supplementary bond

equipotential bond

metal supply pipe
(e.g. gas or water)

current passing alternately along
the phase & neutral (ac)

(A)

Installation Showing No Fault Condition

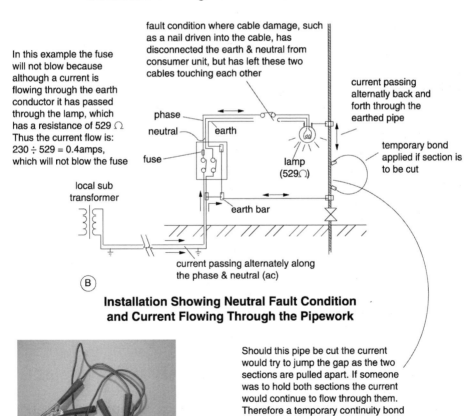

fault condition where cable damage, such
as a nail driven into the cable, has
disconnected the earth & neutral from
consumer unit, but has left these two
cables touching each other

In this example the fuse
will not blow because
although a current is
flowing through the earth
conductor it has passed
through the lamp, which
has a resistance of 529 Ω
Thus the current flow is:
230 ÷ 529 = 0.4amps,
which will not blow the fuse

phase

neutral   earth

fuse

lamp
(529Ω)

current passing
alternatly back and
forth through the
earthed pipe

temporary bond
applied if section is
to be cut

local sub
transformer

earth bar

current passing alternately along
the phase & neutral (ac)

(B)

Installation Showing Neutral Fault Condition
and Current Flowing Through the Pipework

Should this pipe be cut the current
would try to jump the gap as the two
sections are pulled apart. If someone
was to hold both sections the current
would continue to flow through them.
Therefore a temporary continuity bond
<u>must</u> be put in place to bridge the cut.

Temporary Bonding Wire

12 Electrical Work

Safe Isolation

Relevant Industry Document
BS 7671

Under the Electricity at Work Regulations it is illegal to work on live electrical circuits unless the operative is undertaking essential testing work that can only be completed with the supply on. Working on live systems should be regarded as highly dangerous and only be practised by the fully trained individual who clearly knows all the dangers.

All exposed electrical work must be treated as live, and you must never assume otherwise. To check that a supply is dead the following procedures should be carried out:

1. Select an approved test instrument, such as a fused voltmeter or test lamp, as shown. *Note*: Electrical screwdrivers that light up do not comply and should be regarded as unsafe.
2. Check that the test instrument is working correctly by checking on a known supply or proving unit.
3. Locate the means of isolation. Turn off the supply and affix a locking device to the switch to ensure that the power cannot be reinstated without your knowledge.
4. Expose the electrical cables with appropriately insulated tools, taking great care not to touch any exposed conductive parts.
5. Verify that the circuit is dead by taking sample readings between
   - phase–phase (three-phase);
   - phase to neutral;
   - phase to earth;
   - neutral to earth.
6. Re-check that the test instrument is working correctly by testing on a known supply or proving unit.
7. Begin work.

Capacitors And Their Ability To Hold A Charge

Even when everything has been done to check that a circuit is dead, there is still a possibility that a shock could be administered from an electrical component called a capacitor or condenser. These devices, found in several shapes and sizes, although typically cylindrical like a small cotton reel, are designed to store a charge of electricity that can be used for various purposes, such as to give an initial boost to the supply to get a motor turning. It is possible to discharge a capacitor, but generally awareness is all that is required.

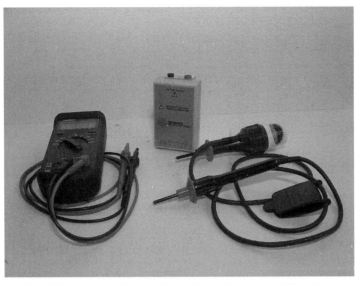

Multi Meter, Proving Unit and an Approved Test Lamp

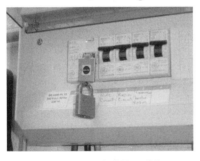

Supply Locked Off at Meter

Supply Locked Off Locally

12 Electrical Work

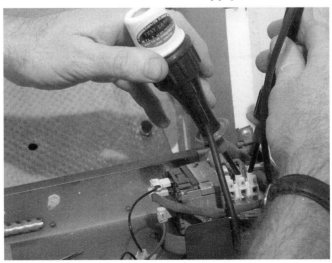

Testing the Supply to Confirm Dead

Inspection and Testing

Relevant Industry Document
BS 7671

Prior to supplying electricity to an appliance for the first time, several checks need to be made. The Gas Installation Commissioning Checklist on page 265 has a section that deals specifically with the preliminary electrical checks. These include:

- checking that conductors are secure;
- checking that earth continuity and bonding is maintained;
- checking that the polarity is correct;
- checking insulation resistance (>0.5 Mohm);
- checking fuse rating.

Prior to undertaking the following tests, the installation must be confirmed dead by following the safe isolation procedure on the previous page.

Conductors secure

This task clearly falls under the heading of inspection. With the cable supplying an appliance having been completed, it should be given a gentle tug to check that it is secure. A check also needs to be made to ensure that in the event of the cable being pulled from its connection, the last conductor to break would be the earth, so this needs to have a little slack.

Earth continuity and bonding maintained

This is a test that is carried out using a multimeter, set on a low ohms setting (Ω). With the earth (circuit protective conductor) connections exposed at the supply connection, i.e. plug of spur connection and appliance, a reading is taken by placing a probe at each point to check the resistance between each. This should be as low as possible, reading no more than possibly 0.1–0.3 ohms. Leaving one probe on the circuit protective conductor, the other probe is then placed on several exposed metal parts to check for a good bond.

Correct polarity

This is a similar test to earth continuity and is designed to check that the phase and neutral connections are wired in correctly. Again using the multimeter and on a low ohms setting, one probe is now positioned on the phase conductor at the supply end. The other probe is then located on the exposed phase connection block at the appliance. As before, a low resistance should be read. This probe is then placed on the exposed earth and phase connections to check that no cross-connection has been made. These two further readings should indicate a high resistance (≥ 1 ohm), going outside the working range of the meter, and proving they are not connected to the phase. Finally, one probe is put on the neutral conductor at the supply and the other at the neutral at the appliance, and the same procedure followed to check that the neutral is going to where it is supposed to and nowhere else.

Insulation resistance (>0.5 Mohm)

This test is carried out to check the condition of the conductors passing to the appliance and will identify any poor cable insulation that would not be detected by the multimeter. Basically, the insulation meter applies a potential 500 V between the conductors (e.g. between phase and earth) aiming to detect a circuit. If the insulation resistance is poor, when the power is switched on the fuse would undoubtedly trip, or a fire may result.

Warning: It is possible to receive a small shock from the exposed metalwork when doing this test, therefore take care. Also, this section needs to be isolated where electronic components are incorporated within the appliance, to prevent damage.

With all switches in the 'on' position (with the exception of the mains supply) the test is completed as follows. One probe is positioned on the phase connection and the other probe positioned on the earth. When in place the button on the meter is depressed, which sends the 500 V through the conductors. Should the current get through from the phase to the earth it would register a reading. Any recorded reading should read greater than 0.5 Mohm. The test is then repeated to check between phase and neutral and finally between neutral and earth.

Fuse Rating

The fuse is designed to protect the cable and appliance. It will not protect you against electric shock, see page 428. The fuse rating can be determined by applying Ohm's law ($V \div R = I$) but you will find it in the appliance's instruction book. For most appliances this is 3 amps, but lower fuse ratings may be required, and these should not be exceeded. Failure to observe this simple rule may lead to appliances becoming live, damaged or cause a fire.

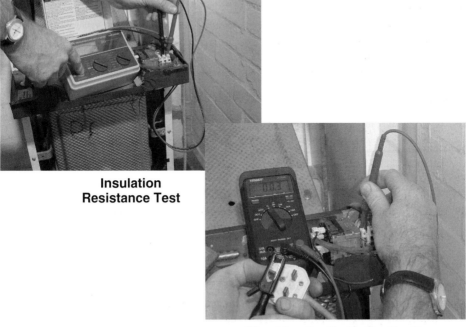

**Insulation
Resistance Test**

Checking for Earth Continuity

Fault Diagnosis of Basic Electrical Controls

With the supply established, it may be that an appliance fails to operate. All the electrical connections need to be inspected to ensure that there are good sound connections and that the PVC insulation has been sufficiently stripped back at the ends, thus exposing the conductor and making a suitable contact. These tests can be made with the power isolated. However, you may need the power on for further fault diagnosis. Again it must be stressed that working with live electrical systems poses a substantial hazard and work must not proceed unless the operative is fully competent. Below and opposite are examples of flow charts showing the logical sequence that needs to be undertaken for fault diagnosis, analysing each step for operation in order to determine a particular fault.

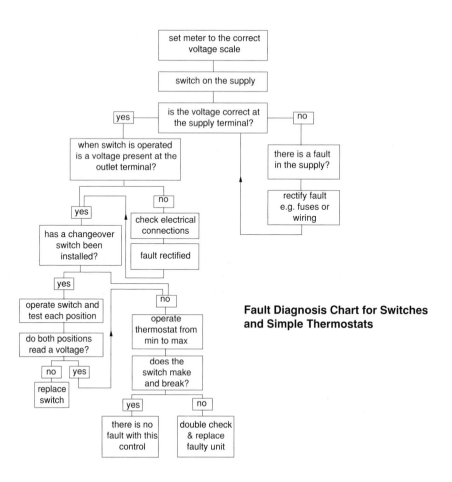

Fault Diagnosis Chart for Switches and Simple Thermostats

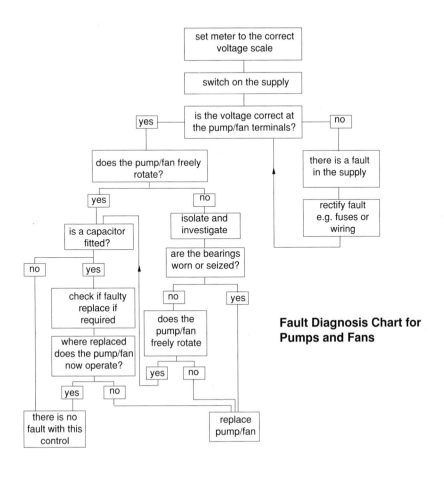

Fault Diagnosis Chart for Pumps and Fans

These two flowcharts are given to provide an illustration of the concept of fault finding and may be limited in the vast number of fault conditions that could occur with a particular appliance. It is very common to find diagrams of this type in a manufacturer's instructions, taking you step by step through the sequential operation of an appliance, and including more than just the electrical controls that may be at fault.

Index

Index

Index